INLAND WATERS
AND
THEIR ECOLOGY

'Mangowa and the round lakes'

In Aboriginal mythology '... the stars that Mangowa tore from their homes, falling to earth, made the circular lagoons that fringe the shores of the coastal lakes of South Australia'; *here* a welcome to limnologists

From the painting by Ainslie Roberts (reproduced in *The Dreamtime*, published by Rigby Ltd)

INLAND WATERS AND THEIR ECOLOGY

I. A. E. Bayly
Reader in Zoology, Monash University

W. D. Williams
Reader in Zoology, Monash University

LONGMAN

Longman Australia Pty Limited
Camberwell Victoria Australia
Associated companies, branches, and representatives
throughout the world

© Longman Australia 1973
First published 1973

ISBN 0 582 71421 4

This book is copyright. Apart from any fair dealing for the purposes of private study, research, criticism or review, as permitted under the Copyright Act, no part may be reproduced by any process without written permission. Enquiries should be made to the publishers.

Text set in Australia by Dudley E. King Pty Ltd
Printed in Hong Kong by
Dai Nippon Printing Co. (H.K.) Ltd.

To Peter and Trevor, and Simon and Richard

preface

Most limnological textbooks are unsuitable for student use in Australia and New Zealand because nearly all biological examples chosen belong to a basically different biota, and there is usually inappropriate emphasis with respect to various physico-chemical features. More comprehensive accounts, on the other hand, are either too large or not in English. Hutchinson's magnificent *A Treatise on Limnology* is too long, too detailed, and too expensive to be anything other than a basic library reference text from the viewpoint of the undergraduate student. Dussart's *Limnologie: L'Etude des Eaux Continentales* has several attractive features and is not as encyclopaedic as Hutchinson's work, but being in French is scarcely suitable as a recommended text in Australasian universities. It will be obvious why the present book has been written.

We make no claims for comprehensiveness and, indeed, throughout most of the book we have deliberately written a rather parochial text. We have made no attempt to provide from a world standpoint a geographically representative or balanced selection of examples. By and large our policy has been not to use an overseas example where a suitable Australasian example sufficed. Frequently, of course, because of the relative infancy of limnology in this region of the world, no suitable example from Australia or New Zealand could be found. Exceptions to this general situation are provided by those parts of the text (e.g. Chapter 2) wherein we deal with subjects which we feel are ill served at present with relatively comprehensive and up-to-date accounts at this level. Considered as a whole, therefore, we hope this book will be of some interest to limnologists wherever they may be.

The treatment of Australian and New Zealand limnology together in the one text may be justified on a number of grounds. It is certainly justifiable on biological grounds, for Australian and New Zealand freshwater biotas have a very close affinity and differ from those of almost all other countries and continents (South America excepted). Even from the viewpoint of physical limnology there is a certain unity about Australian and New Zealand lakes. The overriding importance of warm monomictic lakes and the apparently complete absence of dimictic lakes may be quoted as examples. Finally, for historical, political, and linguistic reasons there is an increasing amount of interchange of biologists and other scientists between the two countries.

We are not insensitive to the fact that to some New Zealanders the adjective 'Australasian' is a source of irritation, but we hope New Zealand limnologists will forgive us for using it fairly freely for conciseness or in the zoogeographical sense. If it is any comfort to them, we might add that in certain areas our current knowledge of 'Australasian' limnology seems to be based more on New Zealand than Australian studies.

Whilst all chapters have been the object of criticism, debate, and contribution by both authors, Chapters 1, 3, 4, 5, and most of 6 and 10 have been written by I. A. E. Bayly, and Chapters 2, 7, 8, 9, 11, 12, and part of 6 (non-planktonic biota) and 10 (nekton and benthic macrofauna, flora, salt lakes as ecosystems) have been written by W. D. Williams.

We thank the following for reading and commenting on portions of the book: Professor H. B. N. Hynes, University of Waterloo, Canada (Chapters 2, 7, 8, 11, and 12); Professor R. G. Wetzel, Michigan State University, U.S.A. (the first part of Chapter 2); and Dr P. A. Tyler, University of Tasmania (the plankton portion of Chapter 6).

For unpublished information we thank Miss Helen Aston, Royal Botanic Gardens and National Herbarium, Melbourne; Australian Atomic Energy Commission, Lucas Heights; Dr G. W. Brand, Monash University; Mr M. Geddes, Monash University; Dr R. B. Glover, D.S.I.R. (N.Z.); Professor J. Illies, Hydrobiologische Station, Schlitz, Western Germany; Mr J. S. Lake, University of Sydney; Dr G. C. Paterson, Mount Alison University, Canada; Mr H. K. Schminke, University of Kiel, Western Germany; the State Electricity Commission of Victoria; the late Mr Hans Steen; Mr B. V. Timms, Monash University; Dr P. A. Tyler; various officers of the State Fisheries Departments of Victoria and New South Wales; Dr K. F. Walker, Monash University; and Mr D. Wimbush, C.S.I.R.O. (Aust.).

Many of the illustrations for the chapters written by I.A.E.B. were prepared by Mrs D. Troon. The remaining illustrations for these chapters were prepared by their author. The figures for the chapters written by W.D.W. were mainly drawn by Mr W. E. C. Ward.

Acknowledgements for photographs or figures appear in the captions except for two plates where it was specifically requested that no acknowledgement be made.

Finally, we acknowledge our gratitude to Mrs Thea McConnell for typing services within and well beyond the bounds of duty.

I. A. E. Bayly, W. D. Williams
Clayton, 1971

contents

PART ONE	General features of inland waters	1
Chapter 1	Chemical features	1
Chapter 2	Biological production	24
PART TWO	Lakes	48
Chapter 3	Preliminary considerations	49
Chapter 4	Productivity	66
Chapter 5	Pnysical and related chemical and biological phenomena	78
Chapter 6	Ecology of the major biological communities	97
PART THREE	Running waters	134
Chapter 7	Non-biological features	135
Chapter 8	Biological features	151
PART FOUR	Other bodies of fresh water	170
Chapter 9	Other bodies of fresh water	171
PART FIVE	Non-marine or athalassic saline waters	192
Chapter 10	Non-marine or athalassic saline waters	193
PART SIX	Man and inland waters	214
Chapter 11	Uses of inland waters	215
Chapter 12	The effects of man on inland waters	232

Bibliography	261
Index of lakes and reservoirs	284
Index of organisms	289
General index	296
Postscript	315

part one
GENERAL FEATURES OF INLAND WATERS

Macquarie perch. *By courtesy of the Victorian Fisheries and Wildlife Department*

chapter one
CHEMICAL FEATURES
Major Inorganic Ions and Related Biological Considerations

THE INDIVIDUAL IONS AND THE IONIC COMPOSITION OF WORLD AVERAGE FRESH WATER AND SEA WATER
Inland or athalassic waters are remarkably varied in their ionic composition. This is undoubtedly one of their most outstanding characteristics despite the fact that many authors have in the past stressed their 'standard composition' rather than their variation, especially with respect to fresh waters. Although the variation in ionic composition is so great, four cations (sodium, potassium, magnesium, and calcium) and three anions (chloride, sulphate, and bicarbonate) account for all but a very small fraction of the total ionic concentration or salinity. The composition of world average fresh water (river water) and of average ocean water is shown in Table 1:1.

In many parts of the world, including much of Europe and North America, fresh waters are usually dominated by calcium and bicarbonate, and world average fresh water (Table 1:1) provides a useful yardstick with which various observed compositions may be compared. In Australia, however, apparently as a consequence of the overriding importance of atmospheric supply of oceanic salts to athalassic waters, sodium and chloride tend to dominate not only saline waters (see Chapter 10), but also fresh waters. In other words, in most regions of Australia the occurrence of standing waters conforming with Rodhe's (1949) 'standard composition' or Conway's (1942) world average fresh water (Table 1:1) is more the exception than the rule. Thus, if in Australia one wishes to compare the observed ionic proportions of an athalassic water with some sort of standard, it is usually more appropriate to compare it with sea water (Table 1:1) than so-called world average fresh water. This is especially true of the saline athalassic waters of Australia.

It may be noted that Conway's original value for the salinity of world average fresh water was 146 mg/l, not 177 mg/l as quoted in Table 1:1. However, in arriving at the former value Conway supposed that the total bound CO_2 is present as carbonate whereas in fact it is represented almost entirely as bicarbonate. Furthermore, Conway included iron and aluminium oxides and silica in his calculation. These are not normally present in ionic form and therefore do not contribute towards salinity when it is defined as below. When nitrate and phosphate are included the salinity of world average fresh water becomes 178 mg/l. These two minor ions together account for about 1·8 per cent of total anion equivalents.

TABLE 1:1
IONIC COMPOSITION OF WORLD AVERAGE FRESH WATER AND OF AVERAGE SEA WATER

	Fresh Water[a]		Sea Water	
	meq %[b]	mg/l	meq %	mg/l
Cations				
Na^+	16	8	77	10,810
K^+	3	3	2	390
Mg^{++}	18	5	18	1,300
Ca^{++}	63	30	3	410
Anions				
Cl^-	10	8	90	19,440
$SO_4^=$	16	18	9	2,710
HCO_3^-	73	105	<1	140
Salinity	—	177	—	35,200

[a] Total ionic concentration is 4·70 meq/l. Some authors use 'total ionic concentration' in the sense of either total cationic concentration or total anionic concentration which, of course, are equal, both having the value 2·35 meq/l. To the present authors, however, the latter usage is confusing.

[b] Sometimes, as here, inorganic ions are reported as the (milli)equivalent percentage of total cations or total anions, and sometimes as the equivalent percentage of the total ionic concentration. The former method allows a slightly more precise presentation of results without resorting to three figures.

This is not primarily a text on methods but it may be noted in passing that all of the major cations may be determined by atomic absorption spectrophotometry. Sodium, potassium, and calcium may also be determined by emission flame photometry. An additional method for calcium and magnesium is complexometric titration with EDTA (also known as sodium versenate) using suitable indicators such as Eriochrome Black T for calcium plus magnesium and Hydroxy Naphthol Blue for calcium alone. Bicarbonate (or bicarbonate plus carbonate, if the pH is greater than 8·3) may be determined by titration with a standard acid until the pH has dropped to 4·5 as determined colorimetrically by a '4·5 indicator' or electrometrically with a pH-meter. Some natural waters have a pH of 4·5 or less and it is assumed that these contain no bicarbonate. Chloride and sulphate may be determined by ion-exchange procedures. In addition, chloride may be estimated by potentiometric titration, and sulphate by determining with atomic absorption spectrophotometry the barium left in solution after precipitating all the sulphate with an excess of barium chloride. Finally it is noteworthy that total anionic concentration may be easily determined by an ion-exchange procedure. This is a useful measure in that it serves to check the accuracy of the determination of major cations; the sum of the cation concentrations in (milli) equivalents should agree with the total anionic concentration if no errors have been made or pH is not exceptionally low.

METHODS OF PRESENTING ANALYTICAL RESULTS

This may be regarded as an almost trivial consideration, yet it is true that the value of some potentially very useful limnological data is considerably reduced because the mode of reporting is inappropriate or ambiguous. An interesting discussion of this topic together with its historical aspects is given by Anderson (1940).

CHEMICAL FEATURES

It was once a common practice to report analyses in terms of hypothetical undissociated compounds. However, a good deal of guesswork and a number of arbitrary decisions are involved in such a method. Indeed, Clarke (1908) described it as 'a meaningless chaos of assumptions and uncertainties'. It should also be noted that the compounds which are deposited on evaporation are not necessarily identical with those that were originally dissolved by the water.

When presenting the absolute amounts of ions in fresh waters it is convenient to use as units either milligrams per litre (which at low concentrations are very nearly the same as milligrams per kilogram or parts per million) or milliequivalents per litre. The latter has an advantage in that it permits a direct and easy check on the accuracy of analysis (Table 1:2), but perhaps a disadvantage in that it is not always easy to think in terms of milliequivalents and the order of the salinity may not be immediately grasped unless it is separately listed.

TABLE 1:2
USE OF MILLIEQUIVALENTS AS A UNIT

Cations	Anions
Na^+ a (meq/l)	Cl^- x (meq/l)
K^+ b (meq/l)	SO_4^- y (meq/l)
M^{++} c (meq/l)	HCO_3^- z (meq/l)
Ca^{++} d (meq/l)	

$$(a+b+c+d)T_1 = (x+y+z)T_2$$

Special attention should be given to the presentation of the amount of (bi)carbonate. In the pH range $4 \cdot 5$–$8 \cdot 4$ ionic inorganic carbon is present almost entirely in the form of bicarbonate and either of the above two units may be used to record its amount. However, where the pH exceeds $8 \cdot 4$ the situation becomes more complex for the reason that an appreciable amount of carbonate, as well as bicarbonate, may be present. Three main systems of presentation are then used and these are shown in Table 1:3. Each has its pros and cons and the preferred method will depend on the circumstances. The first and third are probably most commonly used, but if, for example, work is being carried out on an alga suspected of using carbonate directly in photosynthesis, it would be desirable to know the absolute amount of this ion present; that is, the second method would be used. However, the precise determination of the separate amounts of carbonate and bicarbonate present at the instant of investigation is not an entirely simple matter. Rapid changes in pH may occur after the removal of the sample, especially if in a dilute water the equilibrium has been temporarily displaced by intense photosynthetic activity of algae. If this occurs a subsequent determination may indicate the presence of bicarbonate only, whereas the original field pH value was high enough ($> 8 \cdot 4$) to indicate the definite presence of carbonate. A way to overcome this difficulty is to record the pH (with a glass electrode) and temperature in the field, later determine the total CO_2 present, and then calculate the amount of bicarbonate and carbonate. A good method of determining total CO_2 is given in Scott (1939).

Highly alkaline carbonate lakes are often sufficiently well buffered for very little immediate change to occur despite the high initial pH (sometimes in excess of $10 \cdot 0$).

TABLE 1:3

METHODS OF PRESENTING THE AMOUNTS OF CARBONATE, BICARBONATE, AND OTHER INORGANIC IONS WHEN THE pH EXCEEDS 8·4

Ion	1 Pure meq/l	2 Pure mg/l	3 Mixed meq/l and mg/l
Na^+	(meq/l)	(mg/l)	(mg/l)
K^+	(meq/l)	(mg/l)	(mg/l)
Mg^{++}	(meq/l)	(mg/l)	(mg/l)
Ca^{++}	(meq/l)	(mg/l)	(mg/l)
Cl^-	(meq/l)	(mg/l)	(mg/l)
$SO_4^=$	(meq/l)	(mg/l)	(mg/l)
$(HCO_3^- + CO_3^=)$	(meq/l)	—	(meq/l)
HCO_3^-	—	(mg/l)[a]	—
$CO_3^=$	—	(mg/l)[a]	—
Sum of cations	(meq/l)	—	(meq/l)
Sum of anions	(meq/l)	—	(meq/l)

[a] See note in text.

Below a pH of 7·0 an appreciable amount of free CO_2 is present and at pH 4·5 and less the inorganic carbon system is represented entirely by free $CO_2 + H_2CO_3$ (undissociated). This, however, being non-ionic, does not enter into present considerations.

On the basis of values obtained by titrating samples with standard acid to phenolphthalein and methyl-orange indicator end-points, it has been a common practice to report alkalinity values in terms of parts per million of calcium carbonate. Not only is this an unnecessary convention, it is sometimes highly misleading. Some highly alkaline natural waters are very poor in calcium, the bicarbonate and carbonate having originated by the solution of sodium, not calcium, compounds. Such is the case with Lake Rudolf in the Rift Valley of Africa (Beadle 1932) and Lake Aroarotamahine in New Zealand (Bayly 1963). These lakes have, respectively, about 21·7 and 3·5 meq/l of bicarbonate plus carbonate, but only about 0·25 and 0·05 meq/l of calcium. A similar state of affairs is found in some of the Lahontan Lakes of Nevada. It is therefore best to avoid the above convention and report alkalinity simply as the number of milliequivalents of acid required to neutralize the base present in a litre, or alternatively as milligrams or milliequivalents of bicarbonate (when the initial pH is 8·4 or less) or as milliequivalents of bicarbonate plus carbonate (when the initial pH is more than 8·4).

The noun 'alkalinity' is unambiguous and may be defined as the total quantity of base determinable by titration with a strong acid. The adjective 'alkaline', however, as applied to natural waters, seems to be used loosely to mean high pH, or a large amount of base, or large amounts of the alkali metals sodium and potassium. Sometimes, as with Lake Rudolf for example, all these three occur together, but this is not necessarily so.

Turning from the absolute to the relative aspects of analyses, it would seem that Maucha's field diagrams (Fig. 1:1), a graphical form of presentation, are hard to beat. To construct one of these, a circle of appropriate radius is drawn and within this four diameters are ruled so as to divide it into eight equal sectors. The four sectors to the right of the vertical diameter (the cation field) are used for depicting K^+, Na^+, Ca^{++}, and Mg^{++}, and those on the left for $CO_3^=$, HCO_3^-, Cl^-, and $SO_4^=$. Lines bisecting each sector are drawn and along each a distance proportional to the concentration of that particular ion (in equivalent percentage of the total ionic concentration divided by $8 \cdot 082 \sin. 22 \cdot 5°$) is measured off from the centre of the circle. Each such point is

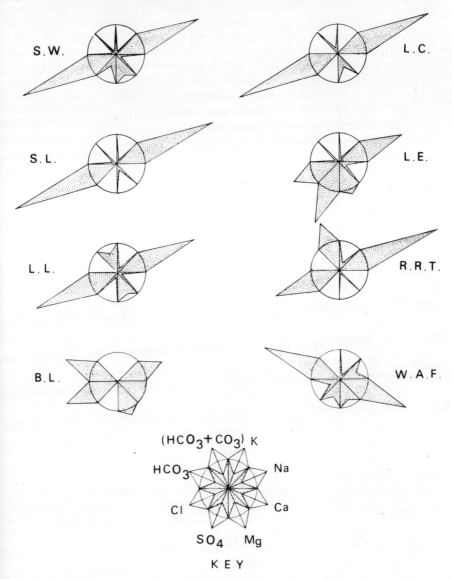

Fig. 1:1 Maucha ionic diagrams for sea water, world average river water, and some Australasian lakes. SW = sea water; LC = Lake Corangamite, Victoria; SL = Salt Lake, Sutton, New Zealand; LE = Lake Edward, South Australia; LL = Lake Leake, South Australia; RRT = Red Rock Tarn, Alvie, Victoria; BL = Blue Lake, Mount Gambier, South Australia; WAF = world average fresh water (river water)

joined to the outer extremity of the radii enclosing it, and the sixteen-sided polygon thus formed is shaded or blacked in. The quadrilateral associated with each sector has an area proportional to the concentration, so that the relative importance of the various ions and differences between various waters are immediately apparent. Brock and Yake (1969) describe a modification of Maucha's ionic diagram that enables one to appreciate immediately differences in total ionic concentration as well as differences in the relative proportions of the various ions.

Although Maucha's field diagrams are undoubtedly very good, their construction takes time—a valuable commodity for most limnologists—and it is often not feasible to use them, especially when a large number of analyses are involved. It is more usual, therefore, to find a tabulation of the various ions as percentages of total cation or anion equivalents (Table 1:1). A convenient way of combining in the same table relative and absolute values for a series of analyses is shown in Table 1:4.

A convenient way of presenting the relative proportions of the major anions is to use triangular diagrams of the type shown in Fig. 1:2. In these the percentage of any one of the anions is proportional to the length of the perpendicular from any point on the side opposite the apex of the triangle marked by the symbol of the anion. The

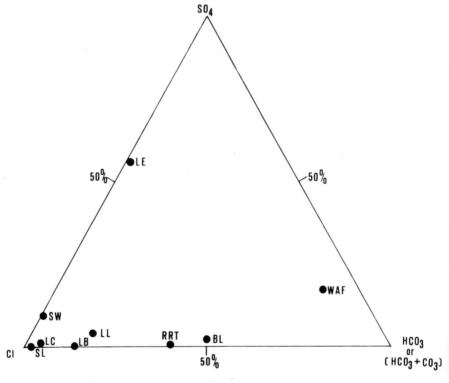

Fig. 1:2 Anionic diagram for sea water, world average river water, and some Australasian lakes. BL=Blue Lake, Mount Gambier, South Australia; LB=Lake Boemingen, Fraser Island, Queensland; LC=Lake Corangamite, Victoria; LE=Lake Edward, South Australia; LL=Lake Leake, South Australia; RRT=Red Rock Tarn, Alvie, Victoria; SL=Salt Lake, Sutton, New Zealand; SW=sea water; WAF=world average fresh water (river water)

CHEMICAL FEATURES

TABLE 1:4
CHEMICAL COMPOSITION OF SOME NATURAL WATERS

Lake or water	Value[a]	Na^+	K^+	Mg^{2+}	Ca^{2+}	Cl^-	SO_4^{2-}	HCO_3^-	$(HCO_3^- + CO_3^{2-})$	Salinity (p.p.m.)
Sea water	A	10,810	390	1,300	410	19,440	2,710	140	—	35,200
	B	77	2	18	4	90	9·3	0·4	—	
Lake Corangamite, Vic.	A								4	12,000–115,000
	B									
Salt Lake, Sutton, N.Z.	A	5,300	230	160	40	8,980	1	430	—	15,180
	B	91	2	5	1	99	<1	3	—	
Lake Edward, S.A.	A	500	63	130	43	544	975	40	—	2,305
	B	60	4	30	6	43	56	2	—	
Lake Leake, S.A.	A	760	33	110	37	1,263	86	—	17	—
	B	73	2	20	4	79	4	—		
Red Rock 'Tarn', Vic.	A	7,350	800	80	<20	7,710	<1	—	43	24,300
	B	92	6	2	<0·3	61	6	196		
Blue Lake, Mt Gambier, S.A.	A	64	4	24	31	114	2	49	—	439
	B	43	2	31	24	49	18	105	—	
World average river water	A	8	3	5	30	8	16	73	—	177
	B	16	3	18	63	10				
Lake Boemingen, Fraser Island, Qld	A	11·8	0·5	1·3	0·4	20·0	4·8	0	—	39
	B	78	2	17	3	86	15		—	

[a] A = absolute amount in p.p.m.
B = relative amount as equivalent % of total cations.

apex lying closest to the ionic symbol thus corresponds to 100 per cent of that particular anion.

SOURCES AND MECHANISMS OF ION SUPPLY

Athalassic waters receive ions either from the weathering of rocks and soils, especially by the action of carbonic acid, or from the atmosphere. A good summary of the processes involved is given by Gorham (1961).

The *action of hydrogen ions* in the weathering of soils and rocks is important and will be considered first. A major source of these ions is carbonic acid. Carbon dioxide is found in normal air at a concentration of $0 \cdot 03$ per cent by volume. This value may be higher in the vicinity of volcanoes, and these have undoubtedly been important throughout geological time in supplying this gas to the atmosphere. On solution in water it forms carbonic acid, but this is strongly dissociated and very little exists as such in natural waters. The importance of carbonic acid in weathering is well illustrated by the high proportion of bicarbonate ions in many river waters (see Table 1:1). Strong acids, such as sulphuric acid and hydrochloric acid, washed down by rain from polluted air, may also be important in some cases. If free acids are absent, hydrogen ions may be provided by the dissociation of water molecules in the process of hydration and hydrolysis. Ordinary *solution* not involving free acids or hydrolysis is important mainly in sedimentary deposits rich in soluble salts. Leaching of marine salt beds may thus produce a lake water enriched in sodium, potassium, and chloride, because these are more soluble than other ions.

A further mode of ion supply from soils and rocks is by *oxidation and reduction*. Iron, manganese, sulphur, nitrogen, and phosphorus are the main elements influenced by these processes, and although all of these are biologically important only sulphur is relevant in a discussion of major ions. Iron sulphides are commonly present in swamps and rocks, and their oxidation may provide a significant source of sulphate. Usually this is in the form of sulphuric acid, and as this is a strong solvent the supply of other ions may also result. Another significant process involves *ion exchange* between the solids and liquids of soil and between lake muds and overlying water. Often an equilibrium may be established because of the limited adsorptive powers of the soil colloids, and in this case the overall ionic composition of percolating water may be little changed. Under non-equilibrium conditions (e.g. recent flooding of previously dry soil), however, the effect of such a process may be quite appreciable. Even the commonest weathering process usually involves some ion exchange with hydrogen ions replacing metal ions incorporated in soil minerals. Finally, there is *complex formation*; ions can be taken from soils and rocks by chelation or other forms of complexing. They are usually joined to a complex organic molecule (the chelator) by more than one valence bond, and are prevented from combining with other ions, but remain in solution. This is the principle of the EDTA method of determining calcium and magnesium, mentioned above; EDTA forms stable un-ionized complexes with calcium and magnesium (and some minor ions).

Athalassic waters may obtain salts from the atmosphere in several different ways. Usually most of the atmospheric salt is washed down by *rain*, but *dry fallout* may also occur since significant differences are found between analyses of rain water from containers left open continuously and of that from containers open only during periods of precipitation. In high-altitude regions *snow* may supply appreciable amounts of ions. Finally, all of the major anions may be cycled through the atmosphere in the

CHEMICAL FEATURES

form of *gases* and dissolved in athalassic waters. Bicarbonate is, of course, derived from carbon dioxide, and sulphur occurs in air either as sulphur dioxide or hydrogen sulphide which may on solution and oxidation yield sulphate. Occasionally chlorine may also be liberated in the atmosphere as gaseous hydrochloric acid which is readily soluble in rain droplets and cloud, giving chloride.

The sources of atmospherically supplied ions are also various. The *sea* is probably most important. Sodium, magnesium, chloride, and sulphate are all supplied to the air in large amounts from sea spray. Studies in Australia (Hutton and Leslie 1958) and elsewhere have shown a marked decrease in chloride in rain water with increasing distance from the sea (Fig. 1:3), indicating that sea spray is a major source of this ion in coastal regions. A second source is *soil*; dust from the ground may contribute a significant amount of ions to rain and snow. This seems to be an important source of calcium and potassium. It also seems worthwhile suggesting at this point that ions such as sodium and chloride, initially supplied to relatively coastal salt lakes from the sea via the air, may later be transported further inland from dried pans. They could be carried directly further inland in dry form by wind alone, or else washed down in rain from dry salt particles temporarily suspended in the atmosphere by wind. The violent fluctuations in salinity observed in some of the small temporary salinas (mainly sodium chloride) in western Victoria (Bayly and Williams 1966a) strongly suggest that a massive removal of salt by deflation between successive fillings can occur. The suggestion is, then, that dry salt pans may be regarded as a special type of soil capable of supplying to the air ions other than potassium and calcium. Such a multiple transfer process would allow salts initially of oceanic origin to find their way far inland. In this connection it may be noted that despite the fact that most of the Lake Eyre drainage basin is remote from the sea (Fig. 10:1) the salt in the lake itself, which is

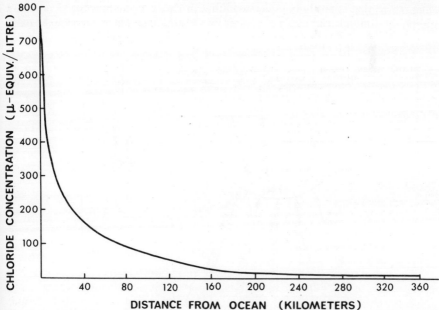

Fig. 1:3 Effect of distance from ocean on concentration of chloride in rain water. *After Hutton and Leslie (1958)*

mostly sodium chloride, was probably mainly atmospherically supplied ultimately, but not necessarily directly, from the sea (see Chapter 10). Wopfner and Twidale (1967) claimed that the theory that the salts in the Lake Eyre Basin have been carried through the atmosphere from the sea is discounted and quoted the work of Hutton and Leslie (1958) in support of this contention. However, to both of the present authors the observations of Hutton and Leslie do not support this conclusion, or at least do not lead inescapably to it. Wopfner and Twidale stated that these salts have 'undoubtedly been derived from the rocks which crop out in the catchment area', but offered no explanation as to how such large absolute and relative amounts of sodium and chloride could have originated from the weathering of these rocks.

In addition to the sea and soil, a few other sources of atmospherically supplied ions are left for mention. Domestic and industrial *air pollution* is today an important source of atmospheric ions in many countries. In Australia areas thus affected are relatively small because of the low overall population density and the extreme centralization of population in State capitals. Nevertheless it can be marked near the major cities. Thus, in the air near Sydney, Twomey (1953) observed droplets which remained liquid at 20 per cent relative humidity and appeared to consist of sulphuric acid. Air pollution may also add significant amounts of hydrochloric acid, calcium, and potassium to rain water. Another source is *organic debris* floating in the air; in some regions a correlation between the amounts of soot and sulphate in rain has been demonstrated. Finally, *volcanoes* are major sources of carbon, chloride, and sulphur which are given out in the form of carbon monoxide, carbon dioxide, hydrochloric acid, hydrogen sulphide, sulphur dioxide, and sulphur trioxide.

The processes of ion supply by the weathering of soils and rocks and by the atmosphere have both been outlined above and are summarized in Fig. 1:4. For some reason that is not fully understood, but probably a consequence of excessive aridity, the atmospheric supply of oceanic ions to athalassic waters seems to be more

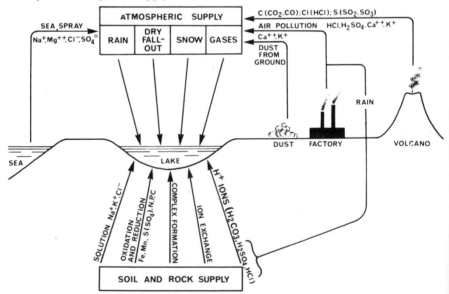

Fig. 1:4 Diagram showing the various processes whereby ions are supplied to an inland water

important in Australia than on any other continent. It is not surprising, in view of its quite marked intensity, that there has long been an awareness of the importance of this process in Australia, whereas most workers in the northern hemisphere have not drawn attention to its significance until fairly recent years (e.g. Gorham 1961). Quite possibly the atmospheric supply of ions is no more effective in Australia than elsewhere but its importance is unmasked because of the diminished significance of supply from terrestrial weathering processes. The latter is probably depressed because of climatic aridity. An interesting illustration of the low intensity of terrestrial weathering in Australia is provided by the comparatively frequent chance discovery of stony and stony-iron meteorites, especially in Western Australia (McCall 1967). In most regions such meteorites disappear rapidly as a result of weathering and are usually recovered only as a result of an observed fall; their long survival in Australia may be attributed to its marked aridity.

Some workers (e.g. Currey 1970 and Venables 1970) dispute the claim that atmospherically supplied oceanic salts are of overriding importance in most Australian inland waters that are dominated by sodium and chloride. These authors, however, provide no explanation of the remarkable ionic homogeneity found throughout geologically diverse regions; if atmospherically supplied ions are as relatively unimportant as they maintain, it is difficult to see why there is nothing like as much ionic heterogeneity as there is in types of soils and rocks.

THE TOTALITY OF INORGANIC IONS AND VARIOUS MEASURES OF IT

Salinity may be defined as the total concentration of the ionic constituents and may be obtained by the determination and summation of the seven ions Na^+, K^+, Mg^{++}, Ca^{++}, Cl^-, $SO_4^=$, and HCO_3^- (plus $CO_3^=$ in some cases). Failure to include nitrates, phosphates, or bromide may result in a small error in exceptional cases.

Total dissolved solids (T.D.S.) can be determined simply by filtering a water sample and evaporating a known volume of the filtrate to dryness at 103°C. When the residue is thoroughly dry the container is cooled in a desiccator and weighed. From this the weight of the empty container is, of course, subtracted. Since the dried residue contains organic as well as inorganic materials the T.D.S. usually exceeds the salinity. It has been found, for example, that for the more saline athalassic waters in Australia the salinity is about 90 per cent of the T.D.S. In highly humified but low-salinity waters the discrepancy between the two measures can be very considerable, salinity being a much lower percentage of the T.D.S. because of the large amount of coloured dissolved organic matter. However, discrepancy in the opposite direction is also possible. Some CO_2 (the *half-bound* CO_2) is lost from bicarbonates on evaporation to dryness, and for this reason the T.D.S. value for highly alkaline waters (e.g. those with large amounts of sodium carbonate) may be distinctly *less* than the true salinity. For most waters, however, agreement between these two measures is good and T.D.S. is the greater.

If following evaporation of a volume of water the dried residue is ignited over a bunsen or in a furnace at 500–600° the *non-volatile solids* (per unit mass or volume) may be obtained. The difference between this and the initial weight of dried sediment is termed the *loss on ignition*, and is due mainly to organic matter and CO_2 (*firmly-bound* CO_2) lost from carbonates. Loss on ignition is a very useful measure in palaeolimnology when dealing with lacustrine sediments.

Specific conductance or conductivity has long been used to estimate both T.D.S. and salinity. Within a restricted range of T.D.S., and under conditions of constant temperature and constant relative ionic proportions, T.D.S. is very nearly directly proportional to conductivity, so that we may write:

$$T = cK \qquad (1)$$

where T = T.D.S., c = coefficient, and K = conductivity. Over a wide range of T.D.S. c itself is subject to some variation.

The conductivity of a given solution increases with increasing temperature and it is therefore customary to correct conductivities to a standard temperature which is usually 18, 20, or 25°C. Correction to 18°C may be made using the formula:

$$K_{18} = \frac{K_t}{[1 + 0 \cdot 025(t - 18)]} \qquad (2)$$

where K_{18} = the conductivity at 18°C, and K_t = the conductivity at the experimental temperature t°C.

It has further been found (Williams 1966) that over a wide range of T.D.S. the value of c (equation (1) above) from Australian athalassic waters (which tend to be dominated by sodium and chloride) is given by the equation:

$$c = \frac{3 \cdot 4 K_{18}}{10^6} + 0 \cdot 666. \qquad (3)$$

The accuracy of this equation is highest where the conductivity exceeds about 5,000 μmhos.

Substituting in (1) we thus obtain:

$$T = \left[\frac{3 \cdot 4 K_{18}}{10^6} + 0 \cdot 666\right] K_{18} \text{ (p.p.m.)}. \qquad (4)$$

Within the salinity range 3–100‰ the value obtained from conductivity via equation (4) usually agrees with that obtained by a gravimetric determination to within ±1‰.

FRESH AND SALINE ATHALASSIC WATERS

In most countries the majority of non-marine waters are fresh; that is, their salt concentration is below that detectable by human taste. In Australia and other arid regions, however, a very significant proportion of the total athalassic aquatic environment is saline. The decision as to where to put the boundary between fresh and saline waters is necessarily a somewhat arbitrary and subjective one (see Bayly 1967b for full discussion). For our purposes, however, we shall define saline waters as those with salinity of more than 3,000 p.p.m. (3‰).

It should be noted that athalassic saline waters not only have a higher salinity than fresh waters, but usually they also have a much more *variable* salinity, and may thus be described as *poikilosaline*. These changes in salinity are substantial and often rapid. Fresh waters usually undergo comparatively small fluctuations in salinity and are thus *homoiosaline*. This difference is important in considering osmotic regulation (see below).

Athalassic saline waters are discussed in more detail in Chapter 10.

TYPES OF OSMOTIC REGULATION

The concentration of the body fluids of most freshwater organisms is significantly lower than that of their marine counterparts, but usually well above that of their external medium. In freshwater animals the osmotic concentration of the body fluids ranges from about 50 mOsm/kg in some bivalve molluscs to about 650 mOsm/kg in some crabs (Potts and Parry 1964), but values in the range 100–350 mOsm/kg are most common. These values may be compared with *ca.* 3 mOsm/kg for world average fresh water. Freshwater organisms are thus *hyperosmotic regulators* and the differential between internal and external concentration can be maintained only by the continuous expenditure of energy. There is a tendency for water to be gained by osmosis and for ions to be lost by diffusion. These tendencies are usually offset by the excretion of a urine hypo-osmotic with respect to the body fluids and the active uptake of ions.

It is instructive to consider what happens when the salinity of the external medium of a typical freshwater animal is gradually elevated either naturally or experimentally. Lockwood (1959), working on *Asellus aquaticus*, showed that up to a certain point the concentration of the haemolymph rose as the concentration of the medium was increased (Fig. 1:5A). At high concentrations the haemolymph approached isosmoticity with the medium but was still slightly hyperosmotic at the highest medium concentration still permitting survival. It is incapable of maintaining the osmotic pressure of its body fluid below that of the ambient medium (i.e. incapable of *hypo-osmotic regulation*) and unable to tolerate very high salinities (the upper limit of salinity tolerance in this case was about 12‰). Beadle (1943, 1959, 1969) pointed out that the upper limit of salinity tolerance in 'freshwater' animals is related to the concentration of body fluids exhibited when in fresh water; the higher the normal blood concentration the higher the salinity to which these animals can be adapted, at least in those species incapable of hypo-osmotic regulation. He has further pointed out that most freshwater animals are unable to hypo-osmoregulate and are usually restricted to salinities below about 10‰. This restriction probably results from irreversible adaptation of the tissues to a dilute body fluid or the inability to obtain sufficient water for excretion from anything other than a hypo-osmotic medium. It

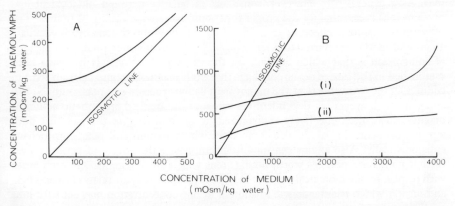

Fig. 1:5 A: hyperosmotic regulation in a freshwater animal (*Asellus aquaticus*). *After* Lockwood (*1959*). B: hypo-osmotic regulation in salt-lake animals (i) *Haloniscus searlei*, (ii) *Artemia salina*. *After* Bayly and Ellis (*1969*)

might be pointed out that, even with this restriction, many, if not most, 'freshwater' animals are not restricted to fresh water.

It seems that the highest salinity to which any unmodified freshwater animal can adapt is about 30‰. Saturated saline waters may have a salinity more than an order of magnitude greater than this and contain living organisms. One special adaptation permitting existence in highly saline athalassic water is hypo-osmotic regulation (defined above). This not only provides the tissues with a concentration low enough for efficient functioning but it also insulates them from the extreme variations in salinity that are especially characteristic of such habitats (see above). Perhaps the most famous example of a hypo-osmoregulator is *Artemia salina* (Fig. 1:5B). This species has a body fluid concentration of slightly less than 1·0 Osm/kg when in a medium with a concentration of 9·5 Osm/kg. The Australian *Parartemia* behaves in a similar manner (M. C. Geddes, unpublished). Another interesting Australian example is the oniscoid isopod *Haloniscus searlei* Chilton (Bayly and Ellis 1969) (Fig. 1:5B). However, it should not be thought that hypo-osmotic regulation is an essential adaptation for existence in highly saline waters; almost complete conformity between body fluids and external medium over a wide salinity range also occurs. Thus calanoid copepods *Calamoecia salina* and *C. clitellata*, which occur in Australian salt lakes, are very eurysaline but are essentially osmo-conformers (Bayly 1969a). The same is probably true of salt-lake ostracods. In such cases it may be presumed that either cellular osmoregulation or exceptional cellular tolerance has been evolved.

BIOLOGICAL SIGNIFICANCE OF IONIC PROPORTIONS (ION ANTAGONISM, IONIC REGULATION, AND IONIC EQUILIBRIUM)

It has been known since the time of Ringer (1883) that solutions containing only a single cation and a single anion are very toxic. Solutions of pure sodium chloride have thus been shown to be injurious to a wide range of organisms, and in general potassium solutions are even more toxic. Pure solutions of calcium chloride seem not to be universally toxic; there are cases in which organisms can function without apparent injury with the presence of no other cation except calcium and, of course, hydrogen ions. As a general rule, however, some sort of balance between monovalent and divalent ions is essential for the proper functioning of a wide variety of tissues. The offsetting of the toxic effects of an excess of one ion or group of ions by an adequate amount of another is referred to as *ion antagonism*. As already suggested, a simple type of antagonism is that which depends on the valency of cations. It has thus been established in the case of sea-urchin eggs that both magnesium and calcium tend to counteract the toxic effects of the monovalent cations sodium and potassium. Not all antagonism, however, is dependent on valency; the marked anaesthetic action of magnesium is often effectively antagonized by calcium despite the fact that both ions are bivalent. There are also some cases where sodium and potassium behave antagonistically. Also anions can modify or emphasize the effects of cations. Two further points are perhaps worth noting: ions which are antagonistic to each other with respect to one biological process may act synergistically in relation to another, and an ion which is antagonistic to another at one concentration may act with it at another concentration.

The ionic proportions found in sea water, diluted sea water (brackish water of estuaries), and concentrated sea water (hypersaline lagoons) are fairly constant and physiologically well balanced. In athalassic waters, however, enormous variation in ionic proportions occurs, and actually or potentially toxic combinations may occur. Thus in some highly alkaline and saline waters found in volcanic regions the ratio of monovalent to divalent cations is very high, and so too is the ratio of bicarbonate plus carbonate to other anions. Such lakes are very probably toxic to some organisms.

Ionic regulation occurs extensively if not universally in aquatic animals (and is probably phylogenetically older than osmotic regulation); even in those organisms which show little or no osmoregulation, such as the majority of marine invertebrates, differences in ionic proportions between body fluid and external medium usually exist. Thus, although sea water may be regarded as a fairly 'balanced' medium, a large number of marine invertebrates maintain the internal concentration of magnesium considerably below that in the medium. The ability to perform ionic regulation evidently permits many organisms to maintain, in a wide variety of athalassic waters, internal ionic proportions that are non-toxic to their cells and tissues, despite wide external variations. Nevertheless the problems of ionic regulation are very probably greater in athalassic than thalassic waters, and there is evidence that toxic situations do occur.

Beklemishev and Baskina-Zakolodkina (1933) investigated the cause of the lower toxicity of the Aral Sea, as compared with the Black and Caspian Seas, with respect to the cladoceran *Daphnia 'pulex'*. These lakes differ, of course, in salinity, but the samples were diluted or concentrated so that isosmotic triplets of solutions were used in the experiments. In addition, the solutions were adjusted to a standard pH value, although this appeared to be unimportant. They considered the antagonism between monovalent and divalent cations, chloride and sulphate, and magnesium and calcium, and concluded that the ratio of magnesium to calcium was of prime importance in accounting for their results. In Aral Sea water this ratio was low ($1 \cdot 34$), but in the Caspian and Black Seas it was distinctly higher ($2 \cdot 51$ and $3 \cdot 12$ respectively). It was thought that Cladocera were absent from Lake Tanganyika and that this was caused by large quantities of magnesium and sodium and small amounts of calcium. Many Cladocera occur in the Lofu River which flows into Tanganyika, but on reaching the lake they apparently die. However, Fryer (1969) noted that cladocerans do occur in Tanganyika but are restricted to a few species and to shallow regions. Beklemishev and Baskina-Zakolodkina also noted that pure solutions of magnesium sulphate were highly toxic to the cladocerans *Daphnia 'pulex'* and *Simocephalus exspinosus* and the calanoid copepod *Diaptomus coeruleus*, but that this was greatly reduced by the addition of calcium chloride. The cladoceran *Polyphemus pediculus*, however, does not find pure solutions of magnesium salts highly toxic and its survival time is not lengthened by the addition of a calcium salt. In this connection it is interesting to note that the Polyphemidae is the only cladoceran family which is widely distributed in the sea (it also occurs in the Caspian Sea) and it is clear that it has evolved an immunity to relatively high levels of magnesium salts.

Boone and Baas-Becking (1931) and Croghan (1958) produced evidence that large amounts of potassium or (bi)carbonate, or both, are toxic to *Artemia salina*. Cole and Brown (1967), on the other hand, showed that this species does occur in some waters with a high concentration of these ions and from which it should be excluded according to earlier work. The apparent conflict in these observations may be explicable on the basis that the taxonomy of *Artemia* is unsatisfactory and that different workers have used different physiological races, or subspecies, or possibly even species. It seems likely, therefore, that excessive amounts of potassium or (bi)carbonate are indeed toxic to certain forms of *Artemia*. Bowen (1964) found that water from Mono Lake, California, which contains large amounts of (bi)carbonate, were lethal to *Artemia* from carbonate-poor waters (Great Salt Lake, Utah, and commercial salterns in California).

Bayly (1969c) discussed the occurrence of the typically freshwater calanoid copepod *Boeckella triarticulata* in saline waters containing large amounts of (bi)carbonate and a high ratio of monovalent to divalent cations, and the exclusion of the halobiont calanoid *Calamoecia clitellata* from the same waters. A number of hypotheses can explain this situation: large amounts of (bi)carbonate *per se* permit an upward extension of the salinity range of *B. triarticulata* but are toxic to *C. clitellata*; within a certain salinity spectrum a medium strongly dominated by monovalent cations is toxic to *C. clitellata* but not *B. triarticulata*; (bi)carbonate is capable of rendering an unbalanced ratio of monovalent to divalent cations non-toxic to *B. triarticulata* but not to *C. clitellata*.

There is some evidence (Hutchinson 1932, Horne 1967) that some freshwater branchipods are eliminated from certain saline waters not because of a low salinity tolerance but because they are intolerant of high magnesium concentrations.

There is a group of species referred to as *natronophils* which, if not limited to athalassic waters with high concentrations of (bi)carbonate, seem to show a marked preference for them. Löffler (1961) includes in this category the following: *Spirulina platensis* (Cyanophyta); *Hexarthra jenkinae*, *H. fennica medica*, *Branchionus novaezelandiae hungaricus* (Rotatoria); *Lovenula africana*, *Diaptomus transvaalensis*, *D. spinosus* (Copepoda); possibly *Cypridopsis inequivalva* (Ostracoda); and *Aedes natronius* (Diptera). It may be noted that the whole of the rotifer genus *Brachionus* is limited almost entirely to alkaline water, and some of its species may be very abundant in extremely alkaline waters. Thus *Brachionus plicatilis* occurs in abundance in several lakes near Alvie, Victoria, all of which have a pH in excess of 9·0 (Bayly 1969c). *Brachionus caudatus* f. *vulgatus* is an important rotifer in the plankton of Lake Aroarotamahine, New Zealand, which always has a pH in excess of 8·0 (Bayly 1962).

Although it is stated above that ionic proportions in diluted sea water are fairly constant, this no longer holds when it is so highly diluted as to be fresh or almost so (say less than 3‰). Thus two adjacent coastal lakes with closely comparable salinity may have marked differences in flora and fauna that are explicable in terms of differences in ionic ratios. Pora (1969) quoted an interesting case involving certain pools and lagoons of the Razelm complex of lakes to the south of the delta of the Danube River. The salinity of the Babadag lakes, and of Big and Little Razelm, is almost the same (in the range 1·0–2·5‰). If the ratios between monovalent and divalent ions are calculated, values of 3·3 for Little Razelm, 3·7 for Babadag, and 5·5 for Big Razelm are obtained. Barnacles are found in Little Razelm and Babadag, but they are lacking in Big Razelm in spite of the large connection that exists between the two Razelms. This difference may be attributed to the difference in ionic ratio, and this in turn to the nature of the connections between Little and Big Razelm, one with fresh Danube water, the other with marine water arriving by the Strait of Portita.

Khlebovich (1968) pointed out that there is a distinct relative increase of calcium, with concomitant ecological significance, in brackish waters at salinities below about 5‰. It might be pointed out, however, that Australian river waters contain smaller relative amounts of calcium than European rivers, so that sea-water ionic

CHEMICAL FEATURES

ratios remain constant to lower values of salinity than this. This is one reason, additional to that of the ignorance of salinity-temperature interaction (Den Hartog 1960), why the Venice System of brackish water classification is not universally applicable.

Dissolved Organic Matter

This has usually been determined by measuring the amount of potassium permanganate (or some other oxidizing agent) required to oxidize the organic matter present under standard conditions of temperature and acidity. A large number of determinations of this type show that about 10–20 mg of oxygen per litre are consumed by matter that does not contribute significantly to the colour of the water (Fig. 1:6). This is the basis for the common practice of considering total dissolved organic matter as consisting of the following two portions:

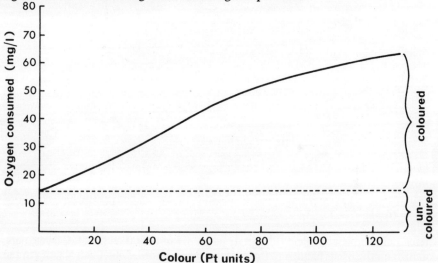

Fig. 1:6 Mean relationship between dissolved organic matter (as measured by the permanganate oxygen consumed method) and water colour expressed in platinum units

1 Uncoloured and derived from the decomposition of plant material, especially phytoplankton, that has developed within the lake or pond. It is thus also *autochthonous* organic matter. This contains about 24 per cent crude protein, and has a carbon to nitrogen ratio of about 12 to 1.

2 Coloured and consisting mainly of derivatives from peat surrounding the lake or pond, or from peat bogs and soils somewhere within the catchment area. It may thus also be described as *allochthonous* organic matter. This contains only about 6 per cent crude protein, and has a carbon to nitrogen ratio of about 45 to 1.

The biological significance of dissolved organic matter is discussed in more detail below (see pp. 74–6).

Hydrogen Ion Concentration

RANGE IN NATURAL WATERS

The pH of the vast majority of natural inland waters lies within the range 4·0–10·0, but even more extreme values have been recorded. Early in 1966 the Mount Ruapehu

Plate 1:1 Mount Ruapehu Crater Lake, North Island, New Zealand. The pH of this lake is sometimes less than 1·0. Ruapehu Peak is shown in the left foreground. *Photograph by New Zealand Herald*

Crater Lake in New Zealand (Plate 1:1) had a pH of only 0·9[1] (R. B. Glover, pers. comm.); on the other hand, a pH of 12·0 has been recorded for Lake Nakura in the African Rift Valley (Jenkin 1932). Small amounts of sulphuric acid seem to be responsible for some bogs and bog lakes having a pH in the range 3·0–4·0. However, in dilute waters containing large amounts of coloured dissolved organic matter, humic acids alone may take the pH down to 4·0 (Bayly 1964) or probably even a little less. Shapiro (1957) showed that an aqueous solution of a purified preparation of humic or 'limno-humic' acid may have a pH as low as 3·6. Values of less than 3·0 can probably be reached only in presence of considerable amounts of strong inorganic acid. On the other hand, readings in excess of 11·0 imply the presence of hydroxide, but values from 9·0–11·0 may be due to large amounts of sodium carbonate which is highly soluble, or to a temporary displacement of the normal equilibrium because of intensive photosynthetic activity. Leaving aside highly acidic or alkaline situations, pH is regulated by the CO_2-HCO_3-CO_3 system and usually lies in the range 6·0–9·0.

---

1. It was subsequently even more acidic but pH was not measured. On 24 January 1967 acidity to pH 3·8 and 9·3 was 320 meq/l and 575 meq/l respectively. On this occasion the major anions and their concentrations were: chloride 10·2‰, sulphate 17·8‰. The lake sometimes overflows the discharges into Whangaehu River and has been known to lower the pH to 2·5 at Tangiwai some 30 km downstream.

CHEMICAL FEATURES

ECOLOGICAL SIGNIFICANCE

Despite the fact that the measurement and recording of pH has long been, and still is, very fashionable with limnologists, the biological and ecological importance of this factor itself seems not to be very great. Hutchinson (1941) seriously doubted whether it had been satisfactorily demonstrated that pH *per se* influences the natural abundance of any aquatic animal with the exception of the protozoan *Spirostomum ambiguum*. This species does seem to be limited to a pH of less than 7·8, and some other species may be restricted in a comparable manner. Most species, however, tolerate a wide range of pH. Thus, for example, Lowndes (1952) pointed out that entomostracans tolerate wide variations of pH and recorded the common cosmopolitan copepod *Mesocyclops leuckarti* from a range of 3·4–9·8. Hutchinson has suggested that some independence of external variations in pH may have been one of the first requirements in the evolution of higher organisms.

It is thus true that pH has been overemphasized as an ecological factor but this does not mean that it is not worthy of measurement; it frequently does give a clue to the existence of other factors that may be highly significant. For example, very low pH waters often contain large amounts of dissolved organic matter and small amounts of calcium and these latter two factors may be significant. Furthermore, when the total alkalinity of a habitat is constant, pH change is proportional to CO_2 change and is therefore a useful measure of the latter. The important thing is not to confuse correlation and causality, and to exercise considerable caution in attributing an observed distribution or abundance to pH *per se*.

Despite the relative insignificance of pH in most ecological investigations, there is some evidence (Baylor and Smith 1957, Bayly 1963) that unusually high pH (9·0 or above) influences the manner in which planktonic crustaceans react towards fluctuations in light intensity. At least two reversed diurnal vertical migrations have been recorded in highly alkaline lakes.

Nitrate and Phosphate

Nitrogen is an important constituent of a wide variety of organic compounds and is especially important in the polypeptide linkage on which the formation of proteins is dependent.

Some nitrate (from the oxidation of nitrogen dioxide) and ammonium nitrogen is present in the atmosphere and is washed down in small quantities in rain water. Despite traditional teachings, lightning now appears not to be of importance in forming atmospheric nitric acid. Appreciable amounts of nitrogen are fixed in the soil by nitrogen-assimilating bacteria, and this not only becomes available for terrestrial plants, but may be leached out of the soils of catchment areas and into lakes. In addition, nitrogen-fixing bacteria and blue-green algae are usually active within the water itself.

The main decomposition product from the proteins of living organisms is ammonia, but in the presence of oxygen this is quickly converted into nitrate by bacteria. Thus, in high-oxygen waters unpolluted by human activity, nitrogen is present almost entirely as nitrate and typically in amounts of less than about 1 mg/l. In eutrophic[2] lakes, however, the amount of nitrate present is often in excess of 1 mg/l, except in the epilimnion[3] towards the end of the summer stagnation period, and in oxygen-

---

2. Defined on p. 64.

deficient hypolimnia[3] where it is common for large amounts of ammonia to be present.

Phosphorus is an important constituent of many proteins including DNA, and of ADP and ATP which play a key role in energy transfer in biological systems. It is also very important in determining the amount of living material that can exist in a lake (see Chapter 4).

Although phosphine is possibly formed in some marshes, phosphates are probably the only phosphorus compounds of limnological importance. Unlike nitrogen, phosphorus is not atmospherically supplied. It ultimately comes from the weathering of phosphatic rocks (mainly fluorapatite, $Ca_{10}(PO_4)_6F_2$) and from the soil. However, part of the supply to aquatic environments is indirect in that the living matter of terrestrial organisms intervenes. Again unlike nitrate, phosphate is strongly held by soil and is not easily leached away.

Although phosphate exists in lake waters in a number of different forms (orthophosphate, polyphosphates, and residual phosphates, each of which may be in the liquid, colloidal, or solid phase) most of these are highly variable quantities, and the sum of the different fractions or *total phosphorus* seems the most fundamental and useful measure. Most uncontaminated lakes contain 10–40 mg/m^3 of total phosphorus in their surface waters. The terminology of the different types of phosphate is discussed by Olsen (1967).

Phosphate, and to a less extent nitrate, are discussed further under lake productivity (Chapter 4). Micronutrients (nitrate and phosphate may be regarded as macronutrients) are also discussed in Chapter 4.

Dissolved Oxygen and Carbon Dioxide

In the terrestrial environment oxygen is not usually an important limiting factor. In aquatic environments, however, it is frequently a limiting factor of major importance. Although oxygen is more soluble in water than is nitrogen, the maximum volume of oxygen that a given volume of water can hold is much less than that present in the same volume of air. Thus, if 21 per cent by volume of air is oxygen, there will be 210 ml of oxygen per litre (cf. 79 per cent or 790 ml per litre for nitrogen). By contrast, a litre of water in equilibrium with air at 20°C contains only 6·4 ml of oxygen (and 12·3 ml of nitrogen). This initial scarcity of dissolved oxygen imposed by physical processes is often intensified by subsequent chemical events. This is particularly true of waters containing large amounts of organic matter. Indeed, it has been mentioned above that dissolved organic matter is usually estimated by the amount of oxygen it consumes from an oxidizing agent such as potassium permanganate.

Dissolved oxygen can be determined with fairly great accuracy by the well-known Winkler method or some modification of it suited to the particular circumstance. Modification of the basic method is particularly necessary when appreciable amounts of nitrites or large amounts of dissolved organic matter are present. If appreciable (bi)carbonate is present, as in certain salt lakes, then the Miller method of determination may be used (Walker, Williams, and Hammer 1970). The solubility of oxygen increases with decreasing temperature (Table 1:5), and also depends on pressure. Limnologists are often concerned not so much with the absolute amount of oxygen present per unit volume of water but with *percentage saturation*. If the actual

---

3. Defined on p. 78.

concentration of oxygen is found by analysis to be O, then the percentage saturation is given by $100 \times O/O_s$. Values of O_s for temperatures not listed in Table 1:5 may be obtained by interpolation. For samples taken from localities that are appreciably above sea level, values obtained for percentage saturation should be corrected for altitude (pressure) by multiplication by the appropriate factor extracted from Table 1:6. It should be noted that if intensive photosynthesis is proceeding when a sample is taken, supersaturation or values for percentage saturation in excess of 100 may be obtained.

Solubility of oxygen is influenced not only by temperature and pressure but also by salt content. In Australia this is not just of academic interest but must be a factor of profound ecological and physiological significance in the numerous highly saline

TABLE 1:5
SATURATION VALUES (O_s) FOR DISSOLVED OXYGEN AT NORMAL PRESSURE (760 MM Hg) IN 1 LITRE OF PURE WATER AT DIFFERENT TEMPERATURES
After Truesdale, Downing, and Lowden 1955

Temp. (°C)	O_s (mg[a])	Temp. (°C)	O_s (mg)	Temp. (°C)	O_s (mg)	Temp. (°C)	O_s (mg)
0	14·18	10	10·92	20	8·88	30	7·48
1	13·77	11	10·67	21	8·72	31	7·37
2	13·38	12	10·44	22	8·56	32	7·26
3	13·01	13	10·22	23	8·41	33	7·14
4	12·67	14	10·00	24	8·26	34	7·04
5	12·34	15	9·79	25	8·12	35	6·93
6	12·03	16	9·60	26	7·98	36	6·83
7	11·73	17	9·41	27	7·85	37	6·74
8	11·45	18	9·22	28	7·72	38	6·64
9	11·18	19	9·05	29	7·60	39	6·55

[a] $1·0$ mg $O_2/l = 0·7$ ml O_2/l.

TABLE 1:6
CORRECTION FACTORS FOR ALTITUDE FOR PERCENTAGE SATURATION OF OXYGEN
After Mortimer 1956

Altitude (m)	Correction factor	Altitude (m)	Correction factor
0	1·00	1,000	1·13
100	1·01	1,100	1·15
200	1·03	1,200	1·16
300	1·04	1,300	1·17
400	1·05	1,400	1·19
500	1·06	1,500	1·20
600	1·08	1,600	1·22
700	1·09	1,700	1·24
800	1·11	1,800	1·25
900	1·12	1,900	1·26

inland waters. Animals such as *Haloniscus searlei*, *Calamoecia salina*, *Parartemia zietziana*, *Diacypris*, and *Platycypris* must be adapted for respiration in very low concentrations of oxygen. Indeed, the upper limit of tolerance of these animals in concentrated waters (Tables 10:2 and 10:3) may be determined more by decreasing availability of oxygen than by increasing salinity *per se*. Kinne and Kinne (1962) working on the desert pup fish *Cyprinodon macularius* showed that although initial data suggested that increased mortality was brought about by intolerance of high salinities, further analysis showed that the real cause was the concomitant decrease in the amount of oxygen. Because of the almost infinite variation in the ionic proportions of inland waters, it is difficult to give universally applicable data on the quantitative aspects of diminution of oxygen solubility with increased salt content. It is probably best to adopt the empirical approach of analytically determining the saturation levels for oxygen at a series of dilutions or concentrations of a particular saline water under consideration. Nevertheless, for the reason that many Australian inland saline waters closely approach sea water in the relative proportions of major ions, data relating to diluted and concentrated sea water (Fig. 1:7) is of interest and utility. It can be seen that the saturation value for oxygen at a salinity of 85‰ and a temperature of 30°C is only 3·1 ml/l, whereas for fresh water at the same temperature it is 5·5 ml/l. A salinity of 85‰ is only about half the upper limit of the range tolerated by several of the species mentioned above. The relationships in Fig. 1:7 cannot be extrapolated linearly above 100‰ as a zero concentration of oxygen would be predicted at *ca.* 180‰ and 15°C, whereas G. W. Brand (unpublished data) has found a saturation concentration of 2·8 ml O_2/l at 189‰ and 15°C for water from a salt lake with an ionic composition not greatly dissimilar to that of sea water.

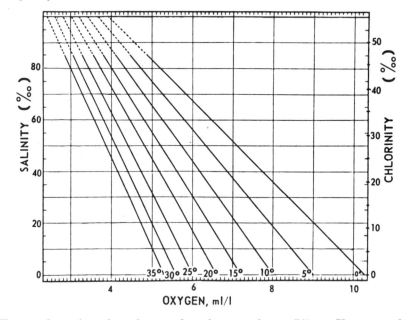

Fig. 1:7 Saturation values of oxygen from dry atmosphere at 760 mm Hg pressure for salt concentrations (diluted or concentrated sea water) up to 85‰ salinity at eight different temperatures. *From Kinne and Kinne (1962)*

CHEMICAL FEATURES

Spatial and temporal changes in the distribution of oxygen in lakes receive further attention in Chapter 5 and oxygen in running waters is discussed briefly in Chapters 2 and 7. Here it will suffice to note that in warm monomictic lakes (the most important type in Australasia [see Chapter 5]) that are fairly highly productive the oxygen content in summer decreases with increasing depth in a manner very similar to the temperature-depth curve (Fig. 5:1A). Such an oxygen-depth curve is said to be *clinograde*. In the same sort of lake in winter, and in relatively unproductive ones in summer, the oxygen content remains almost constant from top to bottom (an *orthograde* curve).

Although present in only small quantities in air, carbon dioxide is extremely soluble in water. In addition to diffusion from the atmosphere, carbon dioxide also finds its way into water from decay processes, and from respiration. Most important, the action of carbonic and other acids on rocks and soils (see Fig. 1:4) results in an influx of bicarbonates and carbonates (which may be regarded as carbon dioxide in ionic form), accompanied by equivalent amounts of mainly divalent cations, into inland waters. From the animal viewpoint, large rather than small amounts of carbon dioxide are likely to act in a limiting manner. However, in that high concentrations of carbon dioxide are usually associated with small amounts of oxygen and large amounts of organic matter, one should be wary of attributing any toxicity stemming from such a combination to carbon dioxide alone. Nevertheless there is evidence that large amounts of free or non-ionic carbon dioxide may be the direct cause of mortality in fish. Increased amounts of carbon dioxide, bicarbonate, and carbonate almost certainly stimulate photosynthesis, but each of these forms would act more or less independently in that a specific group of plants is likely to be able to use one form only. The significance of the availability of carbon in one or other of several different forms (including carbon dioxide, bicarbonate, and carbonate) as a factor influencing lake productivity is discussed at some length in Chapter 4.

The distribution of carbon dioxide and oxygen in inland waters is nearly always of an antagonistic type. Thus in lakes the stratification of carbon dioxide is of the opposite type to that of oxygen; the epilimnia of thermally stratified lakes commonly have large amounts of oxygen and small amounts of carbon dioxide, whilst for hypolimnia the reverse is often true.

chapter two
BIOLOGICAL PRODUCTION

General Considerations

Leaving aside a definition for the moment, there is no questioning the fact that 'production' has become a central (if not *the* central) theme of modern ecology.[1] Insofar as freshwater ecologists, by providing empirical data and refining concepts, have contributed more significantly to this theme than most other sorts of ecologists, a chapter such as the present is not inappropriate, even though part of it deals with principles which are of general applicability to other environments.

Intuitive ideas of production in fresh waters have, of course, been held for a long time by men concerned with amounts of fish produced by a given water body; in general these ideas related the mean weight of fish from a specified water to a defined interval of time. It was not, however, until the 1930s that such ideas were applied on a wider basis to the biota of inland waters. Once they were, their importance was rapidly accepted, and in the next twenty years or so many papers appeared applying, defining, and refining the concept of biological production as related first to inland waters in general, and then, more significantly after elaboration of the ecosystem concept (e.g. by Tansley (1935) and others), to inland waters as more or less discrete, functional, ecological units of the biosphere. The most important contributions during these two decades were made by Thienemann (1931), Strøm (1932), many Russian investigators including particularly Winberg (1936), Borutzky (1939a, b), and Ivlev (1945[2]), Juday (1940), Hutchinson (in Lindeman 1942), Lindeman (1942), and Macfadyen (1948).

The most significant contribution was Lindeman's[3] attempt to relate energy content, trophic structure, and energy transferences within an ecosystem. His ideas on this subject represented a substantial advance in our conceptual approach to ecosystems despite some later criticisms, such as those levelled at his assumptions on

---

1. Limnology itself has even been defined (over thirty years ago!) as 'That branch of science which deals with biological productivity of inland waters and with all the causal influences which determine it' (Welch 1935: 10).
2. For a translation (by W. E. Ricker) of Ivlev's 1945 paper see Ivlev (1966).
3. Although Lindeman is usually given sole credit for the views expressed, it seems to us that some injustice is thereby done to G. E. Hutchinson who clearly played an important role in their formulation.

respiratory and turnover rates, and at balances of his energy budgets (Smith 1955, Slobodkin 1962, Ricker [in footnote in Ivlev 1966]).

Briefly, Lindeman's concept was that ecosystems consist of several trophic levels, the energy content of each being $\Lambda_1, \Lambda_2, \ldots \Lambda_n$. Λ_1 represents the energy content at the autotrophic level. Energy flows unidirectionally between levels, the rate of flow into Λ_1 being λ_1, into Λ_2, λ_2, and into Λ_n, λ_n. Energy is also lost at each level as a result of work. The rate of this loss ($= R_1, R_2, \ldots R_n$) plus the rate of energy passed on to the next trophic level (respectively, $\lambda_2, \lambda_3, \ldots \lambda_{n+1}$) constitutes $\lambda_1^1, \lambda_2^1, \ldots \lambda_n^1$. Thus $\lambda_n^1 = R_n + \lambda_{n+1}$, and the rate of change of energy content at the nth trophic level is

$$\frac{d\Lambda_n}{dt} = \lambda_n + \lambda_n^1.$$

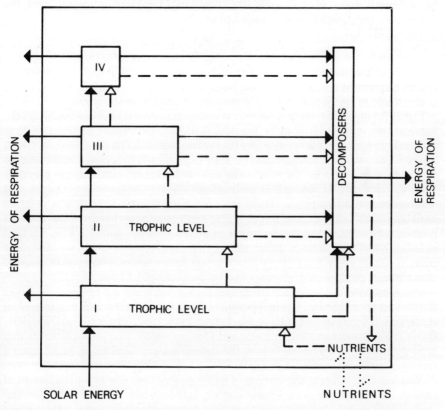

Fig. 2:1 Generalized form of energy relationships and nutrient pathways in an ecosystem. Solid arrows = energy flow; broken arrows = nutrient flow; dotted arrows = potential exchange flow

The energy relationships between trophic levels, together with an indication of nutrient pathways, are expressed in a more general way in Fig. 2:1. Here the amount of energy passed on from a given trophic level constitutes the production of that level. Adhering to the symbolic conventions of Ricker (1968) based on terms approved by the International Biological Programme for trophic levels II, III, IV, and the decomposers (i.e. all heterotrophs):

$$P_s = C - F - U - R$$

where P_s = secondary production, C = consumption (total intake of food by organism in specified period), F = egesta or faeces (food not absorbed into body of organism), U = excreta (material absorbed but then removed from body), and R = respiration (that part of assimilated food used in maintaining life processes by conversion to heat or mechanical energy). Initially P_s may represent merely increase in biomass (i.e. growth) of a given trophic level, but ultimately it represents the amount of energy passed on to the next trophic level and/or decomposers (except in some situations, e.g. peat bogs and even lake bottoms, where organic matter, and hence energy, accumulate in quite large amounts). For the first trophic level, the autotrophs, the equation may be modified to:

$$P_p \text{ (net)} = P_p \text{ (gross)} - R - U$$

where P_p (gross) is the initial amount of light energy fixed. Absolute amounts of energy transferred are, of course, meaningless without being related to time, and it is important to stress that values for production should be so related.

Figure 2:1 also indicates that not only is energy transferred between trophic levels, nutrients are as well. It is quite feasible, therefore, to evaluate material budgets (e.g. carbon, nitrogen, phosphorus) for the various trophic levels of a given ecosystem. There is, however, no nutrient loss (at least theoretically in a closed ecosystem) to an unavailable, degraded form as is the case with energy; nutrients are recycled. Thus $C = P + F + U$, where P = production.

Consideration of empirical results from a natural aquatic ecosystem is deferred until later in this chapter, but it is as well to emphasize here that the relative lack of prominence in Fig. 2:1 of the decomposers as heterotrophs should not be regarded as indicative of their relative unimportance in natural ecosystems; as we shall show later, in certain ecosystems energy flow through decomposers is far more substantial than through the so-called 'grazing chain' of heterotrophs.

Much semantic confusion has arisen over the various terms used by limnologists in discussions of production. It is inappropriate for us to discuss this in detail (interested readers are referred in particular to Macfadyen 1948, 1957; Elster 1954a; Ohle 1956; Davis 1963; Steemann Nielsen 1965; and Dussart 1966). The principal terms that need definition are discussed below, the definitions provided being based upon a variety of modern sources.

Standing crop refers to the amount of living material existing at a given instant of time in unit volume or area. Biomass, stock, standing-crop biomass, and standing stock may be regarded as synonyms,[4] and organomass (after Davis 1963) as the amount of living and non-living organic material at a given instant of time in unit volume or area. Values for standing crop may be expressed in several ways:

---

4. 'Biomass' and 'standing crop' are not always so regarded; for some authors the biomass of aquatic macrophytes includes the above-sediment portion and rooting organs, whereas the standing crop refers only to the biomass above the sediment line. This usage is confusing.

1 as wet weight. It is important to recognize the difference between true wet weight and experimental wet weight.
2 as dry weight. Material is usually dried at temperatures ranging from 60 to 110°C with 105°C the usual one, although freeze-drying or drying over desiccants is preferable. In the former case, temperatures *should* exceed 100°C.
3 as ash-free dry weight. This is important when a significant part of the animal or plant material concerned is inorganic skeletal matter, as in molluscs or diatoms. Ashing should be at 550°C.
4 as organically combined carbon.
5 as estimates from nitrogen, phosphorus, or specific organic constituents.
6 as estimates from pigment (usually chlorophyll) analyses. This method of expression applies particularly to estimates of phytoplankton standing crops even though, following the use of conversion factors, estimates correct to little more than an order of magnitude are to be expected. Chlorophyll estimates are, however, useful *indices* within one locality.
7 as dimensions. Nauwerck (1963) has determined the volume of numerous plankton species.
8 as units of energy (calories or kilocalories), measured by calorimetry (cf. Table 2:1).
9 as numbers.

TABLE 2:1
CALORIFIC EQUIVALENTS FOR SOME AQUATIC BIOTA
Summary data only[a]

Group	Calories per gramme dry weight	Number of values averaged	Calories per ash-free gramme dry weight	Number of values averaged
Algae	4,943	11	5,130	6
Vascular hydrophytes	3,374	10	4,345	5
Microconsumers (*E. coli* only)	5,028	1	5,520	1
Invertebrates (grand mean)	2,933	32	5,490	56
Crustacea	3,115	14	5,504	23
Insecta	5,388	13	5,692	7

[a] Extracted from Cummins (1967). Values based on whole or parts of organisms.

Various conversion factors have been published which permit limited expression of each measure of standing crop in terms of any other. However, the factors cover so wide a range that great caution is needed in their use, and many for most practical purposes are relatively useless.

Methods of obtaining samples for standing-crop determination vary of course with the nature of the standing crop. For plankton a tried and proven, but by no means perfect, method is to centrifuge a large volume of water after it has been filtered through a plankton net to exclude net-plankton. The residue in the net and the centrifugate are then combined. Strictly, this gives a sample of *seston*, not plankton (seston is the total amount of suspended matter, living and non-living, i.e. the living plankton plus organic and inorganic *tripton*). The standing crop of zooplankton, or at least of the larger zooplanktonic species, can be sampled with the Clarke-Bumpus plankton sampler which has a flow-meter in its throat, thus enabling determination of the water volume filtered by the net whilst obtaining the catch. This volume is independent of the degree of clogging of the net by phytoplankton. In sampling, the net is hauled vertically throughout the entire trophogenic[5] stratum (preferably more than once), or towed horizontally at several suitably spaced depths between the top and bottom of this stratum, and a mean obtained. The standing crop of benthos is usually sampled with an Ekman-Birge grab or tube or core sampler. Raw samples may be washed through sieves of decreasing mesh size to retain appropriately sized organisms. Tube samplers, it may be noted, have many advantages over most other sorts of benthic sampling apparatus (Brinkhurst, Chua, and Batoosingh 1969). Details of sampling methods applicable to standing crops of other communities may be obtained by reference to Welch (1948), Schwoerbel (1966), Lagler (1968), and Vollenweider (1969).

Actual values for standing crops in inland waters display a considerable range. Some of the highest relate to saline or near-saline lakes. Thus Rawson and Moore (1944) gave a value of 573 kg ha^{-1} (experimental wet weight) for the benthos standing crop of a Saskatchewan saline lake.[6] And in one area of Cooking Lake, Alberta, Kerekes and Nursall (1966) reported a value of 71·1 mg l^{-1} for the average summer organic biomass of surface-water seston, a value said to be near the possible maximum for a north temperate lake under natural conditions. Reported minimal standing-crop values for benthos are <10 kg ha^{-1}, and for organic seston, <1 mg l^{-1}. Some further values for lacustrine standing crops are given in Chapter 4.

Production and *productivity* present the greatest difficulties of definition; there is no concensus on definitions for them, and imprecision in their use by ecologists persists. Without suitable qualification by authors the imprecision is likely to remain for some time since the terms may be used in connection with several different concepts. A definition (still relatively imprecise) of *production* adopted here which agrees with both current popular usage (cf. *The Shorter Oxford English Dictionary*) and its usage by a majority of ecologists is that production is (a) the process of producing ($\equiv$ process of energy storage), *and* (b) the quantity of material, or biomass, produced ($\equiv$ energy stored). It will usually be clear which of these alternatives is intended.

Primary and *secondary production* refer to production by (of), respectively, autotrophs and heterotrophs. With respect to the former, it is necessary to distinguish clearly between *net* and *gross* primary production. As quantities we have already linked the two by the equation:

$$P_p \text{ (net)} = P_p \text{ (gross)} - R - U$$

which is another way of expressing the fact that net primary production is the amount of stored chemical energy remaining from the total amount of light energy fixed (= gross primary production) after plant metabolic energy-requiring needs (=R) have been satisfied, and energy losses (=U) have been taken into account. It is less usual to distinguish between net and gross primary production in the qualitative sense of referring to the processes involved. Agreeing in principle with Ricker (1968), we consider it better not to use the term tertiary production to indicate fish production. Primary and secondary production are considered in more detail below.

---

5. For definition see p. 64.
6. 573 kg ha^{-1} is an equivalent expression for 573 kg per ha or 573 kg/ha. Similarly, mg m^{-3} day^{-1} is equivalent to mg/m^3/day. The former notation is followed in this chapter.

No attempt is made to define *productivity* with any precision at all and we use it, and suggest its use, in a general way to imply the extent of production. Providing such imprecision is recognized, we suggest, in disagreement with Ohle (1956) and Steemann Nielsen (1965), there is still a case for retaining the term. Concepts of *rate* or potentialities are not involved in the above definitions of production and productivity.

Harvest and *yield* may be regarded as synonymous terms referring to that part of any plant or animal biomass produced (thus, production) which is used by man. *Terminal production* is conceptually the same but refers to heterotroph biomass only. Crop is better not used in this context in order to avoid confusion with the concept of standing crop. In inland waters, harvest, yield, and terminal production apply mostly of course to fish production used by man.

Primary Production

The major energy source for the maintenance of aquatic (and terrestrial) ecosystems is light energy from the sun, the light-fixation process usually involving the use of water as a hydrogen donor in the reduction of carbon dioxide to a carbohydrate. This process is not the only sort of photosynthesis that occurs, and a small number of photosynthetic bacteria may use sources other than water for hydrogen. Some purple and green sulphur bacteria, for example, may use hydrogen sulphide as a hydrogen source, and other purple sulphur and non-sulphur bacteria may use organic as well as inorganic hydrogen donors (Fogg 1968). In addition, there are several sorts of bacteria which obtain energy by a process of *chemosynthesis*, that is by deriving energy for organic synthesis from purely chemical transformations. Nitrite and nitrate bacteria, colourless sulphur bacteria, *Thiobacteria*, and iron bacteria are examples.

Whilst the above-mentioned bacteria are autotrophs, it must be remembered that most bacteria are heterotrophs. As such, however, many aquatic bacteria use both autochthonous and allochthonous organic matter that would otherwise be unavailable to animals, and it has therefore been suggested (Sorokin 1965) that bacterial biosynthesis of this nature in inland waters is to be regarded more nearly as primary than secondary production. It has even been claimed (Kuznetzov 1968) that lake bacteria using organic substrates are frequently more important as energy sources for other heterotrophs than photosynthetic autotrophs. At all events, Nauwerck's (1963) investigations on Lake Erken, Sweden, indicated that phytoplankton is of only secondary importance as a zooplankton food source in that lake. Many algae also, it must not be forgotten, are able to use dissolved organic matter as either sole (in the dark) or supplementary (in the light) carbon sources for growth (e.g. Fogg 1965, Wetzel 1968). However, algal chemo-organotrophy has not been demonstrated for *in situ* populations; all evidence is negative with natural organic substrate concentrations.

With reference to autotrophs *sensu stricto*, three principal groups can be recognized according to their size and habitat: phytoplankton, macrophytes, and periphyton. It is convenient to discuss the groups separately (see below). The contribution of each to the total sum of solar energy fixed within a given ecosystem depends upon several factors. Thus, in swiftly flowing waters, the contribution by the phytoplankton is absent or negligible, whereas in large, deep lakes, phytoplankton is the major contributor. In shallow lakes, on the other hand, sensible contributions are provided by all three groups of autotrophs. Wetzel (1964), for example, found that in saline Borax Lake, California, the total annual mean production for the whole lake on a daily basis was 101·0 kg carbon by the phytoplankton, 75·5 by the periphyton, and 1·4 by the macrophytes (6·6 during the growing season). In a given ecosystem the proportion of energy used that is actually fixed by autotrophs *in situ* and by autotrophs in contiguous ecosystems is also variable. Most running-water ecosystems are maintained mainly by allochthonous energy (Hynes 1963).

PHYTOPLANKTON

A discussion of the phytoplankton that occur in lakes and of their interrelationships is given in Chapter 6. All that is necessary in the present context is to note that although the larger forms may account for most of the biomass at any specific time, the *nannoplankton*, that is that part of the phytoplankton population which would pass through a net with the smallest apertures (normally *ca.* 60μ although nets with apertures of 10μ are now available), may be the most important in terms of primary production. This is substantiated by a number of investigations. Verduin (1956b), working on Lake Erie, North America, found that on average 65 per cent of photosynthesis was carried out by algae smaller than 65μ. A similar situation prevails in Lake Erken, Sweden; Rodhe, Vollenweider, and Nauwerck (1958) concluded that only in late summer and autumn do net-phytoplankton sometimes contribute more than 50 per cent towards total primary production. In spring they found that the nannoplankton contribute as much as 95–98 per cent to total production. On occasion, also, species of the nannoplankton may constitute the most substantial part of the algal standing crop as well as being the most productive element—as does *Synechococcus* sp., a blue-green alga of bacterial size, in Loch Leven, Scotland (Bailey-Watts, Bindloss, and Belcher 1968). An analogous state of affairs exists in certain Australian salt lakes although the algae involved are different of course (K. F. Walker, pers. comm.). In Lake Werowrap, Victoria, for example, the total production for 1970 has been by *Gymnodinium aeruginosum* ($\ll 60\mu$) and smaller forms, to the exclusion of algae $> 60\mu$.

Methods for determining the rates and amounts of phytoplankton production fall easily into two sorts: those which involve incubation of samples and those which do not. Both sorts are to a significant extent applicable to the study of marine and freshwater phytoplankton production, and it is not surprising that some overlap occurs in the usefulness of marine and limnological literature dealing with techniques. Particular reference in this connection is made to Strickland (1960, 1965), Doty (1963), Vollenweider (1969), and National Academy of Sciences (1970).

Briefly, incubation techniques involve enclosure of samples in darkened and clear glass bottles[7] either suspended *in situ* in the natural environment or maintained under standard laboratory conditions for a given interval. Changes which occur during the interval as a result of photosynthesis (clear bottles only) and respiration (clear and dark bottles) are then measured. Such a technique was first used by Pütter (1924) and Gaarder and Gran (1927) who recorded changes in oxygen concentrations before and after incubation and from these calculated production. The basic procedure followed by them is still widely used, although it has a rather low sensitivity and cannot be used when the value of production during incubation is less than *ca.* 20 mgC m^{-3} (the

---

7. Conventionally referred to as 'light and dark' bottles.

practical lower limit of detection is when oxygen evolution is $ca.$ 8 mgO_2 m^{-3} hr^{-1}). A very much more sensitive incubation technique measures the amount of radioactive carbon (^{14}C) that becomes incorporated into plant tissue (or at least is transformed from the inorganic to organic form). It was first used by Steemann Nielsen (1952). A third incubation technique that has not yet, as far as we are aware, been applied to freshwater phytoplankton production involves the use of an electronic particle counter (e.g. a Coulter counter) to detect changes in the size frequency and volume of phytoplankton organisms. This technique has been used in determining marine primary production (cf. Cushing and Nicholson 1966, Sheldon and Parsons 1967), and has been used to study certain population characteristics of freshwater phytoplankton.

There are a number of common and separate difficulties with respect to the various incubation techniques. Major difficulties common to all include: (1) those due to problems associated with sampling heterogeneously distributed algal populations; (2) unnatural algal sedimentation following reduction of turbulence and circulation after sample enclosure; (3) modification of normal algal activity subsequent to experimental manipulation (which may have caused excessive turbulence or overexposure to light); (4) sample divergence from initial sample structure as a result of growth, grazing, and similar phenomena; (5) increased bacterial respiration from bacterial growths upon bottle walls; and (6) differences which develop in the chemical nature of the enclosed sample (e.g. in pH, oxygen, carbon dioxide, and nutrient concentrations) from that of the surrounding water. The last three difficulties can be minimized by incubating samples for only short periods (less than six hours is recommended), but this then creates the difficulty of deriving factors for converting production during short periods to a day basis. Alternatively, extremely large—but thereby manipulatively difficult—enclosures may be used over longer periods. So far such an alternative (e.g. a large plastic sphere) has been tried by marine ecologists only (cf. McAllister, Parsons, Stephens, and Strickland 1961).

Additionally, in using oxygen evolution as a measure of production, bubble formation inside bottles may often be troublesome, it is necessary to have some knowledge of photosynthetic quotients (ratio of oxygen evolved to carbon dioxide absorbed), and slight temperature differences between dark and light bottles may significantly affect respiratory rates in the two bottles. In general, however, the method does provide a measure of gross production and an estimate of net production.

In using the sensitive ^{14}C technique, on the other hand, there is still some uncertainty as to what exactly is measured, there are considerable technical difficulties of calibration of the radioisotope and instruments, and, since the technique conventionally measures only particulate carbon, significant amounts of fixed carbon that leave the plant as dissolved organic matter after fixation of inorganic carbon are not measured. Fogg (1969) noted that as much as 35 per cent of fixed carbon may be lost in this manner in oligotrophic lakes, but less than 1 per cent in eutrophic ones. Recent opinion on what is actually measured (e.g. Goldman 1968) is that it is a value close to net production.

One difficulty in the technique involving the use of a particle counter is caused by suspended detritus. Thus detritus generally constitutes more than half of seston in fresh waters and is highly variable both spatially and temporally; counters cannot distinguish between it and phytoplankton. Orifice-clogging is also a problem with most models of particle counters (R. G. Wetzel, pers. comm.).

Methods of determining production, which in large lakes at least is largely due to phytoplankton, that do not need sample enclosure and incubation mostly involve the direct measurement of environmental changes in oxygen and carbon dioxide concentrations and of pH values during relatively short periods (e.g. a day). The results of such methods are relatively imprecise because of uncertainties concerning, for example, rates of gaseous diffusion and degree of water turbulence, and these methods are much less used than are those involving isolated samples.

Values for rates of phytoplankton production are usually expressed in units of mgC m^{-3} hr^{-1} (or mgC m^{-3} day^{-1}) or gC m^{-2} day^{-1}. The first is an expression of the intensity of photosynthesis in a given volume of water, the second an expression of the total intensity under a given area of the surface of the water body being considered. Clearly, values of the first sort will vary in any body of water with depth, and at any given depth with time according to the extent of light penetration. The shape of the curve relating such values to depth *typically* is of the general pattern shown in Fig. 2:2. Excessive surface light has a depressing effect on photosynthetic rates so that maximum production occurs at some depth beneath the surface, while light attenuation with depth gradually depresses photosynthesis until none occurs.

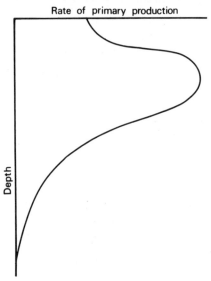

Fig. 2:2 General shape of curve relating rate of primary production to depth

It cannot be too strongly emphasized, however, that photosynthesis-depth curves with other shapes frequently occur, and of course at certain times of the day surface photo-inhibition is absent or negligible. The particular shape of the curve may vary not only between lakes at a given time of day and diurnally within any one lake, but also seasonally within a lake. This is illustrated by Fig. 2:3 (based on Wetzel 1966) which indicates the range of seasonal variations in the shape of the curve relating integrated *daily* production and depth in Goose Lake, Indiana.

Summation of the amount of production at all depths within a water body (in effect, integration of the area enclosed by the photosynthesis-depth curves of the sort shown in Fig. 2:3) provides values for the second sort of expression of phytoplankton

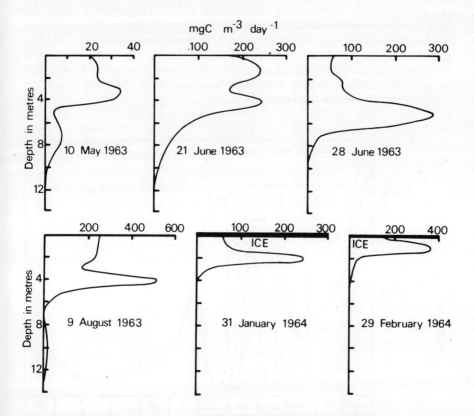

Fig. 2:3 Seasonal variations in the shape of the curve relating rate of primary production to depth in Goose Lake, Indiana. *Redrawn after Wetzel (1966)*

production, total production under unit area of water surface. Such values may of course be expanded to give a measure of absolute phytoplankton production within the whole lake.

The range of rates of phytoplankton production per unit area in lakes is rather considerable. The minimum recorded is 2 mgC m^{-2} day^{-1} (Clear Lake, California, Goldman and Wetzel 1963), the maximum published one, 10 gC m^{-2} day^{-1} (Lake Mariut, Egypt, Vollenweider 1960). Some unpublished values for certain Australian salt lakes exist and exceed this maximum. The value of 41·3 gC m^{-2} day^{-1} for Soap Lake, Washington (Anderson, in Edmondson 1969a), is so much in excess of the theoretical maximum figure of 10–13 gC m^{-2} day^{-1} suggested by Lund (1967) that it cannot be accepted at present without question. Most values applicable to freshwater lakes are less than 1 gC m^{-2} day^{-1}.

In comparing values for production per unit area in different lakes, attention must be given to the depth of the column of water beneath the unit area. It may well be, for example, that values for an apparently productive (eutrophic) but shallow lake are in fact absolutely less than those for a deep and apparently unproductive (oligotrophic) lake (see also Chapter 4); in the latter, light may penetrate to great depths resulting in algal photosynthesis in a very long water column. It is also important to realize that much variation occurs according to season and position within a single lake. Figure 2:4, based on Wetzel's (1966) data for Sylvan Lake, Indiana, clearly shows this. The best comparative values are those for mean annual daily rates.

Factors which govern phytoplankton production are many and varied. Lund (1965) has reviewed nutrient-limiting factors in detail and they and some others are considered briefly later in this book (Chapter 6).

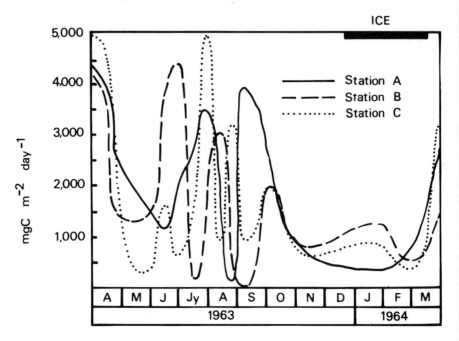

Fig. 2:4   Seasonal and local variations in rates of primary production in Sylvan Lake, Indiana. *Redrawn after Wetzel (1966)*

PERIPHYTON

The periphyton contributes to *in situ* carbon fixation in both still and running waters. In the former, of course, its components (see Chapter 6) occur peripherally and do not constitute the only source of fixed carbon, although, as indicated by Wetzel's (1964) work on Borax Lake discussed previously, they may be an important source. In running waters, on the other hand, the periphyton is frequently a major or the only contributor to carbon fixed *in situ*, and except in deep turbid rivers is not generally restricted to a peripheral position.

Apart from the relative importance of the periphyton contribution within the whole of an aquatic ecosystem, various investigations have shown that periphyton production on a unit area basis must not be underestimated. Pieczyńska and Szczepańska (1966), for example, found that in Lake Mikołajskie, Poland, even within the littoral region covered with emergent macrophytes, periphyton production was more important than phytoplankton production and constituted 23 per cent of total production. And in shallow Borax Lake, Wetzel found that periphyton production per unit area was about three and ten times that of, respectively, phytoplankton and macrophytes. The maximum rate of periphyton production in this lake was determined as $5 \cdot 8$ gC m^{-2} day^{-1}, a value greater than that for maximum phytoplankton production both in this lake and in most other lakes.

In still as well as flowing waters, one method that has been used extensively to estimate periphyton production involves the submergence of artificial substrata, frequently glass and less frequently plastic, for known periods. After submergence the biomass of the adnate material is determined and then related to time. Reviews of this method are given in particular by Cooke (1956), Sládecková (1962), and Wetzel (1964). There are, however, several criticisms that may be raised against a narrow interpretation of the results as periphyton net production. The principal ones (Wetzel 1965), in brief, are that periphyton components have different selectivities towards artificial substrata, they may exhibit different turnover rates, and manipulative and mechanical errors are likely to be large. Moreover, many studies of this sort have entailed positioning the artificial substrata in ecologically unrealistic situations, that is, suspended freely from buoys in open water, and, according to the work of Dumont (1969), have paid too little attention to the effect of orientation of the artificial substrata.

Some results, particularly with respect to lakes, have been obtained using methods similar to those used in studying phytoplankton production. Wetzel (1964) with dark and clear perspex cylinders and ^{14}C methodology studied periphyton production in isolated but *in situ* areas of Borax Lake, and various authors have applied similar techniques but measured changes in oxygen concentration. The greatest drawback in such methods is, of course, the heterogeneous distribution of most periphyton communities. Certain authors (e.g. Goldman, Mason, and Wood 1963) have even suspended detached fragments of the periphyton (small discs of the epilithic algal mat) in conventional light and dark bottles and determined production as for phytoplankton samples.

A completely different approach is provided by those methods which monitor certain physico-chemical diurnal changes, principally in oxygen, at a given locality. This type of method is generally applicable to slow-moving streams only, and the results may and frequently do include contributions from submerged macrophytes. In summary, and assuming oxygen concentration is the parameter being measured, the consumption of oxygen by autotrophs and heterotrophs, its production by autotrophs, and the physical diffusion of oxygen in or out of the system interact to produce a characteristic daily curve of change in concentration (Odum 1956). Analysis of this enables the determination of photosynthetic rates. Usually the concentration changes are measured at two stations, one upstream and one downstream on a stretch of river receiving no run-off and tributaries. In this stretch, if we consider dissolved oxygen concentration, the change per unit of surface area may be expressed as

$$x = \text{ph.} \pm \text{diff.} - r_c$$

where x is the rate of gain or loss of oxygen per unit surface area, ph. is the rate of production of oxygen (photosynthesis), diff. is the rate of gain or loss of oxygen by diffusion, and r_c is the rate of biological removal of oxygen (respiration). A major difficulty is the determination of the rate of diffusion. A consideration of the various ways that have been used, and of the whole technique, is given by Owens (1965).

Under certain circumstances readings at only a single station may also be meaningful (Odum 1956).

Rates of diffusion in swiftly flowing streams are too rapid to permit analysis of the effects of stream metabolism on dissolved oxygen, and other methods of determining production need then to be applied. On such occasions pH measurements as an estimate of carbon dioxide changes may be feasible (Wright and Mills 1967). Another approach is that of McConnell and Sigler (1959). Dealing with a fast-flowing river in Utah, they first estimated periphyton standing crops as chlorophyll (whole rocks supporting algae were immersed in acetone for pigment extraction), and then related chlorophyll to photosynthesis by conventional light and dark bottle methodology. However, since the restriction of water movement greatly affects metabolism in rheophilous periphyton (e.g. Whitford 1960) such a technique is viewed with some reservation (Wetzel 1965).

MACROPHYTES

Like the periphyton, macrophytes may contribute significantly to *in situ* carbon fixation in both still (Straškraba 1968) and running waters, and, with certain important exceptions, in the former mostly occur peripherally and in the latter are not generally so restricted. The exceptions include many free-floating noxious weeds of which some are mentioned in Chapter 12.

On an areal basis, macrophytes are much more productive than phytoplankton communities under comparable conditions, emergent macrophytes even more so than submerged ones. Thus, in terms of annual net production on fertile sites measured as metric tons of dry organic matter per ha, Westlake (1965) quotes values for lake phytoplankton of 1–9, for submerged macrophytes of 4–20, and for emergent macrophytes of 30–85. Many recorded values for daily fixation rates in macrophytes exceed $1 \text{ gC m}^{-2} \text{ day}^{-1}$, a value previously noted as not usually exceeded by freshwater phytoplankton.

The high rates of production in macrophytes, taken with generally low rates of turnover, are reflected by the high values that have been recorded for standing crops; whereas phytoplankton standing crops are rarely above 100 g dry weight m^{-2}, macrophyte biomass in the same terms lies between 200 and 10,000 g. With respect to submerged macrophytes in lakes, the maximum standing crop recorded according to Sculthorpe (1967) applies to an infestation of *Lagarosiphon major* in Lake Rotoiti, New Zealand, where Fish (1963a)[8] recorded $1,000 \text{ g}$ dry weight m^{-2}. In rivers this maximum is 519 g for temperate rivers (River Ivel, England, in September; Edwards and Owens 1960) and is only slightly higher, 621 g, for tropical ones (Odum 1957). Apparently, submerged crops are generally no greater in flowing waters than in still ones, or in polluted than unpolluted ones. Considerably higher values are recorded for the standing crops of emergent macrophytes, particularly for temperate reed-swamps dominated by such genera as *Scirpus*, *Typha*, and *Carex*.

There are considerable difficulties involved in measuring the rate of production in macrophytes, and not least in terms of units which are meaningfully comparable between different forms. The oldest method, 'harvesting', is based upon the determination of biomass changes in relation to time. For certain macrophytes,

---

8. Following Sculthorpe (1967) and for ease of comparison we have assumed that dry weight of *Lagarosiphon major* is about 10 per cent of the fresh weight recorded by Fish (1963a).

especially those exhibiting annual regrowth patterns (as many temperate species), this is a fairly appropriate method since seasonal maximal biomass *approximates* the maximum cumulative net production (Westlake 1965). However, underground production of biomass presents problems as significant proportions of the biomass of both submerged and emergent macrophytes may occur as rhizomes or roots (cf. table 3 in Westlake 1965).

Other methods include those measuring *in situ* changes in various environmental parameters, for example oxygen, carbon dioxide, pH, and those which in principle involve incubation of macrophyte biomass. The former has already been discussed in our consideration of periphyton production, and as indicated is applicable only in certain situations. With incubation methods a whole range of problems arises according to the exact technique used. Thus, if ^{14}C is used (cf. Wetzel 1964), unless the relatively tedious process of complete combustion is carried out, self-absorption is a problem. Moreover, it has recently become clear that after fixation significant amounts of dissolved organic carbon may be excreted as is the case with phytoplankton (see earlier discussion). Little is yet known of the extent of such excretion in macrophytes (Wetzel 1969a, 1969b). If oxygen evolution is used as a measure of photosynthesis, then long periods of incubation may be needed, with consequent increase in attendant errors. Errors may also accrue because of oxygen storage in intercellular lacunae (Wetzel 1965). Some of the difficulties inherent in the study of macrophyte production are of course negated when completely submerged but rootless species are studied—as is illustrated by Carr's (1969) laboratory stream investigation of *Ceratophyllum demersum* from Lake Ohakuri, New Zealand.

Secondary Production

Although, as indicated, direct estimates of primary production are still subject to considerable debate as to accuracy, they are nevertheless far more numerous and comparable than are those presently available for secondary production. This is not surprising of course when the multitude of life-forms, habitats, life-cycles, and trophic relationships of heterotrophs are considered.

The earliest rigorous direct estimates of secondary production in freshwater animals relate to benthic invertebrates—in Lake Beloie, U.S.S.R. (Borutzky 1939a, 1939b), to be discussed later—but undoubtedly more is now known and understood concerning fish production than concerning production by other aquatic heterotrophs. It is therefore appropriate, if seemingly not entirely logical, to discuss fish production first, then production at lower trophic levels.

There are two principal methods for estimating fish production, one numerical, one graphical. Both have also been used to determine heterotroph production other than by fish but were elaborated first on fish populations. The numerical method, as initially expounded by Ricker (1946)[9] and Ricker and Foerster (1948), depends upon the erection of a mathematical model with certain assumptions. Thus, if Ricker's model or formula is used (it is only one of several alternatives that could be), it is assumed that there is no emigration or immigration, that mortality and growth are distributed seasonally in the same manner, and that mortality and growth are

---

9. Clarke, Edmondson, and Ricker (1946), with acknowledgements to G. A. Riley, followed a similar line of reasoning.

exponential (which is unlikely over long periods but is a good approximation over short ones). The rate of change of the biomass of a population of contemporaneous age (or cohort) is then:

$$\frac{dB}{dt} = (G - Z)B$$

or, by integration:

$$B = B_0 e^{(G-Z)t}$$

where B = biomass, B_0 = initial biomass, G = instantaneous rate of growth, Z = instantaneous rate of mortality, e = base of natural logarithms (2·71828), t = time.

The instantaneous coefficients (G, Z) are estimated by:

$$G = \frac{\log_e \overline{w}_2 - \log_e \overline{w}_1}{\Delta t}$$

and

$$Z = \frac{\log_e N_2 - \log_e N_1}{\Delta t}$$

where $\overline{w}_1, \overline{w}_2$ and N_1, N_2 = mean weights and numbers of fish present at times t_1, t_2. The mean biomass, $\overline{B}$, for the year t_0 to t_1 is then:

$$\overline{B} = \int_0^1 B_0 e^{(G-Z)t}\, dt$$

$$= B_0 \frac{(e^{(G-Z)} - 1)}{G - Z}$$

and the annual production, P, is the product of the annual growth rate and the mean annual biomass, that is:

$$P = G\overline{B}$$

or

$$P = GB_0 \frac{(e^{(G-Z)} - 1)}{G - Z}.$$

This applies of course only to a population of contemporaneous age; for the computation of total production (i.e. from all age groups), production for each separate contemporaneously aged population must be summed. Using hypothetical data, Chapman (1968) provides a clear worked example to illustrate the computational procedure involved in the application of such formulae as Ricker's.

The graphical method of estimating fish production was first developed by Allen (1950, 1951) working on the Horokiwi stream in New Zealand. Basically, the number of individuals (N) in a population at successive times is plotted against the mean weight of an individual at the same times. No fixed mathematical model of growth or mortality is needed as only experimental data are used in curve-plotting. An example of such a curve (now usually referred to briefly as an 'Allen curve') is given in Fig. 2:5. Here the number of survivors in a group of contemporaneously aged fish is plotted on the ordinate and the mean individual weight on the abscissa. The curve starts at

BIOLOGICAL PRODUCTION

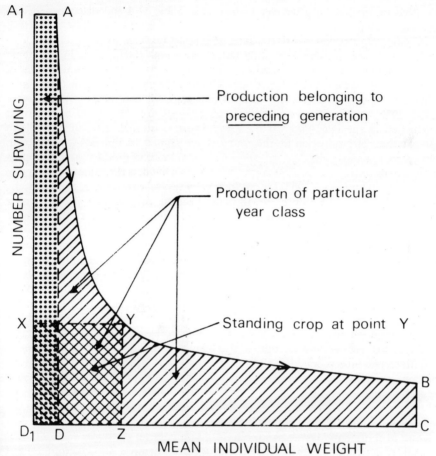

Fig. 2:5 Hypothetical example of an Allen curve. For explanation see text

the upper left corresponding to the time the fish hatch and with continuously decreasing numbers and increasing mean individual weight ends up at the lower right. The area under the curve (ABCD) is equal to the production of fish of a particular age-group (usually designated year I, II, etc.), and may be obtained by planimetry. The area A_1ADD_1 represents the initial biomass of the population or the weight of newly hatched fish. This should of course be assigned to the production of the preceding generation. At any point Y on the curve the standing crop is given by the rectangle $XYZD_1$.

The results of production estimates for fish in natural standing waters in temperate regions indicate a range of rates of *ca.* 2–15 g m^{-2} year^{-1} when only one species predominates (table 1 in Chapman 1967). Ricker and Foerster's (1948) classic study, for example, indicated that the value for *Oncorhynchus nerka*, the sockeye salmon, in Cultus Lake, British Columbia, was *ca.* 5·9 g m^{-2} year^{-1}. Of course production by a given species can be much less than this if the species is merely one of several present and occupies a high trophic level, and very much greater if the water body is fertilized or situated in a warmer climate (see Table 11:4 and discussion of fish farming in Chapter 11).

In temperate streams and rivers, rates of production seem to be generally somewhat higher than in natural standing waters. Le Cren (1969), reviewing salmonid production in small streams, suggested that there is apparently a maximum value of about 12 g m^{-2} year^{-1} even in streams of quite different ecology. However, under other conditions higher values obtain, as indicated, for example, by Allen's work (1950, 1951) on the Horokiwi stream. His estimate for the rate of brown-trout production in the Horokiwi was 54·7 g m^{-2} year^{-1}. Even if criticisms of Allen's computations of production by the youngest age groups are upheld (see Chapman 1967, Le Cren 1969), his data still show a salmonid rate of production higher than elsewhere recorded. The nearest approach to Allen's figure is the value of 42·6 g m^{-2} year^{-1} [10] given by Mann (1965) for production by several coarse fish in the artificially enriched River Thames. Since growth rates of brown trout in Australia are similar to those in New Zealand (cf. Weatherley and Lake 1967), it is unfortunate that no direct production estimates are available for any Australian locality.

The application of methods which have been used to compute fish production to determine production by invertebrates, whilst theoretically feasible, often presents considerable difficulties; invertebrates are, for example, less easy to sample, are more diverse, are frequently multivoltine (have several generations per year), and are usually difficult to segregate into cohorts. It is not surprising that thus far the number of *direct* estimates of invertebrate production are relatively few. However, the subject has important practical repercussions and is being increasingly studied, with notable contributions from eastern Europe, so that the situation may alter reasonably soon.

With certain invertebrate populations, of course, it is quite feasible to apply those methods generally used for measuring fish production, that is when individuals may be aged or are clearly divisible into cohorts and have only few generations per year. The usefulness of empirical Allen curves to determine invertebrate production has been demonstrated, for example, by Nees and Dugdale (1959), who used data relating to chironomid larvae. And, referring to numerical methods, Waters (1966) was able to determine the production of *Baetis vagans* nymphs (Ephemeroptera) in a Minnesota stream by applying Ricker's basic formula:

$$P = G\overline{B}.$$

Similarly, Negus (1966), using shell annular structures as a means of ageing specimens, was able to elucidate growth rates of *Anodonta* and *Unio* (freshwater mussels) in the River Thames, England; she then combined her results with biomass data to give production. A potentially useful and allied numerical approach recently tried (Schindler 1968; but see earlier discussion by Clarke, Edmondson, and Ricker 1946) for zooplankton production—but investigated thus far in laboratory populations of *Daphnia magna* only—uses the basic relationship:

$$B_t = B_0 e^{(A-R)t}$$

where A and R = instantaneous assimilation and respiration rates per unit biomass, B_0 and B_t = biomass at beginning and end of period, t, and e = base of natural logarithms.

---

10. Exclusive of gonadal production.

BIOLOGICAL PRODUCTION 41

The earliest rigorous direct estimates of production by invertebrates were provided, as has been indicated, by Borutzky (1939a, 1939b), who worked on a limited number of species of the benthic fauna of Lake Beloie, Russia. Essentially his method was an iterative procedure involving the successive measurement of standing crops and the determination of losses (ostensibly) due to removal by predators and decomposers; the cumulative losses were regarded as the production. The same basic procedure (summation of mortalities) was used by Anderson and Hooper (1956) and by Tilly (1968) to estimate the production of certain invertebrates in, respectively, a Michigan lake and a spring in Iowa. In such cases, in terms of an Allen curve where the ordinate (number of survivors) begins at N_n and goes to N_0 (D_1 to A_1 in Fig. 2:5) and the abscissa (mean individual weight) from $\bar{w}_0$ to $\bar{w}_n$ (D_1 to C in Fig. 2:5), production is given by:

$$\int_{N_0}^{N_n} \bar{w} dN.$$

The same result is also given by:

$$\int_{\bar{w}_0}^{\bar{w}_n} N d\bar{w}.$$

That is to say, cumulative increases in biomass can be used as a basis for determining invertebrate production.

This latter approach has been made use of by several eastern European workers (e.g. Hillbricht-Ilkowska, Gliwicz, and Spodniewska 1966 [for crustacean plankton], Kajak and Rybak 1966 [for benthos])[11] as well as others (e.g. Andersson 1969). A modification of the method for stream-living animals was used by Waters (1966) to derive production of *Baetis vagans* (Ephemeroptera). He used the formula:

$$P = \Delta B + (D_0 - D_i) + E$$

where P = production rate, ΔB = rate of change in population density, D_0, D_i = biomass of individuals drifting out of and into study area (see Chapter 8 for discussion of stream drift), and E = daily emergence of adults. Results gained with this formula, though higher, compared favourably with those obtained using Ricker's formula (see above).

In general it appears (Nelson and Scott 1962) that iterative methods of determining invertebrate production tend to underestimate true values, and although this must therefore be taken into account when such methods are used it frequently is not. Tilly (1968), working with invertebrates in a spring, used an empirically derived corrective factor as high as 3·2 by which to multiply values gained iteratively.

When populations being investigated are not easily divisible into cohorts or generations, methods different from those described above must be employed. One that has been suggested by Elster's (1954b) and more particularly by Edmondson's (1960, 1965, 1968) work for certain sorts of zooplankton populations involves the

---

11. Reference is not made to several Russian authors because of the relative difficulties in Australia of obtaining copies of their publications.

determination of instantaneous rates of birth and population growth, and estimates of biomass and egg development time. This method has the distinct advantage that, apart from the experimental determination of egg development time, preserved samples can be used. In short, a finite birth rate is first calculated using the relationship:

$$B = \frac{E}{D}$$

where B = finite birth rate, E = number of eggs per female, D = development time in days. This finite rate is converted to an instantaneous one, b, by the relationship:

$$b = \log_e(1 + B).$$

The instantaneous rate of increase, r, is derived by determining population size on successive occasions, N_0, N_t:

$$r = \frac{\log_e N_t - \log_e N_0}{t}$$

where t = time of interval, and, assuming no immigration or emigration, continuous reproduction, and exponential population growth, that is growth is described by:

$$N_t = N_0 e^{rt}.$$

Knowing the instantaneous rates of birth and growth, the instantaneous death rate, d, is easily computed since:

$$r = b - d.$$

This death rate is then combined with biomass estimates to give a value for production.

Amongst others, Hall (1964) has used this method to interpret the population dynamics of *Daphnia galeata mendotae* in a Michigan lake, and it has been used to determine rotifer production in two Polish lakes by Hillbricht-Ilkowska *et al.* (1966).

Cooper (1965), dealing with a freshwater amphipod of which the females during the breeding season produce a succession of broods not clearly divisible into cohorts, estimated mortality in the field as the difference between actual and predicted population sizes. Predicted data were derived from laboratory experiments on growth rates, the intrinsic rate of natural increase, instar duration, and mortality. The product of field mortality and biomass data (involving a variety of instars, temperatures, and sampling times) gave estimates of production. Further estimates were arrived at by a combination of biomass data and information on growth rates derived from field samples.

For stream invertebrates not rigorously divisible into cohorts, Hynes (1961a) attempted to estimate production by summing successive losses in size-groups. Later, however, it was realized (Hynes and Coleman 1968) that growth other than that by the smallest size-groups had not been taken into account, and a modified method was proposed. Using this, it was hoped, quantitative samples collected at intervals could be used to compute approximate annual production estimates. The modification was criticized by Fager (1969) and Hamilton (1969)—each, in turn, critical of each other's criticisms—and further modified by Hamilton. Readers are referred to Hamilton (1969) for a more complete discussion.

BIOLOGICAL PRODUCTION

Because of the difficulties involved in the direct determination of production by invertebrates, several early workers (Juday 1940, Lindeman 1941, Dineen 1953) used the idea of *turnover rate* as a means—albeit an approximate one—of deriving indirect determinations. Thus a figure for annual production was obtained by multiplying the mean standing crop by the number of times yearly this was thought to replace itself, on the basis that the time taken for one replacement was approximately equal to the time of a complete life-cycle. In the past such rates have often been arrived at by relatively arbitrary guesswork, but can be made *somewhat* more precise by growth studies. They must remain, nonetheless, crude underestimates since they do not consider mortality before measurement of the standing crop or potential growth afterwards (Allen 1951, Waters 1966).

An allied approach has been to use the concept of *turnover ratio*, that is the ratio of production to (average) biomass. Here production is initially independently and directly determined, but the ratio, if assumed to be more or less constant, is then used to derive indirect estimates of production from biomass data alone. Although potentially of great use in production studies, insufficient data have yet accumulated to evaluate fully its significance. It is interesting to note, nevertheless, that Waters (1969), in a valuable theoretical examination of the way in which turnover ratio varies according to the shape of the Allen curve, and in a comparison of some original and published empirical data, concluded that for *cohorts* of freshwater benthic insects the ratio is relatively constant, with a modal value of 3·5. In this connection it is of great interest to note that if a turnover ratio is calculated for *Chironomus* from the directly estimated data of Borutzky (1939b) (which, inexplicably, Waters seems to have overlooked), the value 3·4 (74·1/21·6 see Table 4:1) is obtained. The turnover ratios for *Chaoborus* and '*Tanypus*', less important insect components of the Lake Beloie macrobenthos, are 2·8 and 3·8 respectively, lying within Waters's range of 2·5–5·0. The turnover ratio for fish, and probably crustaceans as well, is somewhat higher than that for insects. Borutzky's data suggest that the ratio for oligochaetes (2·3) is less than that for insects. Of considerable interest of course is the ratio for the entire macrobenthos. If this consists almost exclusively of univoltine insect species which spend little time in the egg stage or as emergent adults, then the ratio would presumably ie between 3 and 4. The turnover ratio for the macrobenthos of Lake Beloie from Borutzky's (1939b) data is 2·9 (Table 4:1).

Finally, with respect to indirect methods of estimating secondary production, brief mention may be made of one based on a back-calculation from fish production estimates—sometimes referred to as the 'predation' method (e.g. Allen 1951, Hayne and Ball 1956, Horton 1961). In this a knowledge of fish production (directly derived), feeding efficiency, and metabolic requirements is used to derive an estimate of food consumed by the fish (i.e. *minimal* production). However, considerable discrepancies have so far shown up between food biomass values (benthos) and production rates so computed; the fish seemingly consume an amount of food several times the mean standing crop of benthos (turnover ratios are too high). The method must therefore be regarded at the moment as of somewhat doubtful use. Offered explanations for the discrepancies include biassed estimates of fish feeding efficiency and metabolic requirements, underestimates of the standing crop of fish food, overestimates of fish production, lack of knowledge concerning other sources of fish food. Some recent experiments by J. Illies (unpublished; pers. comm.) involving the enclosure of whole stretches of a stream suggest, however, that production of stream insects may be considerably greater than previous estimates have indicated.

In view of the variety of aquatic invertebrates and their environments, the relative paucity of data on invertebrate secondary production, and the diversity in methods of obtaining such data, little purpose is served in attempting to survey and summarize in isolation published results on this subject. Table 2:2 lists a selection of the more interesting or appropriate values currently available.

So far in our discussion of secondary production we have limited ourselves to a consideration of macroorganisms. But, as mentioned earlier in this chapter, most bacteria are heterotrophs and the extent of their production (and of other microorganisms) should not be underestimated, and nor should the significance of their role as biosynthetic organisms capable of utilizing auto- and allochthonous organic matter that would otherwise be unavailable to larger heterotrophs. Unfortunately, while estimates of their numbers and types are possible, determination of their rates of production in the natural environment is most difficult. It is not surprising that at a recent international conference at which aquatic production processes were discussed in detail (UNESCO/IBP symposium, Warsaw, 1970) it was noted that 'there was an outstanding lack of papers dealing with microbial processes in lakes'.

Briefly, two main approaches are currently used to provide information on microbial activity. One, employed extensively by many Russian workers (e.g. Sorokin 1969), involves the determination of bacterial growth in bottles; the results, though giving direct estimates of production, nevertheless remain open to many criticisms. The other involves the measurement of rates of uptake of labelled organic solutes. These rates, whilst not directly indicating production, may be used to derive information on substrate types and their turnover rates, and on the probable number and activity of bacteria (Hobbie 1969).

Energy Budget and Ecosystem Dynamics

A logical corollary to our discussions of primary and secondary production would be a detailed consideration of the energy budgets of individual aquatic species and of the total trophic-dynamics of those few aquatic ecosystems on which sufficient appropriate data have now accumulated. It is felt, however, that to pursue such a consideration strays too far from our original intentions in writing this book (see Preface). We provide only a brief introduction and readers are referred elsewhere for greater penetration. Mann (1969) has provided a valuable review in this connection.

Early in this chapter we noted that for trophic levels above the first (see earlier for explanation of symbols):

$$P_s = C - F - U - R.$$

The same basic equation applies of course to individual populations within a trophic level. The elaboration of each of the terms as units of energy allows the construction of the so-called *energy budget*. A knowledge of this is important in a number of ways including, for example, the calculation of various ecological efficiencies of energy transfer. In general the construction of a budget requires a knowledge of production, biomass, calorific values of individuals and their products, feeding, egestion, and respiration. To illustrate the procedure, we refer to Comita's (1964) work on *Diaptomus siciloides* (a copepod) in Severson Lake, Minnesota. As well as studying in great detail seasonal fluctuations in field populations, laboratory data on calorific values of the species and its food were obtained, as well as on oxygen consumption and algal grazing rates. Comita's results are summarized in Table 2:3.

TABLE 2:2
INVERTEBRATE PRODUCTION: SELECTED VALUES

Invertebrate group	Production $g\ m^{-2}\ yr^{-1}$	Locality	Source of data
Lotic environments			
Baetis vagans (Ephemeroptera)	9·1[a]	Valley Creek, Minnesota	Waters (1966)
	12·6[a]		
unionid mussels (3 species)	20·5[b]	River Thames, England	Negus (1966)
total benthos	620·2[c]	River Speed, Ontario	
	12·2[c]	Afon Hirnant, N. Wales	} Hynes and Coleman (1968)
Lakes			
Chironomus	7·4		
oligochaetes	4·3	Lake Beloie, Russia	} Borutzky (1939b)
Chaoborus	2·8		
total macrobenthos	14·9		
Asellus aquaticus (Isopoda)	16·7	Lake Pajep Mäskejaure, Sweden	} Andersson (1969)
	31·6	Lake Erken, Sweden	
total zooplankton (main pelagic groups)	296	Lake Mikołajskie, Poland	} Hillbricht-Ilkowska et al. (1966)
	199	Lake Tałtowisko, Poland	
	ca. 220	Lake Tałtowisko, Poland	
	ca. 170	Lake Mikołajskie, Poland	
total benthos (average for whole lake)	100	Lake Sniardwy, Poland	} Kajak and Rybak (1966)
	ca. 20	Lake Lusine, Poland	
	ca. 15	Lake Flosek, Poland	

[a] By two separate methods.
[b] Wet weight, soft tissue.
[c] But see Hamilton (1969) for discussion of computational procedure.

TABLE 2:3
ENERGY BUDGET OF *DIAPTOMUS SICILOIDES*. GENERATION AND ANNUAL VALUES ARE SHOWN
Data from Comita (1964, Table 1)

	Generation					Annual
	1	2	3	4	5	
Energy (calories)						
Input	0·11	0·38	0·69	1·08	0·49	2·75
Respiration	0·12	0·29	0·31	0·58	0·19	1·46
Reproduction	0·03	0·025	0·04	0·08	0·01	0·18
Excretion and egestion	—	0·065	0·33	0·42	0·29	1·11
Percent of input						
Respiration	103	76	45·5	54	38·4	53
Reproduction	26	6·4	6·1	7·2	1·7	7
Excretion and egestion	—	17·1	48	38·8	59·9	40

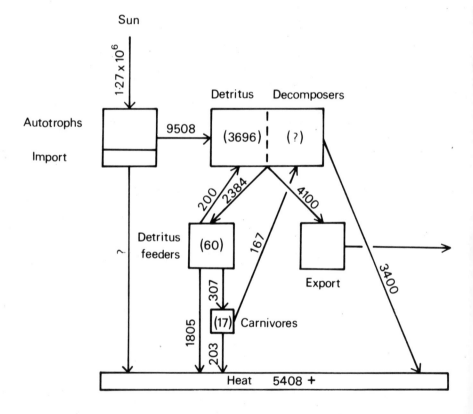

Fig. 2:6 Diagram summarizing principal energy pathways in Cone Spring. Bracketed numbers represent mean monthly standing crop, and all numbers relate to kcal. m^{-2}. Arrows and associated numbers indicate annual flow rates. *Based on Tilly (1968)*

The construction of an energy budget for one species within an ecosystem is difficult enough; when all species are to be considered, the difficulties are almost insuperable and it is not surprising that few attempts have yet been made to do so. We need not emphasize the complexity of most aquatic ecosystems. The first attempt to treat an aquatic ecosystem as a whole in terms of quantitative energetic[12] relationships between component biota was made by Juday (1940), who discussed Lake Mendota, Wisconsin. A year later Lindeman (1941) dealt in much the same way with Cedar Creek Bog, Minnesota. Although much of the actual data of both authors were rather crude estimates and have been criticized subsequently, their approach proved to be a sound conceptual one. A more precise dissection of the total energetics of a freshwater ecosystem did not eventuate for some seventeen years, that is until Odum (1957) discussed energy relationships within a small running-water community, Silver Springs, Florida, and Teal (1957) those within a cold spring in Massachusetts. More recently Tilly (1968) has investigated in a similar fashion a cold spring in Iowa and, largely as a result of the impetus provided by the International Biological Programme, several other comprehensive studies are in progress or their results are shortly to be published (as indicated by the nature of papers delivered at the recent UNESCO/IBP conference in Warsaw, 1970).

Tilly's diagram (modified) summarizing energy flow pathways in Cone Spring is reproduced in Fig. 2:6. It shows that the detritus feeders and decomposers formed an extremely important part of the ecosystem. In other ecosystems, however, they are less important than direct herbivores and carnivores, and it has therefore been suggested that aquatic ecosystems may be characterized by the relative importance of the *detritus food chain* and the *grazing food chain*.

---

12. Total ecosystem studies could be founded—theoretically at least—on the circulation of materials rather than the transfer of energy. Thus far, complete studies so founded are not available.

part two
LAKES

Lake Mulwala, Victoria. *Photograph by Anne Hederics*

chapter three
PRELIMINARY CONSIDERATIONS

INITIAL DISTINCTIONS
Standing bodies of water such as lakes, ponds, and pools are collectively referred to by limnologists as *lentic* environments. In this section, Part II, we are concerned with the larger lentic environments or lakes. The common terms pond and lake can scarcely be subject to rigorous definition, but from the biological point of view the fundamental distinction is one of depth rather than area; a typical pond is shallow enough for rooted vegetation to be established over most of the bottom, whereas a typical lake is deep enough for most of the bottom to be free of rooted vegetation (see also Chapter 9). The fact that most lakes are permanent and many ponds are temporary is a factor, additional to the extent of rooted vegetation, that produces biological differences between the two environments. Obviously organisms without resistant stages in their life history tend to be eliminated from ponds that have only a seasonal existence. Ponds, pools, farm dams, and other water bodies are discussed in Part IV.

Running waters such as rivulets, streams or creeks, and rivers may be collectively referred to as *lotic* environments. These are discussed in Part III.

Usually, for at least part of the year, a lake has an outlet or gives rise to a lotic system. It is then described as an *open* lake. Sometimes, however, there is no outlet at any time and the lake is then said to be *closed*. Closed lakes receive further attention in Part V.

ORIGIN OF LAKE BASINS
As will be shown later, the biological characteristics of a lake may be strongly influenced by the nature and morphology of the lake basin. There is ample justification, therefore, for some discussion of the ways in which lake basins originate. An excellent comprehensive account of lake origins is given by Hutchinson (1957). Many Australasian examples of the different types are also mentioned by Cotton (1945) and Hills (1940). A very brief general account of lakes with special reference to those in New Zealand was given by Allen (1949).

Tectonic lake basins are formed by movements of the deeper parts of the earth's crust. One such type of lake is formed by tilting or folding movements reversing the previous drainage pattern. A good example is provided by the upwarping that formed the Great Basin of South Australia which was occupied during the wet phases of the

Plate 3:1 Lake George, in the Cullarin Range, New South Wales, near Canberra. Photograph taken from the southern end of the lake looking northward. The Cullarin fault-scarp on the western side of the lake is clearly seen on the left. According to one theory it was this fault which betrunked a former westward-flowing river and formed the Lake George basin

Pleistocene by a lake which is now represented by the usually dry Lake Eyre. There was a slight warping of the earth's crust along an east-west line just south of Lake Eyre (Plates 10:1 and 10:2) and extending as far as the Nullabor Plain. The land rose along this line and became slightly depressed to the north where the Lake Eyre basin sank below sea level. This slight elevation was sufficient to dam the waters of the ancestral south-flowing rivers and form a large lake. Deflation following the onset of aridity during the Pleistocene has also contributed, and is largely responsible for the fact that part of the lake bed lies below sea level. According to one theory Lake George near Canberra (Plate 3:1) is of similar origin although in this case the original drainage would have been in a northward direction and impoundment was presumably caused by warping along a line (the Norwood Warp) to the north of the lake. Lakes Buchanan and Galilee in Queensland and Lake Disappointment in Western Australia also probably originated as a result of tectonic warping. The Great Lake in Tasmania was formerly believed to be of glacial origin but more recent opinion is that it too was probably formed by tectonic movements. However, the origin of the Great Lake is still a contentious issue. It may be noted that the lakes referred to are all large; indeed the largest lake in Queensland, New South Wales, South Australia, and Tasmania in each case very probably belongs to the tectonic class. But not all lakes of this class are large. Lake Edgar in the southwest of Tasmania is a small lake which is fairly definitely of tectonic origin. Some tectonic lakes, including the oldest and deepest in the world (Lake Baikal in Siberia), are associated with fault scarps. Lake Torrens in South Australia, though it now contains only a little water, is an example of a *fault-scarp lake*.

PRELIMINARY CONSIDERATIONS

Plate 3:2 The Mt Gambier Lakes, South Australia. In the left foreground is Browne Lake behind which lies Valley Lake. Blue Lake is in the background to the right. A small lake, Leg of Mutton Lake, is tucked away in a depression between Valley and Blue Lakes. All these lakes are of volcanic origin. Blue Lake at least may be classed as a maar. *Photograph by D. Darian Smith*

Lake Omeo, a shallow lake in the eastern highlands of Victoria, was also probably formed as a result of the damming of a stream by faulting.

A second important group of lakes is those associated with *volcanic activity*. Maars[1] are formed by a violent eruption caused by hot lava coming in contact with ground water or by degassing of magma. They are usually subcircular and often contain lakes of great depth. Lakes Barrine, Eacham, and Euramoo on the Atherton Tableland, north Queensland, are fine examples, the first two being very deep (> 100 m) and one having an exceptionally large area for this class of lake. Hutchinson (1957) quotes 1·6 km as being the approximate upper limit for the diameter of such lakes but Lake Barrine has a diameter of 1·9 km. Tower Hill in Victoria with a diameter of 3 km is probably the largest known maar in the world (Ollier 1967) but the several shallow lakes it contains occupy only a small proportion of its total area. Maar lakes are common on the western volcanic plains of Victoria, for example Lakes Gnotuk, Bullenmerri, and Keilambete (saline), and Elingamite and Purrumbete (fresh). The Blue Lake at Mount Gambier, South Australia (Plate 3:2), is a spectacular lake of volcanic origin and superficially has the appearance of a small caldera. It is thus

---

1. The original meaning of maar is a lake. However, since the water has no direct connection with the volcanic formation of a maar, most geomorphologists do not regard the presence of water as an essential part of the definition of the landform. Ollier (1967) gives the following definition: 'Maars are landforms caused by volcanic explosion, and consist of a crater which reaches or extends below general ground level and is considerably wider than deep, and a surrounding rim constructed of material ejected from the crater.'

considered by earlier authors, but Ollier and Joyce (1964) and Ollier (1967) regard it as a maar in which a small lava flow occurred before the explosive, maar-forming eruption. The average diameter of Blue Lake is only *ca*. 1 km. New Zealand lakes of this type include Lake Pupuke near Auckland, and Lakes Ngapouri, Okaro, and Rotongaio in the Rotorua district. There is a cluster of seven lakes collectively known as the Red Rock Lakes, near Alvie, Victoria (see Bayly 1969c and Plate 10:3). These are regarded by Ollier (1967) as occupying the depressions of a complex maar produced by closely spaced multiple eruptions.

Volcano crater lakes constitute a further class of volcanic lakes. These are usually (see Hutchinson 1957) called explosion craters but, in that 'explosion' is just as descriptive of the origin of a maar as it is of a crater at the top of a volcano, we prefer the present terminology. Such lakes are found near the summit of distinctly raised, and usually extinct (but see Plate 1:1), volcanoes in which the orifice has usually been enlarged during the final stages of eruption. Mt Eccles near Macarthur, Victoria, contains a deep crater lake. Mt Quincan on the Atherton Tablelands is a young scoria cone with a crater about 0·8 km wide which contains a single small lake. Mt Le Brun near Gayndah, Queensland, has two distinct craters each containing a small, almost perfectly circular lake; these are the Coalstoun Lakes. Lake Rotokawau near Rotorua, New Zealand, is another example (Allen 1949). Lake Rotomohana, also in the Rotorua district, is a multiple crater lake occupying a series of confluent craters instead of a single one.

Another type of volcanic lake occupies a *caldera*. These are formed by the subsidence of the central portion of a volcano when it is left unsupported following the extrusion of a large amount of magma. The result is usually a more or less circular cavity with precipitous sides. Calderas are larger than maars, usually having a minimal diameter of about 5 km. Sometimes calderas are considerably modified by the upwelling of rounded domes from the floor. These transform the circular shape of the space in which a lake can form, and diminish the possible area. Lake Aroarotamahine on Mayor Island in the Bay of Plenty, New Zealand, is one such *modified caldera* lake. Here secondary domes, old and new, have pushed the possible lake-containing space to one side (the eastern) so that only a small portion of the area enclosed by the caldera cliffs is occupied by water (Brothers 1957). In the Rotorua district of New Zealand lakes Tikitapu, Okataina, Rotoiti, Rotoehu, and Rotoma are also modified caldera lakes representing subdivisions of the floor of Haroharo Caldera (Healey 1963). That is, these lakes occupy hollows formed by the extrusion of lavas from the floor of a large caldera.

Some volcanic lakes originated by a process somewhat like caldera collapse formation but on a much larger scale; so much material was ejected, and so much empty space created underground, that not only the central portion of a volcano but extensive tracts of surrounding land subsided. Such subsidence usually occurs along pre-existing fault-lines. It is often difficult to assess the relative role of tectonics and volcanism in these collapses so that the term *volcano-tectonic* is employed. Lake Taupo (Plates 3:3 and 11:2), the largest lake in New Zealand, and Lake Corangamite (Plate 10:3), the largest lake in Victoria, both probably originated in this manner despite the fact that one is deep and the other shallow. Currey (1970), however, considers that Lake Corangamite was formed when lava flows blocked a former course of the Barwon River. Grange (1937) is adamant that, because of the flatness of its floor, warping and faulting alone are not sufficient to account for all of the features of Lake

Plate 3:3 Lake Taupo, North Island, New Zealand. This, the largest lake in New Zealand (area 370,000 ha), was formed by volcanism and tectonics. Kuratau Spit on the west shore of the lake appears on the right and the volcanic cone Pauhara on the left. *Photograph by Whites Aviation*

Taupo. He states that it is quite certain that great explosive eruptions have taken place from the lake particularly in the northeastern parts. To the northeast of Lake Taupo, Lakes Rotorua and Tarawera may also be regarded as originating by volcano-tectonic collapse.

Yet another kind of volcanic lake is produced by the *collapse of lava flows*. It is not uncommon for a surface of a newly formed lava flow to cool and form a crust while the lower layers are still fluid and moving. An unsupported crust may then collapse, producing a lake basin. In western Victoria many small lakes have been formed by depressions of this sort extending below the ground-water level.

The last type of volcanic lake to be considered is produced by *volcanic damming*. Lake Omapere in the North Island of New Zealand was dammed by a basaltic lava flow which reversed the hydrographic pattern of the surrounding landscape. The upper Waitangi valley (now covered by Lake Omapere) formerly drained to the east, but a lava flow dammed the Waitangi River to produce Lake Omapere and reverse the drainage to the west. Several lakes in Victoria, including the Cockajemmy Lakes near Mt Abrupt, and Lake Condah near Mt Eccles, were formed by volcanic dams.

The crater at Mt Hypipamee on the Atherton Tableland, Queensland, contains what may be regarded as an extraordinary volcanic lake (assuming the essential difference between a lake and a pond is one of depth rather than area). The crater is

Plate 3:4 Lake Pedder, a Tasmanian lake formed by glacifluvial outwash from the Frankland Range (background). Photograph taken from the northeast corner of the lake looking south southwest. The eastern quartzite beach is clearly seen on the left. This beautiful lake, which lies in a national park, is the home of several endemic species. It is seriously threatened as a result of the short-sighted policies of Tasmanian governments and the Hydro Electric Commission of that State. The Premiers of the governments implementing these policies are Mr W. A. Bethune and Mr E. E. Reece, and the Hydro-electric Commissioner is Sir Allan Knight. *Photograph by Vern Reid*

like a large well with quite sheer walls and technically is known as a *diatreme* or *gas maar*. The diameter of this natural well is about 60 m and its depth 140 m. The depth of the water itself is about 70 m. In this lake the morphometric ratio (average depth)/(maximum depth) must be almost unity and the ratio of area to mean or maximum depth must be exceptionally high.

Some lakes are formed when a *landslide* fills a floor of a valley and dams a stream. Such lakes are often short-lived since once the lake rises and flows over the top of the dam the outflowing water may carry away the unconsolidated materials. Lake Waikaremoana in New Zealand (N.I.) is a good example of a lake held by a rockslide dam. Like many artificially dammed lakes, this lake occupies a system of branching valleys and has a strikingly dendritic shape. It is held up by a great rockslide through which water leaks but very little overflows. Lake Minchin in Canterbury is another example (Gage 1959). Within a circumscribed region of unglaciated and non-volcanic mountains, landslide lakes may be unique features. Lake Tarli Karng near Mt Wellington, Victoria, is a case in point; it is Victoria's highest natural lake (altitude 915 m) and its only highland lake of any appreciable depth (51 m).

An important group of lakes is formed by *glacial activity*. These are especially important in the South Island of New Zealand; Gage (1959, 1969) estimated that out of forty-six sizeable lakes in Canterbury twenty-four are of glacial origin.

PRELIMINARY CONSIDERATIONS

Plate 3:5   Lake Wakatipu, a piedmont lake in the South Island of New Zealand. The photograph is taken from the head of the lake and looks south southwest along the upper arm of the lake. The township of Glenorchy is in the foreground. *Photograph by Whites Aviation*

Kettle lakes are formed by the delayed melting of large ice-boulders detached from receding ice-fronts. Usually they are small. Lake Marymere in Canterbury, New Zealand, is a kettle lake and there are doubtless many more in the South Island. Many of the smaller lakes and ponds in the western part of the Central Plateau of Tasmania—Australia's 'Land of Ten Thousand Lakes'—have been formed in this way by *ice-blocks melting amongst morainic material*. These have a maximum dimension of up to about 100 m. Other small lakes in the same region are completely surrounded by solid rock, having been gouged out of the dolerite by *small-scale glacial corrasion* (Jennings and Ahmad 1957).

Cirque lakes are found at the bottom of amphitheatres formed by ice action at the heads of glaciated valleys. The excavation of a cirque is usually attributed to frost-riving or the action of freezing and thawing on a rocky concavity of a mountainside. When deglaciated many cirques contain lakes that are held either by a rock lip (true rock basin) or a moraine. They are generally small and comparatively shallow, but some are very deep relative to their area. Numerous cirque lakes are found in formerly glaciated parts of Tasmania: in the Frenchman's Cap National Park region (Peterson 1966), in the Frankland Range south of Lake Pedder (Plate 3:4), in the western Arthurs Range and other parts of southwest Tasmania. On the Australian mainland there are four such lakes restricted to a small area on the Kosciusko plateau of New

South Wales (Peterson 1968). The deepest of these, Blue Lake, has a maximum depth of about 29 m, and owes its existence mainly to glacial gouging and only to a small extent to a moraine. The other three, Lakes Cootapatamba and Albina and Club Lake, are shallower and are held entirely by terminal moraines (Dulhunty 1946). Numerous examples also occur in the heavily glaciated Fjordland district of southwestern New Zealand. Thus Lake Quill occupies a huge hanging cirque and overflows as the Sutherland Falls into the Arthur Valley. Actually Lake Quill is the lower member of a cirque stairway; the upper cirque is still completely ice-bound, but may eventually contain a lake.

Cirques occur neither greatly above nor below the permanent snow-line which, of course, oscillates to some extent according to climatic changes. Valley glaciers, however, may descend considerably lower and can produce rock basins by corrasion. Where glaciers occupied long valleys at low elevations, *piedmont lakes* may be produced. These lie in narrow, greatly overdeepened, U-shaped valleys. Lake Wakatipu in southern New Zealand (Plate 3:5) is a good example. This lake has a maximum depth of 378 m and the lowest part of its floor is 70 m below sea level. It is held mainly by a rock rim and to some extent by a moraine. The lowest part of the rim of the rock basin is at least 305 m above the deepest portion and the moraine adds about 70 m on top of this. Lakes Te Anau and Wanaka may also be classified as piedmont lakes. Lake St Clair in Tasmania (Plate 3:6) is the only Australian example of a true piedmont lake. Lake Manapouri in Southland, New Zealand (Plate 3:7), is classed because of its association with a much indented coastline as a *fjord lake*, but this is merely a special case of a piedmont lake. With a maximum depth of 445 m this lake is easily the deepest in Australasia and it ranks fifteenth on the world list of deep lakes.

The Laurentian Great Lakes of North America which, except for the Caspian Sea, include the largest (with respect to volume and area) lakes in the world, represent *large rock basins produced by continental ice*. They represent the aftermath of glacial scouring on an enormous scale, and differ from piedmont lakes in the low profile of the mountains around their upper ends, in addition to size.

Sometimes the terminal moraines of valley glaciers persist and dam the stream that replaces the glacier, giving rise to *lakes held by morainic dams*. Some lakes in the western part of the Tasmanian Central Plateau—Clarence Lagoon, for example—have been impounded in this way by an end moraine. Other examples occur in the South Island of New Zealand.

In limestone country water may dissolve away localized patches of material and form funnel-shaped hollows. The bottoms of these may then become choked with lumps of rock and fine insoluble material washed in from the surface so that, after rain, water accumulates, and a *solution lake* or pond temporarily comes into existence. Sometimes the bottom of these sink-holes extends below the water-table, so that it is unnecessary to assume some degree of impermeability in the floor for a lake to exist. In this event they are likely to be of a more permanent nature. Small solution lakes are found near Lake Tyrrell in the Mallee district of Victoria. Lake Tyrrell itself may

Plate 3:6   Lake St Clair, a piedmont lake in Tasmania, from the northern end. The photograph looks somewhat east of south. *Photograph by Vern Reid* (top)
Plate 3:7   Lake Manapouri, a fjord lake in the South Island of New Zealand. This is the deepest lake in Australasia and fifteenth deepest lake in the world. The depth before artificial alteration was 445 m. *Photograph by Whites Aviation* (bottom)

have been formed, at least in part, by solution. Other examples are found in the Coorong district of South Australia, and in the Waitomo, southern Wairarapa, and Nelson districts of New Zealand.

Some lake basins have been produced by fluvatile action or processes associated with rivers. The most important kind of *fluviatile lake* is the *lateral lake*. These are formed when a large river builds up its bed at a faster rate than its lateral tributaries are able to, by the deposition of levees along the sides. This is quite a common situation since smaller tributaries usually have much less opportunity to collect sediments. The side branches thus tend to become obstructed and eventually drowned by the lateral sediments deposited by the main river. A large number of lateral lakes are found bordering the rivers and anabranches[2] of the Murray-Darling drainage system; they are especially common along the Darling River in western New South Wales. Lakes Wairarapa and Waikare provide New Zealand examples (Allen 1949). In the South Island, Lakes Pearson and Lyndon also originated from the deposition of alluvial fans (Gage 1959).

Another type of lake is formed by the meandering of rivers. Whenever a small change in topography or geological structure produces the slightest bend in the course of a river greater turbulence occurs on the concave side, and it is here that erosion begins. The concavity is thus accentuated and sediment is deposited from the slower running water on the convex side. This process may continue until eventually whole loops are cut off and isolated. Part of the abandoned channel is usually over-deepened and may persist as a shallow crescentic lake which in Australia is usually called a *billabong*. Elsewhere they are termed *oxbow lakes*. Many examples occur along the Darling and Murray Rivers, a few also occur along the Goulburn River in Victoria. Many of these are not permanently filled with water. Examples are also found in the Wairau and other mature valleys in New Zealand.

Some lakes are formed solely or partially as a result of *wind action*. These are of two main types; *dune lakes* and *lakes in rock basins produced by deflation*. Existing classifications of dune lakes, including that proposed by Hutchinson (1957), seem unsatisfactory, and as an attempt at improvement subdivision into the following four types is proposed:

(i) Dune barrage lakes (lakes behind sand dams) In the simplest case these are formed by dunes moving inland to block a river valley draining towards the coast. They are typically sub-triangular in plan with the deepest part close to the dune dam. Some lakes on King Island, in Bass Strait, are essentially of this simple barrage type (Jennings 1957). The history of many dune barrage lakes, however, is more complex than this; sometimes the two sides of the valley consist of quite different material, and quite commonly one or both sides of the valley themselves consist of dunes distinctly older than the barrage itself. Such is the case with some of the dune lakes on King Island and on the west coast of the North Island (N.Z.) (Cunningham *et al.* 1953) where there are two main series (Northland and Patea-Otaki). At least two of these New Zealand lakes (Swan Lake and Lake Heaton) conform to the simple type in having a roughly triangular plan and the greatest depth near the barrage.

Sometimes little or no inland movement of dunes is involved in the barrage formation; it may be simply a low foredune (e.g. Lake Ainsworth, northern New

---

2. The name anabranch is apparently an abbreviation of anastomosing branches, first proposed by Jackson in 1834, and has been adopted only in Australia.

South Wales) or, in regions which have undergone eustatic emergence, even a former wave-built off-shore bar. The latter is evidently true of a former lake on King Island (Jennings 1951). Sometimes wave action and deflation contribute almost equally to the formation of the obstruction that produces impoundment.

As indicated briefly below, dune barrage lakes may occur in desert or inland as well as coastal regions.

(ii) Lakes between the trailing arms of parabolic dunes (impervious layer if present not of organic origin) If a large amount of deflation occurs so that little or no sand is left on the floor between the two arms, and an impervious rock floor is exposed or only just covered, a lake may develop. Many of these lakes seem to owe their existence to deflation alone, and some undergo a slow but almost continuous change of position. They are not usually as permanent as those of the next category. In some cases at least they appear to have only a seasonal existence. Thus those near Cape Flattery on Cape York Peninsula have a seasonally fluctuating water level and in the dry season deflation tends to deepen and extend the lake basin (Bird 1964a). It is not clear whether these particular lakes are dependent upon the exposure or near-exposure of impermeable rock. It is conceivable that they tolerate continuous loss of water by seepage during their periods of existence. Parabolic dune lakes are also found in the southern wallum country of Queensland and on King Island.

Some dune lakes originate by the formation of a dune barrage between the arms of a parabolic dune; that is, they represent a combination of the above two types.

(iii) Lakes resting on impervious organically bonded sand-rock in dune depressions Such lakes are usually perched amongst relatively old and high siliceous dune systems stabilized by vegetation which may even be in the nature of rain forest. Their origin depends upon organic accumulation in addition to deflation. They are commonly quite permanent. The water usually has a very low salinity with (in coastal regions) ionic proportions closely similar to those of sea water, and is often acid-humic in nature with pH's as low as $4 \cdot 0$. The best examples are almost certainly found on Fraser or Great Sandy Island (Bayly 1964, 1966), but good examples also occur in the northern wallum of Queensland (localized portions of Cape York Peninsula) (Brass 1953) and on King Island. Some dune lakes on the west coast of the North Island (N.Z.) which are stated (Cunningham *et al.* 1953) to be 'formed in the hollows of the older consolidated dunes' probably belong to this category.

Some of these lakes are quite large. Lake Boemingen on Fraser Island, for example, has an area of about 260 ha.

(iv) Lakes in permeable depressions extending below a common extensive water-table Lakes of this kind probably occur in low-lying dune areas of Australia or New Zealand, or both, but no definite examples can be cited. Presumably such lakes would not uncommonly occur in localized clusters. Furthermore, the existence of identical or very closely similar levels in a cluster of dune lakes would justify their interim assignment to this class.

Before the construction of O'Sullivan Dam in southeastern Washington, U.S.A., several hundred 'pot-holes' (small lakes and ponds) existed amongst the desert dunes lying south of Moses Lake (Harris 1954). These evidently represented exposures of the water-table. (It may be noted that Moses Lake itself was a dune barrage lake [Russell 1893]—an inland example involving the damming of a canyon.)

A series of five desert dune lakes, Toghraklik-köl and its four daschis or salt pools, adjacent to the Tarim River north of Tibet provides another inland example of this type of lake (Hedin 1904). The lakes in the more extensive series of parallel dune lakes in this same region are quite independent of each other, and this is possibly true also of the parallel dune lakes in Nebraska, U.S.A. In these two instances, although the lakes are completely surrounded by sand, it is probably underlying rock basins, rather than the sand itself, which determine their existence. Despite this, such parallel dune lakes may still have originated as a result of wind action in that the undulations in the underlying bedrock may have been produced by deflation (see below).

Lakes in rock or clay basins produced by deflation The conditions which permit the excavation of a basin by wind action are by no means fully understood, but one of the necessary conditions is an arid climate such as occurs throughout most of Australia. In 'Salinaland', or that part of Western Australia surrounding Wiluna and Kalgoorlie, there are a large number of irregular basins which after rain are occupied by ephemeral saline lakes (Jutson 1934). There are about thirty such basins over 14 km long and a very large number of smaller ones. They are often elongate and sometimes lie in rows as if they were part of a drainage system. The floor of these lakes may be solid rock that is very smooth and level. Sometimes, however, it is covered with fine silt which is usually very thin so that the bedrock is exposed in places. There is usually only a small deposit of salt left when the lake evaporates.

It has usually been supposed that most of the more elongated basins are the remains of river valleys blocked with drifting sand. Some workers, however, do not accept this and believe that the elongated shape of the basins is due to westward migration, the western ends being deflated while sand tends to accumulate at the eastern end. They believe that rapid erosion after rare rainstorms is of considerable importance.

The clay-pans near Broken Hill, New South Wales, represent small deflation basins; wind has removed the sandy surface soil and exposed a hard clayey floor (Collins 1923). Many of these intermittently contain water.

Finally we consider *lakes associated with coastlines*. It is possible for a lake to form, by the process discussed below, along a lacustrine shoreline, but here discussion will be limited to maritime shores. The most usual way these lakes are formed is by the growth of a bar across some marine inlet. If any longshore current is moving along a strait coast indented by a bay, the inertia of the current will tend to carry it across the mouth of the bay. If the current is carrying sedimentary material this will be deposited as a bar following the coastline. Unless erosive action prevents it, the bar can eventually cut off the bay or inlet as a coastal lake or lagoon.[3] In many instances, however, the volume of the river discharge and the strength of tidal currents prevent complete separation. In Victoria the Gippsland Lakes and lakes further to the east are associated with a coastline of submergence, and represent marine-flooded plains and drowned river valleys (Bird 1964b). They are now almost completely cut off from the sea by a long off-shore bar. Lake Bunga in the same region is completely cut off by bar formation. Several lakes between Beachport and Robe in South Australia, on the other hand, have been formed by coastal emergence and lie in a depression between two parallel dune-systems representing stranded sea-beaches.

---

3. In Australia the term lagoon has lost its usual significance; it may be applied to almost any substantial body of water irrespective of its distance from the sea and its height above sea level. Some of the 'lagoons' on the Central Plateau of Tasmania have an altitude of almost 1,000 m!

PRELIMINARY CONSIDERATIONS

In New Zealand, Lake Ellesmere near Christchurch has been cut off from the sea by sand and gravel spits.

MORPHOLOGY OF LAKE BASINS AND MORPHOMETRIC PARAMETERS

The *bathymetric map* is the standard means of recording the morphology of a lake, and the first thing needed for this is an accurate outline map. Unless a special map exists, which has been prepared for the express purpose of accurately showing the shoreline, a vertical aerial photograph is usually the best means of obtaining the outline. It is usually not difficult to obtain from the appropriate authority special maps complete with depth contours for large artificial lakes, but for natural lakes the best available maps are often unsatisfactory for limnological purposes. Recently, however, the Fisheries Research Division of the New Zealand Marine Department commenced publication of a superb series of bathymetric maps for the major lakes in that country. Unfortunately nothing comparable has been undertaken in Australia; nor is it likely to be, so it seems, for a long time. Aerial photographs are available for most parts of Australia and all of New Zealand. These are taken in an overlapping series, and other things being equal the photograph in which the lake is most nearly central should be chosen to minimize distortion. The stated scale of the aerial photograph should be checked against direct field measurement with a surveyor's tape of the distance between two prominent objects that can later be recognized on the photograph.[4]

Stereoscopic pairs of photographs may be used to advantage to obtain an idea of the catchment area of a lake when this is not available from contoured topographical maps.

Once the outline has been obtained, the bathymetric map can be constructed if the position at which soundings are made is fixed by one or more of the following methods: using angulation data recorded by a shore observer; from the lake by the alignment of two prominent objects along each of two different directions; by the counting of oar strokes along a fixed course across the lake under calm conditions (early morning is usually best). With the latter method the arbitrary intervals can be later converted to absolute intervals by reference to the outline map. Where a small lake is involved, it may be feasible to stretch a graduated rope across, and sound at regular intervals along it.

From a suitably contoured bathymetric map the following parameters may be calculated:

1 Area (A) This is best determined by planimetry.[5] If no planimeter is available the outline may be traced on to a piece of paper, cut out, and weighed. A square piece of paper representing a known area would, of course, also have to be weighed. Alternatively, the counting-of-squares method could be used.

2 Volume (V) The best way of obtaining this is to determine with a planimeter the area enclosed by all contours, then to construct a graph of area against depth. The area below this graph represents the volume and can be obtained easily with a

---

4. One of us has on at least two occasions been supplied with aerial photographs for which the stated scale was clearly incorrect.
5. For more details of this, and the construction of bathymetric maps, see Welch (1948).

planimeter (see Fig. 3:1). Alternatively, calculate the volumes between the planes of the successive contours $\left(\text{e.g. } {}_0V_x = \frac{x}{3}(A_0 + A_x + \sqrt{A_0 A_x})\right)$ and sum these volumes. When:

$$\text{total volume} = {}_0V_x + {}_xV_{2x} + {}_{2x}V_{3x} + \ldots$$

where x = distance between contours, and A_0 = area of lake (A).

3 Mean depth ($\bar{z}$) $\frac{V}{A}$.

4 Maximum depth (z_m).

5 Maximum length (l_m) The shortest distance between the two most remote points on the lake shore.

6 Mean width (w) $\frac{A}{l_m}$.

7 Maximum width (w_m) The maximum distance between shores at right angles to the maximum length.

8 Shoreline development (D_s) The ratio of the *length of the shoreline* (s) to the length of the circumference of a circle of area equal to that of the lake; that is:

$$D_s = \frac{s}{\sqrt{2\Pi A}}.$$

This ratio may be regarded as an index of the potential importance of littoral influences on the lake. Obviously the shoreline development cannot be less than 1·0, but in more or less circular lakes, such as many crater, maar, and caldera lakes, it is only slightly greater than 1·0. On the other hand, a high value would be obtained for a dendritic lake such as Lake Waikaremoana. Marked elongation, however, is more important than a high degree of sinuosity alone in producing extremely high values. The highest value recorded is about 20.

9 Volume development (D_v) and the ratio mean to maximum depth ($\bar{z}: z_m$) Volume development may be used to express the form of a basin, and is defined as the ratio of the volume of a lake to that of a cone of basal area equal to the area of the lake and height equal to the maximum depth of the lake; that is:

$$D_v = \frac{A\bar{z}}{\frac{1}{3}A z_m} = 3\bar{z} : z_m.$$

The ratio of mean to maximum depth ($\bar{z} : z_m$) is thus just as good an index of form as volume development and is generally used instead.

For most lakes this ratio has a value between 0·33 and 0·50. Values greater than 0·50 are found in shallow lakes with flat bottoms, calderas and other crater lakes (e.g. for Lake Aroarotamahine on Mayor Island, New Zealand, it is 0·55), flat-bottomed graben lakes, and piedmont and cirque lakes.

Although the ratio $\bar{z} : z_m$ is a useful measure of the form of a lake basin it gives no indication of the way in which deep regions are distributed. In many lakes there is

PRELIMINARY CONSIDERATIONS

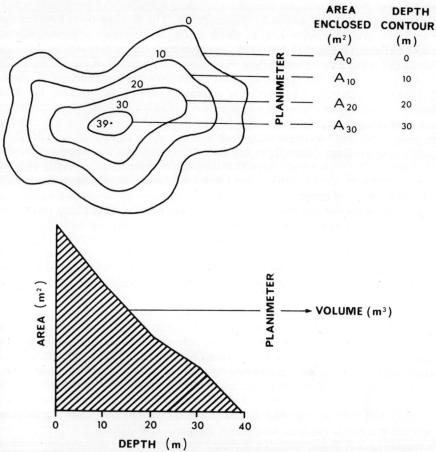

Fig. 3:1   Diagram showing a method of determining the volume of a lake. The first step (above) consists of determining with a planimeter the area enclosed by all contours. The second step (below) involves the construction of an area-depth graph in which the area below the graph (representing the volume) is again determined by planimetry

only one, approximately central, deep area, but the existence of more than one deep area with submerged ridges between them is by no means uncommon. Thus Lake Baikal, the deepest lake in the world, has three distinct basins which probably differ in geological age. Where several distinct basins occur, their deeper portions may differ somewhat in physical and chemical characteristics especially during periods of thermal stratification. In such cases *submerged depression individuality* is said to exist.

TERMINOLOGY OF THE MAJOR ECOLOGICAL REGIONS AND BIOLOGICAL COMMUNITIES OF A LAKE

The open water of a lake above the immediate vicinity of the bottom, and away from the obstructions of any rooted vegetation near the shore, is usually called the *limnetic* region (Fig. 3:2). The corresponding oceanographic term, *pelagic*, is sometimes used for lakes as well. Except perhaps for very shallow lakes, the limnetic region may be regarded as divisible into two vertical layers, the *trophogenic* and *tropholytic*, which have markedly different metabolic characteristics. The dividing plane between the two is the *level of compensation*, at which depth the constructive photosynthetic activity of plants is exactly balanced by their destructive respiratory activity. In other words this is the level at which there is zero net accumulation of organic matter in producer organisms. The exact level of compensation varies continuously of course. During the night it can be regarded as coincident with the lake surface, but with increasing intensity of illumination it sinks lower and lower. Turbidity, temperature, extent of cloud cover, and many other factors are also of importance in determining its precise location. Attempts have been made to relate its approximate depth to Secchi-disc readings (since these are easy to obtain), but the empirical factors obtained display such a wide range that they can be regarded as generally of local applicability only. Verduin (1956) used a factor of 5; other authors report values only slightly in excess of unity.

In some thermally stratified lakes (see p. 78) the lower limit of photosynthesis is determined, at least for a time, by the temperature and other concomitant changes (i.e., by the *thermocline*) rather than illumination. This is equivalent to saying that the region of rapid temperature change may occur above the potential level of compensation. Thus in some nutrient-poor or *oligotrophic* lakes with a well-developed thermal stratification there may be two photosynthetic maxima (one near the surface and one near the thermocline), and phytoplankton is lacking in the deepest waters not so much because of inadequate illumination, but because temperature or chemical conditions are unsuitable (Findenegg 1965). In nutrient-rich or *eutrophic* lakes, on the other hand, the transmission of light may be so poor because of crowding of algae near the surface, that the well-lit or *euphotic* zone is quite shallow and located well above the thermocline. Brown humified lakes also have only a narrow euphotic zone.

The second major region is the bottom or *benthic* region. Complex terminologies have been used to subdivide this (Chapter 6), but only two regions need consideration at this point; these are the shore or *littoral* and the *profundal* regions. The division between the two is the line along which the plane of compensation intersects the bottom (Fig. 3:2). Some workers recognize a transitional zone, the *littoriprofundal* or *sublittoral*, between the two main regions. It is probable that some photosynthetic bacteria and small photosynthetic protistans extend below what constitutes the compensation level for macrophytes.

The limnetic region of a lake is occupied by two major groups of organisms; the *plankton* and *nekton*. The term *plankton*, which comes from Greek, means literally that which was made to drift or wander, and refers to organisms, most of which are small, that drift about in a largely or wholly passive manner. In the case of non-motile

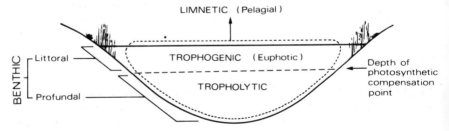

Fig. 3:2 Diagram showing the major ecological regions of a lake

PRELIMINARY CONSIDERATIONS

forms the direction of movement and position within a body of water is completely dependent upon the movements of the water. The plankton may, of course, be divided on a nutritional basis into *phytoplankton* and *zooplankton*. It may also be divided on the basis of size; thus the larger forms which can be caught with a net are called *net-plankton* and those that are so small that they pass through the finest nets that are readily available[6] are called *nannoplankton* (nanno = dwarf). The latter consists mainly of small flagellates and diatoms and is 'captured' by the centrifugation of water samples. *Nekton* includes the powerful swimmers that are completely independent of water turbulence and are able to move 'at will' in any direction. Many members of this group are capable of making substantial horizontal migrations even against strong currents. By far the most important group of nektonic organisms is fish, but the actively swimming insects, such as the notonectids, and natant shrimps are also included. In fact many organisms usually described as planktonic, for example the planktonic Crustacea, are not completely dependent on turbulent movements of the water and are more accurately described as *nektoplanktonic*, implying a degree of dependence on turbulence that is intermediate between that of a non-flagellate phytoplankton cell and an adult fish.

Organisms living in or on the bottom materials are collectively called *benthos*. The terms *phytobenthos* and *zoobenthos* may also be used with obvious meaning.

The German term *Aufwuchs* is often used to designate all organisms that are closely associated with or attached to a submerged surface but do not penetrate into it. This term does not exclude some organisms that may be described as benthos as defined above, although it does exclude rooted macrophytes and organisms entirely contained within bottom materials. It does, however, include some organisms which are not usually regarded as benthic, such as minute plants and animals attached to the stems of large plants. The definition is further considered in Chapter 6.

The word *pleuston* is used to refer to organisms that are associated with the air-water interface.

Finally, the term *psammon* may be employed in collective reference to the community living at the lake margin in sand, including the minute interstitial spaces between the grains of sand. This constitutes a well-developed community in lakes with extensive beaches of quartzite sand. By far the best Australian example of such a lake is Lake Pedder in Tasmania (Plate 3:4). Unfortunately this lake will very shortly be flooded out of existence by a hydro-electric dam being built on the Serpentine River. A brief preliminary account of this lake including its psammon is given by Bayly, Peterson, Tyler, and Williams (1966).

---

6. With the old bolting silk the smallest aperture size available was about 60µ. Modern nylon bolting cloth goes down to about 10µ. Many nannoplankton species are smaller than this, some being less than 5µ in length. It should be noted that in a given volume of water the nannoplankton not only greatly outnumber the net-plankton in most cases, but also often exceed them in mass.

chapter four
PRODUCTIVITY

We have previously discussed in a general way the phenomenon of biological production (Chapter 2). The present chapter is concerned specifically with the production that occurs in lakes, and since emphasis is placed upon the *extent of production* the use of the term *productivity* is appropriate here. Readers are referred to Chapter 2 for a discussion of our definitions of the terms production and productivity. Some of the 'factors' discussed below may not themselves have any fundamental influence upon lake productivity, but are easily identified or measured, and are correlated with factors that are of more fundamental significance. Most of them are briefly discussed by Bayly (1969b).

THE INFLUENCE OF MORPHOLOGY
Several important hyperbolic relationships exist between both the standing crop and productivity of lacustrine communities and mean depth of lakes (Rawson 1955). The mean standing crop of plankton expressed as numbers or biomass per unit volume decreases with increasing mean depth in the manner depicted in Fig. 4:1(a). This curve represents of course a mean relationship or generalization; it is not something to which all lakes rigidly adhere. There is no reason why a small but appreciable mean standing crop, thus expressed, should not exist in an extremely deep lake, and this is the meaning of 'K' (Fig. 4:1(a)) or the fact that the graph does not meet the abscissa. Figure 4:1(a) is not true, however, of the productivity or mean crop of plankton beneath unit surface area.[1] Considered in these terms there may be very little difference between the crops of deep (morphologically oligotrophic) and shallow (morphologically eutrophic) lakes; there is simply greater dilution of the planktonic material in the deeper lake. This is equivalent to saying that the trophogenic region of an oligotrophic lake is thicker than that of a eutrophic lake. This is clearly shown by the well-studied lakes in Wisconsin; in Lake Mendota ($\bar{z} = 12 \cdot 1$ m) the amount of

---

1. Rawson (1955) claimed that it is true of area-based mean standing crop. However, if two somewhat peculiar lakes are eliminated from his plot the relationship is only doubtfully hyperbolic. The consensus is that planktonic crops expressed in these terms are largely independent of depth.

seston was found to be 2·4 mg (dry organic matter)/cm², whilst in Green Lake ($\bar{z} = 33\cdot1$ m) it was 2·7 mg/cm². In this case the mean area-based standing crop of the deeper lake was slightly higher. Also Verduin (1956) showed that the primary production values for nine of ten lakes varying quite widely in the depth of the euphotic zone (and presumably mean depth as well) lay within or only just outside the range 1·8–2·4 gC/m²/day.

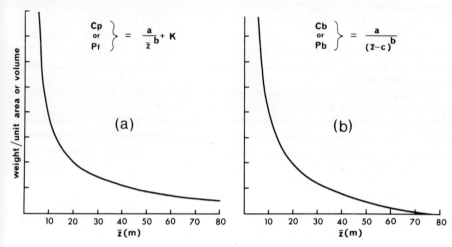

Fig. 4:1 Generalized graphs showing the type of relationship existing between the mean depth ($\bar{z}$) of a lake and (i) the mean standing crop of plankton per unit volume (Cp), (ii) fish productivity per unit area (Pf), and (iii) the mean standing crop (Cb) and productivity (Pb) of benthos per unit area. (a, b, c, and K are constants which have, of course, different values depending on whether Cp, Pf, Cb, or Pb is under consideration.) *Based on Rawson (1955)*

The relationship between fish productivity and mean depth is essentially similar to that for planktonic standing crop (Fig. 4:1(a)). Again there is a 'K', or a small but significant production in very deep lakes, which may be interpreted as production based on purely planktonic feeding from the limnetic region and the existence of insignificant numbers of bottom-feeding fish. In view of the fact that most adult fish are mainly bottom-feeders, and that the relationship between mean depth and mean benthic standing crop is of the type shown in Fig. 4:1(b), the relationship between standing crop of fish and mean depth is also very probably hyperbolic. Most of the useful fish data, however, relate to long-term productivity assessed from commercial activity; information on standing crop is not available except for very shallow lakes (z usually < 5 m). It might be argued that commercial fisheries data represent merely the harvest, not the productivity. This is true, but in large lakes (e.g. the Laurentian Great Lakes) where there has been a sustained fishing effort over a long period of time the harvest probably tends to be maximal or to approach closely the productivity.

Within a given lake district there is a definite tendency for the mean standing crop of benthos to decrease with increasing depth, and some workers (see Bayly 1969b) hold that a hyperbolic relationship (Fig. 4:1(b)) is a reasonable generalization. This time we have elimination of K from the equation. This may be taken to mean that in an infinitely deep lake, if such could exist, the standing crop of benthos would be zero.

In very deep lakes with precipitious sides, for example some caldera lakes, this may not be far from the truth. Referring to the two Wisconsin lakes mentioned above, we find that in Lake Mendota ($z = 12 \cdot 1$ m) the mean standing crop of benthos is $0 \cdot 52$ mg (dry weight)/cm^2 compared with only $0 \cdot 27$ mg/cm^2 in Green Lake ($z = 33 \cdot 1$ m). It is not difficult to account for this fall-off of benthos with increasing depth; the greater the depth, the greater the likelihood that organic matter descending from the upper layers will be mineralized before reaching the bottom, and the less the formation of ooze for benthos to browse upon. If lakes from various parts of the world (or even different lake districts of the same country) are considered together, the sort of relationship shown in Fig. 4:1(b) may be obscured (Hayes 1957). This can be attributed to differences in the edaphic and climatic factors operating in the various lake districts.

Although the number of direct estimates of benthic productivity are very few, it may be assumed that Fig. 4:1(b) is also true of it. For deep lakes comprehensive data even on mean standing crop are scarce enough but exist for Lake Nipigon, Ontario (Table 4:1). It is a safe assumption that the annual turnover ratio for the entire macrobenthic fauna of this lake is not more than about 3, so its productivity would not be more than about 12 per cent that of Lake Beloie, an intensively studied lake near Moscow (see Table 4:1). In some respects the apparent comparisons provided by Table 4:1 are unfair; for example, Lake Nipigon is very cold (<4°C) most of the year

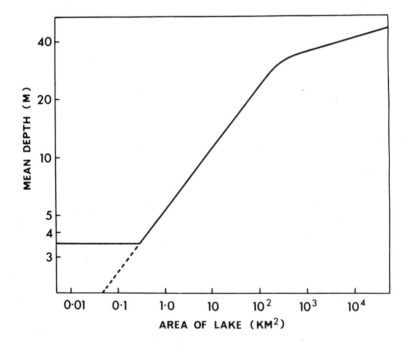

Fig. 4:2 The average relationship existing between the mean depths and areas of lakes, from data on 500 lakes in various parts of the world. The horizontal portion of the graph is probably an artifact reflecting neglect of small bodies of water by limnologists. The true relationship in this region is probably the dotted extrapolation of the middle limb. *After Hayes (1957)*

TABLE 4:1
COMPARISON OF MACROBENTHOS PRODUCTIVITY ESTIMATES FOR TWO SHALLOW AND TWO DEEP LAKES
(kg/ha dry weight)

	Lake Beloie[a]		Linsley Pond[b]		Lake Nipigon[c]	Great Slave Lake[d]
Mean depth (m)	5·6[e]		6·7		59	62
	Productivity (direct estimate)	Mean crop	Crop at time (Nov.) of max. total macrobenthos crop	Mean crop	Crop at time (Jan.) of max. total macrobenthos crop	Mean crop	Mean crop
BROWSERS							
Chironomus	74·1	21·6	18·7	31·5	44·8	0·4	
Oligochaetes	43·0	18·7	29·0	2·7	7·2	0·5	
Amphipods	—	—	—	—	—	1·5	
Molluscs	—	—	—	—	—	2·7	
PREDATORS							
Chaoborus	28·1	10·0	17·6	11·5	19·8	—	
'Tanypus'	3·4	0·9	1·8[f]	1·3[f]	—	—	
Totals	148·7	51·6	67·1	47·0	71·8	5·9[g]	2·5
Multiplication factors (annual turnover ratios)			2·9[h]			☆ 3	☆ 3
Assumed approximate productivity			136			☆ 18	☆ 8

[a] From Borutzky (1939a).
[b] From Deevey (1941) using the Juday factors 0·16, 0·09, and 0·18 for Chironomus, Chaoborus, and oligochaetes, respectively to convert wet to dry weights.
[c] Mean depth from Rawson (1955), biological data from Adamstone (1924).
[d] From Rawson (1953).
[e] Maximum depth (13 m (Borutzky 1939a)) divided by 2·35 (world mean ratio maximum to mean depth (Hayes 1957)).
[f] Including Tanytarsus and other minor constituents of the benthos.
[g] Includes 0·7 kg for insect larvae restricted to shallow regions. [h] From data for Lake Beloie.

whereas Linsley Pond becomes quite warm during the summer and, furthermore, these lakes have very different surroundings. Nevertheless the qualitative indication that, other things being equal, the productivity of a deep lake is substantially less than a shallow one is almost certainly valid.

There is good reason to believe that nektonic and benthic productivity do not increase indefinitely with decreasing mean depth (it has already been indicated that planktonic productivity is largely independent of mean depth). Study of a large number of lakes indicates that on average the photosynthetic region extends down to a depth of about 5·9 m; that is, to the bottom of a lake with a mean depth of about 2·5 m (Hayes 1957). In other words, a pond with a mean depth of 1·0 m is unlikely to be any more productive than a lake with a mean depth of 2·5 m.

For the reason that over the greater portion of the range there is a linear relationship between mean depths and areas in a logarithmic plot (Fig. 4:2), hyperbolic relationships essentially similar to those discussed above would be obtained if productivity or standing crop was considered in relation to area rather than mean depth (see Rounsefell 1946). There is every reason to believe, however, that mean depth is the more fundamental parameter.

Mean depth is an important parameter not only when present-day lacustrine populations are under consideration, but also in palaeolimnology. An inverse relationship between mean depth and quantity of cladoceran remains in bottom sediments has been found for a series of Wisconsin lakes (Frey 1960). In these lakes *Daphnia* decreases and *Bosmina* increases in relative abundance with decreasing mean depth. Chydorids also increase in relative abundance in the sediments of the shallower lakes, and since nearly all of these are strictly littoral forms,[2] this increase probably reflects the progressive shift in importance from limnetic to littoral production. Since chydorids fossilize more readily than nearly all other crustaceans, it would probably be this increasing importance of the littoral region, rather than an increase in planktonic productivity, that would account for the quantity of cladoceran remains increasing with decreasing mean depth. At all events, Harmsworth and Whiteside (1968) have shown no correlation between measurements of primary production and the numbers of cladoceran microfossil remains in two series of lakes, one from Denmark, the other from Indiana. Copepods, which typically account for most of the zooplankton biomass in living communities, decay and disintegrate before reaching the bottom or very shortly afterwards (Deevey 1964).

Since the supply of nutrients to a lake is associated with the formation of soils in the catchment area and the extent to which they are leached, the relative importance of erosion and leaching of these soils influences the potential productivity of a lake (Mackereth 1966). Intensive erosion prevents the formation of soils and results in the deposition of nutrient elements still locked in mineral particles in bottom sediments where they are unavailable to lacustrine organisms. Subdued erosion allows the

---

2. *Chydorus sphaericus* is an exception. Bayly (1962) found this living a true limnetic existence in a New Zealand lake which is eutrophic, high in sodium and (bi)carbonate, and exhibits persistent blooms of the blue-green alga *Anabaena*. This in agreement with Frey's (1960) observation that *C. sphaericus* is limnetic in Lake Mendota and that its abundance seems positively correlated with *Anabaena*. However, Fryer (1968) regards the suggestion that this species directly uses blue-green algae for food as probably incorrect. He considers that the explanation of its occurrence in open water, suggested long ago by Scourfield, is simply that it often takes an algal filament between its carapace lobes and clambers along it.

accumulation of soil in which mineral particles are held in a condition that allows leaching to bring the nutrients into solution in a biologically available form. Periods of severe and subdued erosion would thus correspond, respectively, with low and high lake productivity. With a given set of climatic conditions the productivity of a lake would be partially determined by the topography of the drainage basin since this largely determines its erosional characteristics. It is quite possible, therefore, that the basis for the above relationships between productivity and mean depth is the association of deep lakes with areas of severe relief and high rate of erosion. Shallow lakes are usually to be found in flat or rolling country, and deep ones more often than not are associated with rugged, broken country. We may note, however, that the correlation between nutrient supply and lake productivity is probably not quite as simple as might be assumed; low productivity, for example, may not be so much a function of low nutrient supply as the greater tenacity of mineral substances in binding the nutrients and thus decreasing the rate of their re-use (Frey 1969, Livingstone and Boykin 1962).

Apart from the derivation of inorganic nutrients from the catchment area, mention should also be made of the fact that some lakes receive large amounts of allochthonous organic matter. This is often in dissolved form from peats and swamps, but sometimes it is in the form of wind-blown pollen and other terrestrial plant material. The latter process may be especially important in some salt lakes, and such lakes have been described by Hutchinson (1937) as *anemotrophic*. Steinböck (1958) has called allochthonous wind-borne material *empneuston*, and referred to lakes in which the energetic basis of the food chains lies mainly outside the lake proper as *allotrophic*.

It seems appropriate to comment at this juncture that Forbes's (1887) oft-quoted essay on the lake as a microcosm, whilst stimulating formation of the concept of an ecosystem, has almost certainly retarded appreciation of the profound influence of the nature of the catchment area on the productivity of a lake. Forbes stated that 'The animals of such a body of water [a lake] are, as a whole, remarkably isolated—closely related among themselves in all their interests, *but so far independent of the land about them* [the italics are ours] that if every terrestrial animal were suddenly annihilated it would doubtless be long before the general multitude of the inhabitants of the lake would feel the effects of this event in any important way'. However, the importance of lakes in ecological studies is not that the processes and organisms occurring within them are *independent* of the surrounding land, for they certainly are not,[3] but stems from the fact that the events occur within a restricted space, not a vast continuum, and are thus easier to study and comprehend.

PHOSPHORUS

This element is commonly regarded as being more important in determining the amount of living matter in a lake than any other single factor. It is certainly very important, but is sometimes overemphasized, and the existence of other almost equally significant factors should not be overlooked; the importance of the availability of carbon (in one or other of several different forms) and trace elements such as molybdenum and cobalt is stressed below.

---

3. Many palaeolimnological studies show that changes in lacustrine communities as represented especially by chydorid and *Bosmina* remains tend to mirror changes in terrestrial vegetation as represented by fossil pollen.

As pointed out earlier, most uncontaminated lakes contain less than about 40 mg/m^3 of total phosphorus in their surface waters. However, in cases where sewage or a substantial amount of agricultural drainage enters a lake, much higher values of total phosphorus are found, and this is assumed to be the most important factor contributing to the excessive eutrophy that usually characterizes such lakes (see Chapter 12).

It appears that when large amounts of soluble phosphate are artificially added to a lake it is rapidly absorbed by the phytoplankton and then sedimented. The productivity is increased but only for a limited time. There is no permanent effect even if the treatment is repeated two or three times. We thus have a self-regulating system which returns to a state of equilibrium fairly rapidly following a disturbance. Despite this buffering of fluctuations within a given lake, quite wide variations in phosphate level occur between different natural lakes, and as a general rule those with high values are more productive than those with low amounts. This inter-lacustrine variation is probably explicable mainly on the basis of the nature of the muds and the relationship of the volume of water to the surface area of mud.

Turning from variations between lakes to temporal variations of productivity within the same lake, it may be noted that the presence of algal blooms is probably mainly determined by the onset of rapid decomposition in the littoral region during high temperatures, and a subsequent rapid output of phosphorus into the epilimnion. Despite the great increase in total phosphorus at such times, soluble phosphate may be undetectable because it is taken up so rapidly by the expanding algal population.

It is noteworthy that the addition of phosphate to a lake is probably also equivalent to an addition of, and at least a temporary increase in, nitrate, since phosphate is a nutrient for nitrogen-fixing organisms. Another point is that the concentration of organic nitrogen may be more than 100 times that of nitrogen in inorganic forms such as nitrate and ammonia.

MICRO-NUTRIENTS: MOLYBDENUM AND COBALT

Molybdenum and cobalt are essential constituents of certain plant enzymes. Molybdenum plays a part in the formation of the enzyme nitrate reductase and is also involved in nitrogen fixation. It has been shown that the blue-green alga *Anabaena cylindrica* requires molybdenum when the nitrogen source is nitrate or nitrogen gas but not when it is ammonia. Cobalt is also involved in a large number of enzyme systems and in addition it is an essential element of the vitamin B_{12} or cyanocobalamine molecule.

That the amount of molybdenum and cobalt may limit production is demonstrated by the fact that their addition stimulates photosynthesis (as measured by the ^{14}C method) in *in situ* cultures of certain lake waters (Goldman 1964, 1965). Several lakes in the South Island of New Zealand including Lakes Lyndon and Coleridge (Mb < 0.07 p.p.b., Co < 0.4 p.p.b.) have been found to respond in this way to the experimental addition of molybdenum and cobalt. In the case of cobalt, maximum response is usually reached at about 20 µg/l. The effect of adding molybdenum to Castle Lake in California, the first lake for which a deficiency of the element was demonstrated, is illustrated in Fig. 4:3.

Zinc may also act as a micro-nutrient limiting factor. Thus its addition to Lake Coleridge water (Zn < 1.3 p.p.b.) increased photosynthetic carbon fixation more than 40 per cent over the control, but a similar addition to Lake Lyndon water (Zn 9·3 p.p.b.) produced no effect. On occasion iron, manganese, and even magnesium (Goldman 1961) can also play a limiting role. Magnesium is, of course, an essential element in the chlorophyll molecule.

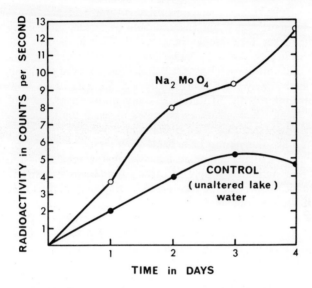

Fig. 4:3 The stimulating effect of adding molybdenum (as Na_2MoO_4) to a lake deficient in that element. *After Goldman (1965)*

CALCIUM AND THE AVAILABILITY OF CARBON IN ONE OR OTHER OF SEVERAL DIFFERENT FORMS

The German limnologist Ohle once proposed the classification of lakes into 'poor', 'medium', and 'rich' on the basis of calcium content as follows:

<10 mg Ca/l   poor
10–25 mg Ca/l   medium
>25 mg Ca/l   rich.

There has been a tendency on the part of some limnologists to assume that calcium content has a fundamental bearing on lake productivity and to use Ohle's designations as if they were productivity terms. The importance of calcium as a limiting factor in the distribution of freshwater molluscs with relatively large calcareous shells appears to have been adequately demonstrated (e.g. Macan 1961). Molluscs are divisible into hard-water species typically found in waters containing more than about 20 mg/l, and those which can tolerate less than this. However, there appears to have been an unwarranted extrapolation and application of this finding to other taxonomic groups, for example Crustacea, and from the benthic to planktonic and nektonic communities. The inability of freshwater crustaceans to satisfy their calcium requirements when only small amounts are present in the external medium does not appear to have been satisfactorily demonstrated. Bayly (1964) showed that all of a series of sixteen coastal dune lakes in southeast Queensland contained less than 1·0 mg/l of calcium yet they were inhabited by quite an abundance of a large species of the crayfish *Cherax* and the smaller shrimp *Caridina*. Similarly, Nygaard (1955) reported *Astacus fluviatilis* thriving in a lake containing only 1·8 mg/l of calcium.

In the case of planktonic crustaceans there is some evidence that carbonate and bicarbonate are more significant than calcium as factors influencing abundance. Certain alkaline lakes, for example some African lakes in the Rift Valley and Lake Aroarotamahine in New Zealand, contain large amounts of carbonate and bicarbonate but very little calcium. In such lakes the population density of planktonic crustaceans may be very high (Bayly 1962, Worthington and Ricardo 1936). Worthington and Ricardo pointed out that the African lakes Bunyoni, Naivasha, Edward, and Rudolf form a series along which alkalinity increases to reach the highest value in Lake Rudolf. They also stated that 'The richness in numbers of the zoo-plankton also increases from Lake Bunyoni to Lake Rudolf'. It seems reasonable to suppose that large amounts of (sodium) carbonate increase phytoplankton production by providing an almost unlimited supply of carbon for photosynthesis. In these cases carbonate and bicarbonate can be considered to have an important indirect effect on planktonic crustacean abundance. Fogg (1968:16) stated that there is no evidence of any plant being able to make use of the carbonate ion. However, the fact that some algae thrive in highly alkaline lakes in which much more carbonate than bicarbonate is present makes it highly probable that it is used, if not directly, then after some sort of transformation. On the basis of what is known of the chemistry of *Spirulina* habitats, together with its capacity for very high rates of photosynthesis (Wood, in press), it would be surprising if this genus were incapable of utilizing carbonate. Hutchinson (1967:309), quoting work by Felföldy, pointed out that some algae may be presumed to use the carbonate ion.

A considerable amount of attention has been paid to the distribution and abundance of freshwater higher crustaceans such as amphipods in relation to calcium, and significant correlations between calcium concentration and presence or absence have been obtained. However, since in dilute inland waters there is a high positive correlation between calcium and bicarbonate, similar correlations may be obtained with bicarbonate. It is surprising, therefore, that bicarbonate has not been given at least as much stress as calcium as a possible factor in determining crustacean distribution and abundance, especially in view of the well-established importance of carbon in photosynthesis and as a basis for organic compounds. For some reason calcium correlations have been more fashionable.

It is important to realize that in low-salinity (<100 mg/l) lake waters as much or even more carbon may be present in dissolved organic matter as in the form of carbon dioxide and bicarbonate. This provides a basis for understanding the following situation: it sometimes happens that in a circumscribed area there exists a series of dilute, ionically homogeneous, and bicarbonate-poor lakes (a series like this may be found on granite or amongst siliceous dunes) which vary strikingly, however, in the amount of coloured dissolved organic matter. Within such a series the abundance of zooplankton is usually greatest in the darkest lakes because of their greater supply of carbon. The utilization of carbon by zooplankton in these humified, bicarbonate-poor lakes may be largely independent of phytoplankton; there is often a striking lack of correlation between zooplankton and phytoplankton abundance. There is increasing evidence that highly humified waters usually have large population densities of bacteria (see e.g. Andronikova 1965) and that these play a major role in converting organic matter in solution into cell substance which is then immediately available as food for zooplankton. Hayes (1963) found that in lakes with an alkalinity of less than 10 p.p.m. counts of sediment bacteria correlated better with colour alone than with alkalinity, but that in lakes with an alkalinity of more than 10 the reverse was true. Parenthetically it may be noted here that another situation in which zooplankton may feed more on bacteria than phytoplankton is meromictic lakes. A characteristic of these lakes is the development of high population densities of green or purple sulphur bacteria in the region of the chemocline. The work of Bicknell and Bunt (1952) and Jackson and Dence (1958) suggests that in Sodon Lake and Green Lake, bacteria constitute the chief food for zooplankton. Saunders (1969) recognized that such habitats constitute a special case in which bacterial feeding is of major importance, but he made no reference to highly humified waters in which the metabolism of carbon compounds is more important than sulphurous ones.

Dissolved organic matter (D.O.M.), as carbamino carboxylates, may also be of significance as a source of carbon in non-humified alkaline waters (Smith, Tatsumoko, and Hood 1960). These authors even suggest that phytoplankton may use complexed carbon dioxide (carbamino carboxylates) in preference to inorganic forms.

To summarize, the view is taken that the availability of carbon, which in total we might represent as [CO_2 + (HCO_3^- + $CO_3^=$) + (C in D.O.M.)] is important in lake productivity and more so than is calcium.

Mention is made above of the role of bacteria in the indirect utilization of dissolved organic matter by zooplankton. In some cases this utilization may be more direct. As discussed in Chapter 2, dissolved organic matter is produced not only by decay but directly by perfectly healthy phytoplankton and it may include components that can be directly assimilated by plants or animals. Compounds which escape from living algal cells include simple organic acids, carbohydrates, amino-acids, and polypeptides. The inclusion of nitrogenous compounds such as the latter fits in with our earlier claim that organic nitrogen, as well as organic carbon, may considerably exceed their respective inorganic forms, especially in dilute waters. The existence of this array of extracellular products means that Pütter's hypothesis that dissolved organic matter is important in the nutrition of zooplankton must now be given serious consideration, despite earlier rejection. Even if direct assimilation of dissolved organic matter by aquatic animals proves not to be of major significance, another possibility exists; in recent years attention has been directed to the fact that particulate organic matter can be produced from dissolved organic matter by the purely physical process of bubbling (Sutcliffe, Baylor, and Menzel 1963). These authors suggested that wind-induced bubbles and circulation of windrows in oceans and lakes should be considered a source of particulate organic matter and of absorbed phosphate, and that this process might provide new insight into the problem of the re-cycling of non-living organic matter. Ragotzkie and Bryson (1953) reported that the abundance of *Daphnia* in Lake Mendota was higher in windrows than elsewhere, but whether this was due to increased abundance of food is not clear. However, the fact that particulate material produced by bubbling has nutritive value has been shown by Baylor and Sutcliffe (1963). On the other hand, Saunders (1969) considers it very unlikely that the mechanisms proposed by Sutcliffe *et al.* (1963) for the production of organic detritus is a general phenomenon in fresh waters.

SALINITY OR TOTAL DISSOLVED SOLIDS

Rawson (1951) stated that 'It is possible that our preoccupation with individual elements [such as Ca, N, and P] has led us to neglect a simple and useful clue to the conditions which lie behind productivity'. It might perhaps have been a fairer statement if Rawson had said that it is convenient to use easily determined T.D.S. values because there is a significant positive correlation between these and certain individual elements which not only are more difficult to estimate, but have a more fundamental influence on productivity. It is very doubtful whether salinity *per se* has any fundamental bearing on productivity. Nevertheless Rawson, and more recently Larkin and Northcote (1958), are quite justified in regarding T.D.S. or salinity as a very useful index of lake production. They take the view that T.D.S. summarizes aspects of both the substrate and the climate as well as their interaction and that it is probably the best single indicator of a large portion of the environmental setting. However, some of the above workers admit that there is a tendency for high carbonate and bicarbonate values to be associated with high T.D.S. content. This being so it would seem that one of the important bases of correlation between T.D.S. and production is the availability of carbon already discussed above.

WATER RENEWAL

The magnitude of influence of the flow of water through a lake will depend on the volume of the lake, the extent of its catchment area, and the amount of rainfall on this area. Thus the effect will be greatest in a small shallow lake with a large catchment area receiving high rainfall, and will become less significant the bigger the lake and the smaller the catchment area and rainfall.

One index of the rate of water renewal is the *replacement quotient* which was defined by Brook and Woodward (1956) as the volume of water contained in a lake divided by the amount of water passing through it during one day. It has been shown that where the rate of water renewal is low, planktonic population densities tend to be highest and most stable and *vice versa*. Rapid outflow could obviously bring about a significant loss of planktonic organisms, and sometimes the rate of replacement may be a factor of overriding importance, producing not only quantitative but also qualitative differences in planktonic communities.

It is often a difficult matter to determine with accuracy the daily amount of water passing through a lake, and not uncommonly the volume is unknown. Thus a useful but apparently less accurate index of the rate of water renewal is given by the ratio (area of drainage basin) : (area of lake). Both of these parameters are readily obtainable from scaled aerial photographs or contoured topographical maps. Using this ratio, Ravera and Tonolli (1956) also observed that as the rate of water renewal increases the planktonic productivity of a lake decreases. If the volumes of the lakes are known it would seem that a better index would be obtained if volume is substituted for area. However, Timms (1968), working on a series of Queensland lakes, experimented with several different indices of water renewal and found that the best correlation with calanoid egg number was obtained using the same ratio as Ravera and Tonolli.

We have already warned, when discussing calcium, against extrapolation from benthic to planktonic and nektonic communities, and clearly water renewal and outflowing water have a more profound influence on plankton than benthos. Perhaps, therefore, we are not fully justified in speaking of lake productivity as if it were monolithic. Strictly speaking, planktonic and benthic productivity should be considered separately, although undoubtedly the correlation between the two is high.

INTERACTION OF FACTORS INFLUENCING PRODUCTIVITY

We can be sure that a shallow lake with large amounts of phosphorus, nitrogen, molybdenum, cobalt, and carbon, and a low rate of water renewal, will be highly productive. Likewise a deep lake in which these factors are reversed will certainly have a low productivity. In nature, however, there is at least some measure of independence in the way that these factors may vary, and at the moment we have no means of quantitatively describing the resultant productivity given the intensity of each. In addition, there is always the possibility that significant factors additional to those discussed remain to be discovered.

Despite these difficulties it is worthwhile discussing one interesting investigation in which some attempt was made to assess the relative importance of different factors influencing productivity. Hayes and Anthony (1964) derived an equation relating an index of fish productivity to lake dimensions and alkalinity. Their equation accounted for 67 per cent of the variability in production index, of which 20 per cent was due to area, 29 per cent to depth, and 18 per cent to alkalinity. An alternative interpretation is that 49 per cent was due to morphology and 18 per cent to the availability of carbon. Although these workers sometimes substituted 'water chemistry' for 'alkalinity' it is clear that only the latter was intended; their investigation of water chemistry was by no means comprehensive and factors such as phosphates, nitrates, and micronutrients such as cobalt and molybdenum very probably contributed towards the 33 per cent of variability that was unaccounted for. They did mention water renewal as a factor contributing towards the uncounted variability. Hayes and Anthony interestingly pointed out that alkalinity had no meaningful relation to productivity when considered by itself, but acquired considerable significance when their triple regression technique was employed. The dominance of morphology was evidently a major factor in masking the effect of alkalinity alone in the single regression approach.

chapter five
PHYSICAL AND RELATED CHEMICAL AND BIOLOGICAL PHENOMENA

Thermal Stratification

DEFINITIONS AND THE DIFFERENT TYPES OF MIXING If temperature is measured at a series of depths in a temperate, sub-tropical, or tropical lake during mid-summer the type of curve shown in Fig. 5:1A will usually be obtained. When this condition exists the lake is said to be *thermally stratified*, and it is divisible into several regions which may be defined as follows (also see Fig. 5:1B).

Thermocline—the *plane* of maximum rate of change in temperature. That is, the thermocline is a plane having no vertical depth.

Metalimnion—the whole region in which the rate of change of temperature with depth is comparatively rapid. It thus has vertical depth—from the 'knee' to the 'ankle' of the temperature-depth graph or more exactly from the upper plane of maximum graph curvature to the lower plane of maximum inverse curvature.

Epilimnion—the body of water lying above the metalimnion.

Hypolimnion—the body of water lying below the metalimnion.

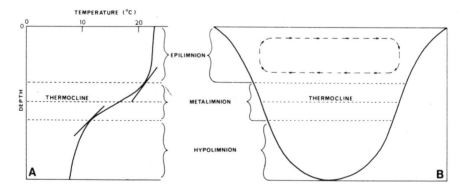

Fig. 5:1 A: a typical temperature-depth curve for a temperate or subtropical lake in mid-summer. B: the terminology for the various regions of the lake corresponding with temperature-depth curve shown in A

In a thermally stratified lake the epilimnion circulates within itself but does not mix with the hypolimnion. However, this incomplete mixing is only a temporary condition and the majority of lakes mix completely at least once a year. A *dimictic* lake is one that circulates completely twice a year. In a *monomictic* lake complete circulation occurs only once a year. Hutchinson and Löffler (1956) divide the monomictic type into *warm monomictic* lakes, in which complete mixing or *holomixis* occurs in winter at a temperature above 4°, and *cold monomictic* lakes with holomixis in summer at a temperature below 4°. The latter are found chiefly in the Arctic and Antarctic circles. Next there are *polymictic* lakes, which either never stratify or have no persistent thermal stratification; heating and temporary stratification may occur during the day but at night much of the heat is lost and holomixis takes place. According to Hutchinson and Löffler this occurs only in the tropics and especially in large shallow lakes exposed to dry winds or in high-altitude lakes. However, it is clear that, in the southern hemisphere at least, polymictic lakes can occur at substantially higher latitudes and lower altitudes than those indicated by Hutchinson and Löffler (1956, fig. 1). Thus most of the large shallow lakes on the Central Plateau of Tasmania (lat. 43°S, altitude *ca.* 1,000 m) including Great Lake, Arthur's Lakes, and Lakes Sorell and Crescent are polymictic (P. A. Tyler, pers. comm.). Lake Dobson (lat. 43°S, altitude 1,030 m), another Tasmanian lake, is also polymictic (Weatherley and Nicholls 1955). Lake Hiawatha in northern New South Wales (lat. 30°S, altitude 17 m) is another example of this type (Timms 1969). This lake has a high transparency, a maximum depth of about 10 m, and is rather exposed. Several shallow lakes (fresh and saline) in the Western District of Victoria are polymictic (Hussainy 1969). Lakes Rotoehu and Rotorua (lat. 38°S, altitude 300–400 m) in New Zealand are also polymictic (Irwin 1968, Jolly 1968). These two lakes have mean depths of 8·3 m and 11·0 m respectively. In addition, many small exposed lakes in the South Island are of this type; Stout (1969a) mentioned the absence of a temperature gradient in summer as being characteristic of the smaller Canterbury lakes and attributed this mainly to the mixing action of strong winds. That polymictic lakes can occur in temperate as well as tropical regions is correctly indicated by Dussart (1966). Evans (1970) showed that polymictic lakes can exist even in the Subantarctic; Lake Prion (A = 36·3 ha, z_m = 32·2 m, $\bar{z}$ = 17·5 m) on Macquarie Island (lat. 54° 30′S) is polymictic as a result of its exposure to high winds and the slow rate of warming during the cool summers. Its annual temperature range is about 1–9°C. Next there are *oligomictic* lakes. These are mainly tropical and are usually small or very deep, or both. They are warm at all depths, but a small temperature difference is adequate to maintain a very stable stratification, and holomixis occurs at rare irregular intervals when abnormally cold and windy weathers occur. Lakes Nkugute, Bunyoni, and Edward in the Rift Valley, Africa, may be cited as examples (see Beadle 1966). It seems highly probable that Lake Kutubu in the Southern Highlands of New Guinea is oligomictic; there are reports that at irregular intervals there is a discoloration of the lake accompanied by bad odours and a mass mortality of fish (Bayly, Peterson, and St John 1970). Finally, there is the *amictic* lake. This is restricted to very cold regions like the Antarctic and is characterized by having a permanent ice-cover.

By far the most important type in Australia and New Zealand is the warm monomictic lake, accounts of which are given by Jolly (1952, 1957a, 1958, 1966, and 1968), Bayly (1962), Irwin (1968), and Timms and Midgley (1969). Although dimictic lakes are important in Europe and North America, there is no known example of this type

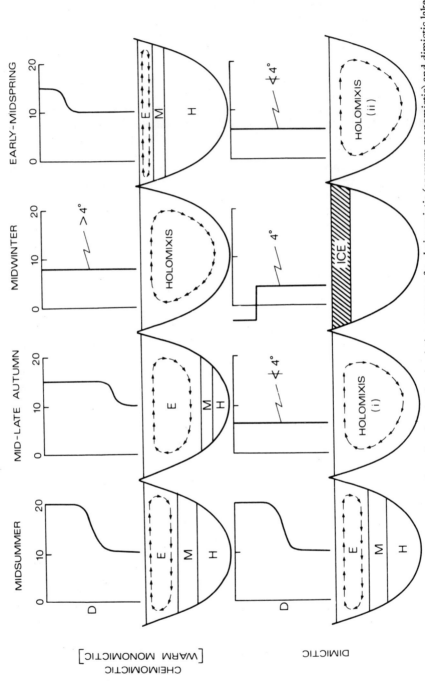

Fig. 5:2 A summary of the seasonal changes that take place in the circulation patterns of a cheimomictic (or warm monomictic) and dimictic lake. The temperature (°C) is shown on the abscissa and the depth (D) on the ordinate. E=epilimnion, M=metalimnion, and H=hypolimnion

PHYSICAL AND RELATED CHEMICAL AND BIOLOGICAL PHENOMENA 81

in Australia or New Zealand. The seasonal cycle of important events in warm monomictic and dimictic lakes is summarized in Fig. 5:2.

It is noteworthy that the thermal classification proposed by Hutchinson and Löffler (1956) does not accommodate some Australian and New Zealand lakes. These are ones that freeze over in winter and which circulate completely in summer at a temperature considerably above 4°C. As such they are only monomictic but do not fulfil the requirements of either a warm monomictic or a cold monomictic lake. It thus appears that the monomictic type would be better divided into *cheimomictic* lakes, which have a winter holomixis at temperatures above 4°C, and *thereimictic* lakes, which have a summer holomixis. The latter would be further divided into *cold thereimictic* lakes, with a temperature below 4°C at the time of holomixis, and *warm thereimictic* lakes, with holomixis at a temperature above 4°C. Lake Nicholls, a cirque lake in Mount Field National Park, Tasmania, is an example of the latter type; in the summer of 1966 (a 'normal' one) it was found to be isothermal at 15°C from top to bottom (Bayly, Peterson, and Williams, unpublished), yet it usually freezes over in winter. The maximum depth of Lake Nicholls is 11 m and its altitude is 976 m. Many other small highland lakes in Tasmania, and some of the Mount Kosciusko lakes on the Australian mainland, probably belong to this category as well. So, too, do many of the smaller highland lakes in the South Island of New Zealand. Thus Lake Lyndon (altitude 890 m) often freezes over in winter but is not stratified in summer and has a temperature greater than $4 \cdot 0°C$ (up to 18°C recorded) (Percival 1949, Stout 1969b). Lakes Ida, Ackland, and Roundabout, all in Canterbury, provide even better examples of the warm thereimictic type of lake in that they almost invariably freeze (Stout 1969a). Some lakes alternate between being polymictic and warm thereimictic, depending on the severity of winter, but lakes that are almost permanently of the warm thereimictic type also exist. Warm thereimictic lakes would of course circulate completely in spring and autumn as well as in summer.

This type of lake probably also occurs in southern South America, especially at high altitudes. If so, it would be a distinctive feature of the limnology of Tasmania, New Zealand, and South America and reflect the fact that they are all relatively narrow peninsulas or islands strongly subject to the climatic influence of the ocean; it becomes just cold enough for winter freezing, but they are not subject to strong continental heating in summer such as occurs in the broader northern continents. High summer winds may be another factor preventing stratification at this time.

The proposed modification to the Hutchinson-Löffler thermal classification of lakes will be clear from the following outline:

1 Amictic
2 Monomictic
  a Cheimomictic (= warm monomictic *sensu* H. & L.)
  b Cold thereimictic (= cold monomictic *sensu* H. & L.)
  c Warm thereimictic (a new type)
3 Polymictic
4 Dimictic
5 Oligomictic
[Meromictic (see below)]

As indicated above, warm thereimictic and polymictic lakes are somewhat similar and may occur together in the same region, the smaller lakes belonging to the former type and the larger ones to the latter. Some lakes may alternate between the two, depending on the severity of winter.

It may also be noted that oligomictic and meromictic lakes have several features in common.

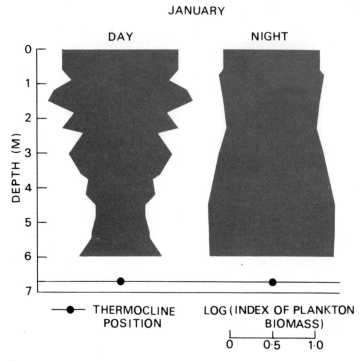

Fig. 5:3 The mean day and 'night' (more correctly post-dusk to midnight) vertical distribution of total zooplankton in Lake Aroarotamahine, Mayor Island, New Zealand, over a twelve-day period in January 1958. Note the sharp cut-off in distribution that occurs some distance above the thermocline and close to the upper limit of the metalimnion. The results appear anomalous because of the occurrence of reversed diurnal vertical migration in the *Daphnia* population. After Bayly (*1962*)

THE CHEMICAL AND BIOLOGICAL CONSEQUENCES OF THERMAL STRATIFICATION Once thermal stratification is established, secondary chemical stratification usually ensues. Even if thermal stratification *per se* were to have no striking biological effect, then chemical stratification probably would.

When in mid- or late-spring the circulation of dimictic and cheimomictic lakes becomes restricted to the epilimnion, the amount of oxygen in the hypolimnion starts to decrease, and at the height of summer stagnation may fall to zero. At the same time the amount of carbon dioxide in the hypolimnion rises above and the pH falls below that in the epilimnion. Considering the situation at the height of summer, the biological consequences of stratification depend on its intensity. If the temperature gradient is very steep, and the hypolimnion very deficient in oxygen or completely deoxygenated, then the vertical distribution of most organisms (anaerobic bacteria

PHYSICAL AND RELATED CHEMICAL AND BIOLOGICAL PHENOMENA 83

and *Chaoborus* excepted) may be sharply cut off at or close to the upper limit of the metalimnion or the thermocline. This has been shown for two New Zealand lakes, Lake Aroarotamahine (Bayly 1962) (Fig. 5:3) and Lake Pupuke (Green 1967) (Fig. 5:4). If this happens, then the epilimnion in itself constitutes almost the entire productive volume of the lake at this time of the year. It may be noted at this point that, other factors being equal, the smaller the area of the lake the closer the thermocline to the surface (Table 5:1). Small sheltered lakes have very shallow thermoclines, and in summer most of their volume is unproductive. In Lake Aroarotamahine, for example, the summer productive volume may be only about 30 per cent of total.

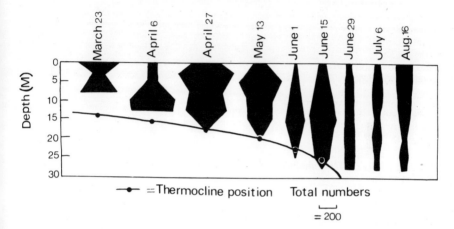

Fig. 5:4 Seasonal vertical distribution of total zooplankton in Lake Pupuke, Auckland, New Zealand. *Redrawn from Green (1967)* and somewhat modified in accordance with text comments by that author to the effect that on two occasions there was a malfunctioning of the sampling device

If, on the other hand, the temperature gradient is not too steep and the hypolimnion not very depleted in oxygen, then certain organisms may even 'prefer' or be limited to the metalimnion or hypolimnion rather than the epilimnion. Thus Lake (1957), working on Lake Canobolas near Bathurst, New South Wales, found that trout (*Salmo trutta* and *S. gairdneri*) preferred the temperature range of 13–15° in the metalimnion to that of the epilimnion, where temperatures were as high as 22°. Similarly, Dendy (1945) found that in thermally stratified North American waters the sauger (*Stizostedion canadense*) selected the deepest and coldest hypolimnetic water which still contained enough oxygen. Bērziņš (1958) work on the rotifer genus *Polyarthra* in a stratified Swedish lake showed that three species, *P. euryptera*, *P. vulgaris*, and *P. remata*, lived in the epilimnion, one, *P. major*, in the metalimnion, and another, *P. longiremis*, in the partially deoxygenated hypolimnion.

In a thermally stratified lake nitrogen and phosphorus may be several times more abundant in the hypolimnion than in the epilimnion by the middle of summer. This results from a steady fall of dead planktonic organisms down into the hypolimnion, and the fact that incomplete circulation prevents the redistribution of these elements. Diminishing amounts of nitrogen and more especially phosphorus may become limiting and retard diatom and other algal production in the epilimnion. The winter overturn in cheimomictic lakes, and the autumnal and spring overturn of cimictic lakes (see Fig. 5:2), are of fundamental importance in bringing about a uniform redistribution of these essential elements.

TABLE 5:1
COMPARISON OF MORPHOMETRIC PARAMETERS AND METALIMNION POSITION

Feature of Lake	Tikitapu[a]	Aroarotamahine[b]
Latitude (°S)	38	37
Altitude (m)	413	3
Area (sq km)	1·55	0·103
Maximum length (km)	1·3	0·720
Average width (km)	1·2	0·142
Maximum depth (m)	24	22·3
Depth range of metalimnion (m)		
February 1956	15–18	—
January 1957	—	3·0–6·0
January 1958	—	5·4–8·9

[a] Data after Jolly (1957a), who used 'thermocline' in the sense of 'metalimnion' in the present work.
[b] Data after Bayly (1962).

If the oxygen content of a hypolimnion falls to almost zero, the level at which the redox potential[1] is 0·2 volts may migrate upwards through the sediments, past the mud-water interface, and into the hypolimnion. Under these conditions, which are usually also accompanied by the accumulation in the hypolimnion of a considerable amount of carbon dioxide, the ferric iron in the bottom sediments is reduced and goes into solution as ferrous bicarbonate (Mortimer 1941, 1942). The iron content of such a hypolimnion thus continuously increases throughout summer stagnation. During this period the mud is a black colour. However, as soon as complete circulation sets in and oxygen is introduced, redox potentials in excess of 0·2 volts migrate downwards to reach the mud-water interface where an insoluble ferric hydroxide complex is precipitated. The surface of the mud now turns an orange or brown colour. The brown layer seems to act as a barrier that prevents a large number of ions from moving into the open water.

The amount of iron in surface waters is very small, but it is not entirely absent— some is present in suspended or complex form but usually there is none in true ionic form. Manganese behaves very similarly to iron except that it reduces more easily and is harder to oxidize. Dissolved manganese can thus be found in higher concentrations of oxygen than can ferrous iron.

As already mentioned in Chapter 1, a considerable amount of ammonia may be formed in oxygen-deficient hypolimnia, but it is usually absent from oxygenated epilimnia.

---

1. Students not familiar with the concept and theory of reduction-oxidation potential should consult a good elementary text on physical chemistry.

It may be noted that in dimictic and warm and cold thereimictic lakes significant photosynthesis may still occur in winter beneath the ice-cover. Sauberer (1950) showed that clear ice transmits light almost as well as distilled water, so that even quite a thick layer of it has little or no effect on the utilization of light by phytoplankton. However, if the ice contains a lot of air bubbles or is covered by a layer of snow, then very rapid attenuation of light occurs and much of the plankton, especially obligate autotrophic algae, may die off. It is for this reason that winter is commonly the time of maximum nitrogen and phosphate values in a dimictic lake. However, even if snow cover produces complete darkness in the water some zooplankton may persist and grow, and it seems likely that they are nourished by very small heterotrophic algae (Rodhe 1955).

Meromictic Lakes

These may be defined as lakes in which some water remains partially or wholly unmixed with the main mass of water at the normal circulation period or periods (cf. *holomictic* lakes discussed above). In other words, they have a permanently stagnant layer of water on the bottom. Such a lake has an upper *mixolimnion* which usually circulates completely, and a lower stagnant *monimolimnion*. The boundary between these two regions is called the *chemocline*. Typically there is a marked difference in the salinity of the mixolimnion and the monimolimnion, and the chemocline may be defined as the plane of maximum rate of change in salinity. Some oligomictic lakes (see above) may be regarded as having a strong tendency towards meromixis.

Soap Lake, a closed saline lake in the lower Grand Coulee region of central Washington, U.S.A., provides an example of a meromictic lake (Anderson 1958a). The mixolimnion has a depth of about 16 m and the monimolimnion stretches from about 20–28 m. Before recent dilution the average salinity of the mixolimnion was about 30‰ while the lowest 4 or 5 m had a fairly uniform salinity of somewhat more

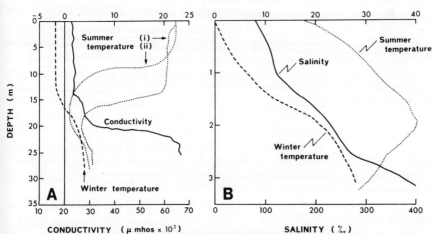

Fig. 5:5 Temperature-depth and salinity (or conductivity)–depth graphs for two meromictic lakes. A: Soap Lake, Washington, U.S.A. *After Anderson (1958a)*. See text for further explanation. B: Hot Lake, Washington, U.S.A. *After Anderson (1958b)*

than 110‰ (maximum of 144‰ recorded) (Fig. 5:5A). The thermal behaviour varies from year to year. In some summers the 'mixolimnion' becomes strongly stratified thermally as in a holomictic lake of temperate regions, so that the term temporarily becomes a misnomer for the lower portion of that region ('summer temperature (ii)' in Fig. 5:5A). During other summers the mixolimnion is kept mixed and almost isothermal. Because of the slow rate of diffusion and the fact that the steep salinity gradient minimizes turbulent exchange, the salinity of the monimolimnion changes only slowly. It is even suggested that the recent dilution of the mixolimnion should protect the monimolimnion against turbulent exchange by steepening the density gradient. An additional mechanism probably also operated to preserve meromixis. Very large crystals of mirabilite ($Na_2SO_4.10H_2O$) are found on the bottom in contact with the monimolimnion, and it is possible that this compound is precipitated at higher levels in winter but undergoes solution in summer, producing dense water which trickles down to the bottom. This would help to maintain the high salinity of the monimolimnion.

Another well-investigated meromictic lake in the United States is Green Lake near Syracuse in New York State (Eggleton 1956). This lake has a depth of about 60 m and the chemocline is located at a depth of 15–25 m; oxygen is always present above 15 m but there is never any below 25 m; conversely hydrogen sulphide is always absent above 15 m but always present below 25 m. The mixolimnion has a specific conductivity (at 20°C) of about 1,800 μmho whilst that of the monimolimnion is about 2,500 μmho.

The abandoned excavation for a third set of locks for the Panama Canal filled with sea water from Panama Bay during high tides and on top of this came freshwater run-off. The result was the formation of a meromictic lake (Miraflores Third Locks Lake) in which the temperature at the chemocline is about 6°C higher than at the surface (Bozinak, Schonen, Parker, and Keenan 1969).

Hot Lake, a small shallow lake (area 1·3 ha, maximum depth 3·3 m) in Washington State, is another meromictic lake with some unusual features (Anderson 1958b). The major ions are magnesium and sulphate, and epsomite ($MgSO_4.7H_2O$) was mined from it for a short time close to 1920. It appears that the lake was re-flooded after mining operations and that as fresh water entered from surface run-off it became stratified and meromixis established. At first sight it seems surprising that meromixis should be maintained in such a shallow lake. However, the lake is well protected from the wind, and each spring fresh water runs off on to the surface to steepen the salinity gradient, and add to the stability of the system. In addition, winter freezing causes precipitation of salts and this compensates for loss of ions from the monimolimnion by diffusion into the mixolimnion. As can be seen from Fig. 5:5B separation into mixolimnion and monimolimnion is not as distinct as in Soap Lake. As the name might suggest, this lake first attracted attention because of local reports that subsurface temperatures were unusually high. In summer the surface temperature is usually about 20°C whilst that at 2 m is about 45°C. Down to this depth the water is fairly clear, but the monimolimnion consists of dark brown water containing hydrogen sulphide. Just below the two-metre level there is a thin layer of opaque green-coloured water containing a high population density of sulphur bacteria. Light is freely transmitted through the transparent mixolimnion, but is rapidly absorbed and converted to heat in the bacterial zone and dark monimolimnion. This heat cannot easily escape because of lack of circulation between the two layers. The mixolimnion, on the other hand, although it can gain a considerable amount of heat from light energy and from atmospheric transfer, also loses it at night and during winter. The most important means of heat escape from the monimolimnion is by conduction to the mixolimnion and to the bottom, but this is a comparatively slow process. It may be noted that in Hot Lake in winter, temperatures of more than 20° may exist on the bottom while there is ice on the top! Lakes similar to this are known in Hungary and temperatures up to 56°C have been recorded at depths. Another one near Elat on the Sinai Peninsula has recently been described by Por (1968). It is noteworthy that considerable experimentation has been carried out in Israel on the trapping of solar energy in artificially constructed meromictic or 'solar' ponds (Tabor 1966). The principle of these is the same as that outlined for Hot Lake. Indeed, it was a natural Hungarian 'hot' lake that inspired Block to propose the artificial solar pond as a means of harnessing the sun's energy on a large scale. From the theoretical viewpoint the idea has great potential.

Meromictic lakes have been reported from the Antarctic. Lakes Bonney and Vanda on the west side of McMurdo Sound are both meromictic with highly saline monimolimnia (Wilson and Wellman 1962, Armitage and House, 1962, Angino, Armitage, and Tash 1965, and Hoare 1966). These lakes have a permanent ice-cover, but temperatures as high as 25°C have been recorded at the bottom of Lake Vanda (Fig. 5:6). The latter situation apparently arises from the trapping and storing of solar energy. However, this is disputed by Angino et al. (1965) who consider that geothermal heat and hot volcanic springs together account for the high temperature of the monimolimnion. The depth of Lake Vanda is 66 m, and the chemocline lies between 55–60 m. It should be noted that Lakes Bonney and Vanda are both amictic and meromictic.

Lake Miers, another lake on the west side of McMurdo Sound, is also meromictic with a bottom temperature of about 5°C (Bell 1967), despite the fact that it is in an area with a mean annual air temperature of about $-20°C$. Baker (1967) recorded a surprising variety of planktonic algae from depths of 15–18 m in this lake, and raised the question of whether this phytoplankton is adapted to photosytheis at very low light intensities or is able to live heterotrophically. He pointed out that these algae could survive low summer light intensities and the six months of winter darkness by heterotrophically assimilating dissolved organic matter produced by summer photosynthesis of surface and littoral plants.

Recent accounts of meromictic lakes include those of Duthie and Carter (1970), Eckstein (1970), and Merilainen (1970). The only known meromictic lake in Australia is West Basin Lake, an unusually well-sheltered lake in western Victoria (B. V. Timms, unpublished). None appears to have been reported from New Zealand.

Dichothermic lakes are those which have a minimal temperature some distance up from the bottom. That is, the temperature rises both above and below this level. Dichothermy is usually associated with meromixis, and more specifically the temperature minimum is usually located at or near the chemocline. Both Soap Lake (Fig. 5:5A) and Green Lake, New York State, are dichothermic, at least in summer.

Optical Phenomena

THE RECEPTION AND REFLECTION OF SOLAR RADIATION AT THE SURFACE, AND ITS PENETRATION THROUGH WATER Solar radiation provides the most important means whereby lakes are heated (geothermal heating is only rarely of major significance[2]). It is also the major source of energy for synthesis of organic compounds (chemosynthesis is normally unimportant), the sole source of energy for photosynthesis, and an important triggering factor in the vertical migration of plankton. The nature and intensity of solar radiation is thus of fundamental importance in limnology.

Fig. 5:6 Temperature-depth, conductivity-depth, and density-depth graphs for Lake Vanda, Antarctica. *After Hoare (1966)*

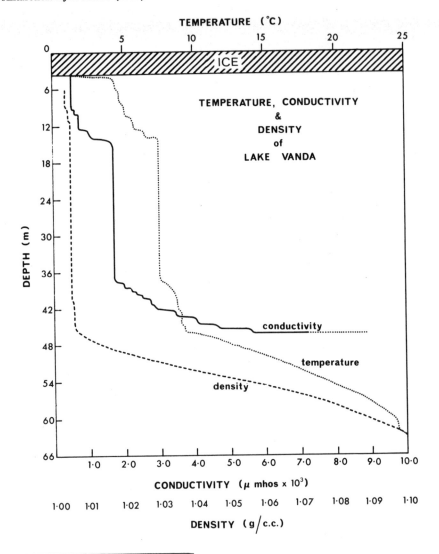

---

2. Yet Jolly (1968) raises the interesting possibility that geothermal heat is responsible for the temperature of Lake Rotomahana being about 3° higher on the average than its near neighbour, Lake Tarawera.

PHYSICAL AND RELATED CHEMICAL AND BIOLOGICAL PHENOMENA

Figure 5:7 shows the spectral composition of the solar radiation that is received above the earth's atmosphere and the way in which this is modified by its passage through the atmosphere to become the solar radiation incident on the earth's surface. Infra-red radiation accounts for about half the total radiation reaching the surface of a lake.

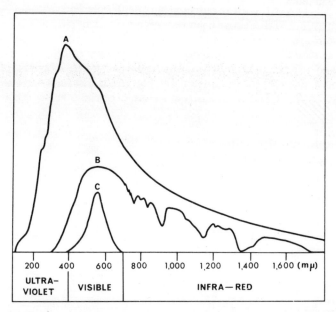

Fig. 5:7 The spectral distribution of radiant energy: (A) outside the earth's atmosphere and (B) at the earth's surface (typical distribution only). (C indicates the spectral sensitivity of the human eye.) *After Hutchinson (1957)*

The total radiation received at a lake surface consists not only of radiation received directly from the sun, but also diffuse radiation from the whole sky. With a cloudless sky the amount of direct radiation received depends on latitude, season, time of day, and altitude. All of these influence the thickness of atmosphere through which the radiation has travelled. The quantity of direct radiation is of course also influenced by the transparency of the atmosphere. The amount of direct radiation received at sea level at the height of summer is of the order of 500–600 cal/cm^2/day. It is a somewhat surprising fact that the maximum daily radiation that can be received at the north and south poles is greater than that received at any other latitude. Thus on 15 June the direct radiation that potentially can be received at the north pole is estimated at 667 cal/cm^2/day, whereas the corresponding value for latitude 30°N is 625 cal/cm^2/day (Hutchinson 1957). This situation arises from the long daily period of potential sunshine in polar regions. The amount of diffuse light from the sky is highly variable but is usually about 20 per cent of the total radiation. This component is most important in the early morning and late afternoon when the sun is very low.

An appreciable portion of the radiation impinging on the surface of a lake is reflected and lost. Considering direct radiation, the fraction reflected depends on the angle of incidence. Where this angle (measured from the perpendicular) is 60° or less, less than 10 per cent is reflected. However, at 70° about 13 per cent is reflected, and at 80° about 35 per cent. The indirect or diffuse radiation strikes the lake surface at all angles and the amount reflected is largely independent of the height of the sun. On the average about 7 per cent of the indirect solar radiation is reflected, providing the horizon is not obscured. With an obscured horizon this percentage would be smaller. It has been estimated that about 6 per cent of total incident radiation is reflected or scattered during a summer day and about 10 per cent during a winter day.

The solar radiation that is not reflected and penetrates into the water is rapidly absorbed and turned into heat. In fact, even in distilled water, about half of the total solar radiation is absorbed and converted to heat in the first metre. If the radiation falls normal to the surface, its intensity at a given depth (z) is given by the formula:

$$I_z = I_0 . e^{-\varepsilon z}$$

where $I_0 =$ the intensity at the surface, $e =$ the base of natural logarithms (2·7183), and $\varepsilon =$ the extinction coefficient.

TABLE 5:2
EXTINCTION COEFFICIENTS FOR DIFFERENT WAVELENGTHS IN PURE WATER

Wavelength ($m\mu$) (colour)	Extinction coefficient
820 (infrared)	2·42
680 (red)	0·455
620 (orange)	0·273
580 (yellow)	0·078
520 (green)	0·016
460 (blue)	0·0054
400 (violet)	0·0134
380 (ultraviolet)	0·0255

The extinction coefficient varies with wavelength (Table 5:2). In other words, absorption is a differential process. Figure 5:8 shows the percentages of the different wavelengths in the visible portion of the spectrum that are transmitted through 1 m of distilled water. It can be seen that below 550 mμ transmission is high but above this value it decreases rapidly and becomes quite low for red light (and the near infra-red). Natural waters always contain chemical impurities and frequently suspended materials, both of which affect the transparency. Thus, while nearly 50 per cent of total radiation will pass through 1 m of distilled water, only about 40 per cent is transmitted through the same distance in the most transparent lakes. In some lakes this value is less than 5 per cent. It should be noted that different lakes vary not only in allowing the passage of different total amounts of light but also in their differential alteration of spectral composition (see Fig. 5:8). In very transparent lakes maximum transmission (or the colour of the light reaching deep water) is in the region of blue light, while in dark brown or highly humified lakes it is near the red end of the spectrum. Crater Lake in Oregon appears to be the most transparent lake known, and has maximum transmission at a wavelength of 450 mμ (Tyler 1965).

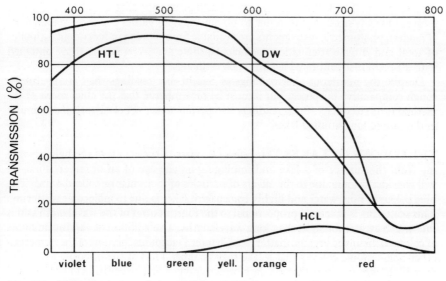

Fig. 5:8 Differential transmission of light of different wavelengths through 1 m of water. DW = distilled water, HTL = highly transparent lake, HCL = highly coloured lake

In thermally stratified lakes transparency within a freely circulating epilimnion is largely independent of depth. However, there is usually a definite minimum in transparency immediately beneath the epilimnion and another close to the bottom. In some cases it appears that additional horizontal strata of characteristic transparency and quite restricted vertical distribution extend across the hypolimnion. This micro-stratification of transparency within the hypolimnia of lakes probably results from a complex stratification of bacteria, dead seston, and dissolved or colloidal colouring matter of different kinds. In hypolimnia that retain a considerable amount of oxygen, stratification of non-bacterial living plankton may also be involved.

THE SECCHI DISC The Secchi disc has long been used to measure, in a rough sort of way, the transparency of lakes. It is a very simple device consisting of a circular white plate, usually with a diameter of 20 cm, which is first lowered with a cord until it disappears, then lowered a little further, and finally hauled upwards until it reappears. The depths at disappearance and reappearance are recorded (the supporting cord should be quite vertical for these observations), and the average depth is referred to as the Secchi-disc transparency. The determination is always made beneath shade to avoid interference from surface reflections and is best made using an underwater telescope. The area of shade, however, should be fairly small.

Unfortunately some of the physical features that influence the value obtained in a given situation have not been satisfactorily standardized or specified. Thus some workers have used a disc diameter of 10 cm or 30 cm instead of 20 cm, and some workers have used alternating white and black quadrants rather than pure white. More important, as pointed out by Tyler (1968), the reflectance of the disc has rarely if ever been specified and this must be maintained at a known value for consistent results. Tyler showed that the Secchi-disc transparency can be used to calculate the sum of the total and diffuse attenuation coefficients, but that some other form of measurement is necessary in order to obtain separate values for these parameters. If modern photometric instruments are available for obtaining independent estimates of total and diffuse attenuation coefficients, these are preferable to those obtained with a Secchi disc calibrated against such estimates.

Despite the numerous shortcomings of Secchi-disc readings they are useful for rough comparative purposes, but it must be remembered that the chief merit of the method is its simplicity. Such readings also permit some sort of historical continuity in the case of long-studied lakes.

THE COLOUR OF LAKES This refers of course to the colour of the light emerging from the surface of a lake and impinging on the eye of an observer. First we will consider colours due to the effects of particles of molecular or colloidal size. The blue colour of pure water and highly transparent lakes is due to molecular scattering. This scattering is inversely proportional to the fourth power of the wavelength and is thus much greater for short than long wavelengths. The addition of varying amounts of coloured dissolved organic matter (see Chapter 1) accounts for most of the numerous other non-sestonic colours that a lake may assume. Thus at the opposite extremity from blue lakes are those that are dark brown and peaty. The addition of only a small amount of humic matter to very transparent water changes the colour from blue to green or some greenish hue. Thus if a lake has a greenish appearance it is not necessarily due to the chlorophyll of its phytoplankton.

The colour of lake waters is usually determined by colorimetry; the colour of the lake water, as viewed by transmitted light, is matched against a series of dilutions of an arbitrary colour standard similarly viewed. The standard is usually prepared by dissolving $1 \cdot 246$ g of potassium chloroplatinate and $1 \cdot 000$ g of crystalline cobaltous chloride in 100 ml of concentrated hydrochloric acid and making up to 1 l with distilled water. Such a standard is said to have a colour of 500 platinum units. The very clearest (blue) waters give a zero value on the platinum scale, whilst highly humified waters may have a colour in excess of 300 units (see also Fig. 1:6).

In a thermally stratified lake there is commonly an increase in colour with depth and sometimes a distinct maximum in the metalimnion or upper hypolimnion. Usually there is also an increase in colour in the lowest part of the hypolimnion. This may be accounted for by organic materials derived from sediments but, as pointed out by Hutchinson (1957), it is probable that in some cases this is due to oxidation of ferrous iron to ferric hydroxide as the former diffuses upward from the bottom mud.

Sometimes the colour of a lake is determined largely or wholly by the light reflected from relatively large suspended particles; that is, by the colour of the seston. This is often the case with highly productive (and frequently saline) lakes. It is thus not uncommon for saline lakes, such as Lake Corangamite in the Western District of Victoria, to be coloured deep green as a result of a bloom of the blue-green alga *Nodularia spumigena*. Some saline lakes in the same area sometimes have a rusty red or brown coloration as a result of immense concentrations of the rotifer *Brachionus plicatilis*. The same effect is sometimes produced by high concentrations of dinoflagellates. Some highly saline lakes are coloured pink by high population densities of the flagellate *Dunaliella salina*. Such is the case with the Pink Lakes at Linga in Victoria (Cane 1962). The Australian salt-lake calanoid copepods *Calamoecia salina* and *C. clitellata* frequently have a bright red coloration which stems from carotenoids, but usually they have little influence on the water colour. Quite apart from saline lakes, highly productive freshwater lakes are often coloured green by blooms of blue-green algae such as *Anabaena*. As mentioned above, meromictic lakes may be coloured by purple or green sulphur bacteria.

Many shallow temporary lakes in the arid parts of Australia are often highly turbid with a brown or grey colour imparted by suspended inorganic particles. The milky appearance of glacial lakes such as Tekapo and Pukaki in the South Island of New Zealand is due to a suspension of fine glacial silt. Needless to say there is minimal penetration of light in such lakes.

Seiches and Wind-Generated Currents

It is not difficult to imagine that a strong wind blowing across a lake would result in a heaping-up of the water on the leeward side. If this wind suddenly stops a current will flow back to the former windward side as a result of the hydrostatic head that accumulated. The momentum of this current will carry it past the horizontal, thus causing a heaping-up on what was previously the windward side. The process will then repeat itself so that the lake surface rocks back and forth. Such oscillations are called *seiches*. In the simplest case seiches are uninodal with no displacement occurring at a point half-way along the length of the lake. However, there is a tendency for harmonics to form, so that binodal, trinodal, etc. seiches also occur.

The period (in seconds) of the nth nodal seiche for a rectangular basin of uniform depth (z) is given by the following expression:

$$T_n = \frac{1}{n} \frac{2l}{\sqrt{gz}}$$

where l is the length of the lake in metres and g is the acceleration due to gravity (9·81 m/sec/sec) [or $T_1:T_2:T_3:T_4 = 100:50:33\cdot3:25$].

Of course such a basin never occurs in nature and a more complex expression, such as that applicable to a symmetrical concave parabolic or prismatic or more complex basin, would be better. However, if the maximum depth of a lake (z_m) is substituted for z in the above expression, and providing that l is much greater than z_m, an approximation of the periodicity for the various nodes can be obtained. The following theoretical expression for a symmetrical concave parabolic basin would usually give a closer approximation:

$$T_n = \frac{\pi l}{\sqrt{(n[n+1]gz_m)}}$$

[or $T_1:T_2:T_3:T_4 = 100:57\cdot7:40\cdot8:31\cdot6$].

The most convenient apparatus for the empirical investigation of seiches consists of a pivoted lever, from one end of which is suspended a float, whilst the other end terminates in a stylus which etches a smoked drum or the like revolving at a constant speed. Some form of damping is required to eliminate local disturbances. Such an apparatus is known as a *limnogram*. In some lakes a very regular trace may be obtained with one of these, but usually a complex pattern, resulting from the superposition of several different nodes, is observed. Russell (1886) appears to have been the first worker to have investigated an Australasian lake from this point of view. He found periods of 131 min and 72 min for the uninodal and binodal seiches respectively

($T_1 : T_2 = 100 : 55 \cdot 0$), in Lake George[3], New South Wales (Plate 3:1), at a time when its length was 30 km and its mean depth $5 \cdot 5$ m.

The most thoroughly investigated Australasian lake from the viewpoint of seiches seems to be Lake Wakatipu (Bottomley 1956a, 1956b, Cox 1965). This lake has long been known to pulsate and according to a Maori legend this is due to the beating heart of a buried giant. Bottomley found periods of 52 min and 27 min for the uninodal and binodal seiches respectively ($T_1 : T_2 = 100 : 51 \cdot 9$) (Fig. 5:9).

Fig. 5:9 Portions of traces taken simultaneously with limnograms at the northern and southern extremities of Lake Wakatipu. Both show a pattern that results from a combination of uninodal and binodal seiches. *Redrawn from Bottomley (1956)*

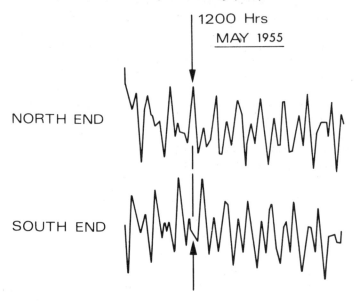

It is the longitudinal seiche that has been under consideration in the above discussions and it has been assumed that the nodal lines are at right angles to the long axis of the lake. Transverse seiches also occur, but are of less significance. Cox (1965) found at a station about half-way along Lake Wakatipu a seiche with a period of only $2 \cdot 86$ min. This was almost certainly a transverse seiche as the period agreed very well with the theoretical value of $2 \cdot 80$ min.

Although the surface seiche discussed above is a very obvious and sometimes spectacular thing (it has probably more than once led to at least the temporary disappearance of a small boat not dragged sufficiently far out of the water!), it is biologically relatively unimportant. Of much greater significance from this point of view is the *internal seiche*. This occurs in thermally stratified lakes when the different density layers oscillate relative to one another. An internal seiche may of course be

---

3. This lake is astatic and has dried up completely several times since 1818, the first year for which reliable information exists. It was completely dry during the following years: 1838–39, 1845–49, 1859, 1904, and 1932–46 (see Russell 1887 and Jennings, Noakes, and Burton 1964). During floods in 1823 and 1874 it reached a maximum depth of $7 \cdot 5$ m.

PHYSICAL AND RELATED CHEMICAL AND BIOLOGICAL PHENOMENA

initiated by a surface seiche; the major factor in the generation of an internal seiche is the wind stress across the surface of a lake.

The period (in seconds) of a uninodal internal seiche in a rectangular basin of uniform depth is given by:

$$T_i = \frac{2l}{\sqrt{\dfrac{g(\rho_h - \rho_e)}{1/z_h + 1/z_e}}}$$

where ρ_h and z_h are, respectively, the density and depth of the hypolimnetic water, and ρ_e and z_e the density and depth of the epilimnetic water.

The period and amplitude of the internal seiche are much greater than those of the surface seiche.

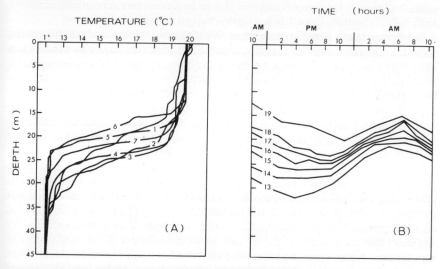

Fig. 5:10 Evidence for an internal seiche in Lake Rotoiti, New Zealand. A: a series of seven temperature-depth graphs taken at different times. The thermocline is seen to fall (1-3), rise (4-6), and fall again (7). B: the change in position of several isotherms with time. The 16°C isotherm may be taken as indicating the extent of oscillation of the thermocline. *After Green, Norrie, and Chapman (1968)*

The only observational work on the internal seiche of an Australasian lake is that of Green, Norrie, and Chapman (1968) on Lake Rotoiti in New Zealand. Their work was carried out in February 1968 when quite extensive oscillations of the thermocline were detected (Fig. 5:10). The observed period in this case was 19·6 hours, which compared quite favourably with 21·2 hours, the period calculated from the above equation. Figure 5:10B is of further interest in that during the earlier part of the oscillation the metalimnion was much thicker than subsequently. This is explicable on the basis of three layers rather than two being involved in the seiche; the metalimnetic water was squeezed alternately from one end of the lake to the other with the same periodicity as the oscillations in the other two layers. Linden (1968) made a theoretical investigation of the internal seiches that can exist in Blue Lake, Mt Gambier (Plate 3:2), on the basis of temperature (and then density) profiles taken at different times of the year. However, it is only in summer, and then only the first- and perhaps second-order modes, that are likely to attain measurable amplitudes. The theoretical period for the primary summer oscillation was found to be about 2 hours 43 min.

Internal seiches cause quite strong currents (of the order of 2–3 cm/sec) so that turbulent movement rather than laminar flow is involved. Thus any piece of water, at all levels in the lake, will suffer random displacement as a result of turbulence. The net result is the mixing of 'contact water' (water touching the bottom) with free water. This is very important because contact water has had the opportunity to dissolve material including nutrients from the sediment surfaces. It is now clear that internal seiches are responsible for substantial horizontal transport of materials and heat even in very deep and permanently stratified lakes. Although the amount of vertical movement is small in comparison with the horizontal, some does occur. The horizontal and vertical movements together play an important part in returning nutrients back into circulation, especially in oligomictic and meromictic lakes.

Seiches are clearly periodic phenomena. We shall now briefly consider non-periodic water movements or *currents*, the most important of which are caused by wind. When wind blows over a lake it exerts a stress at the air-water interface and moves water downwind. Because of the rotation of the earth it is also moved (in the southern hemisphere) to the left. Where there is an unrestricted water surface with a great depth below it the resulting surface current sets 45° to the wind direction. Where the water is homogeneous, reducing the depth tends to reduce this angle, but reduction is appreciable only in quite shallow lakes (less than about 10 m deep). The velocity of the wind drift is usually about 2 per cent of the wind velocity.

In a stratified lake the upper layers of the epilimnion move in the general direction of the wind and a return current is established in the lower epilimnion. Little is known concerning the hypolimnetic currents but they are probably due mainly, if not entirely, to internal seiches. In deep unstratified lakes there is a distinct return current upwind. This current depends on the extra hydrostatic pressure exerted by the water that accumulates at the lee end. Since the influence of the wind decreases rapidly with depth while the velocity of the return current is largely independent of depth, there is a plane at which the two exactly balance, and below this a return upwind current. Sometimes this return current is simple and sometimes it spirals beause of geostrophic effects.

In wide shallow lakes the pushing of water against the lee side produces return currents that usually travel around the sides of the lake and converge in the windward half of the lake surface.

chapter six
ECOLOGY OF THE MAJOR BIOLOGICAL COMMUNITIES
Plankton

Phytoplankton
SOME GENERAL ASPECTS OF DISTRIBUTION AND BROAD GEO-CHEMICAL CORRELATIONS The importance throughout most of the Australian mainland of saline or near-saline waters, dominated in the cations by sodium and usually alkaline, is stressed in Chapters 1 and 10. In agreement with Hutchinson's (1967:351) statement that high levels of sodium seem to be conducive to large myxophycean or blue-green algal populations, this group is undoubtedly of major importance in these waters, especially in summer when temperatures and light intensities are high. Thus in the Lake Corangamite region of Victoria blue-green algal blooms are a feature of salt lakes with a salinity of up to about 60–70‰. Lake Corangamite itself, for example, is characterized by having a permanent bloom of *Nodularia spumigena* (Hussainy 1969). In still more saline lakes of this region the green alga *Dunaliella salina* becomes dominant. In New Zealand some of the lakes in the Rotorua district have an abundance of blue-green algae but this is partly a consequence of comparatively recent cultural eutrophication (Fish 1969).

At the other end of the scale, very dilute natural waters are comparatively scarce, or at least restricted in area, on the Australian mainland (Bayly 1967b: 91). However, such waters occur in the following regions: the sandy coastal country of Queensland (known locally as 'Wallum') and similar regions in other Australian States (see Bayly 1964a, Coaldrake 1961); elsewhere in the southwestern tip and eastern border of the Australian mainland where rainfall exceeds evaporation, especially in granitic regions and in high-altitude localities in the Great Dividing Range (Williams, Walker, Brand 1970); in most of Tasmania; and in New Zealand, especially in the glacial lakes of the South Island. As pointed out by Hutchinson (1967), there can be no doubt that the occurrence of large numbers of species of desmids is usually correlated with dilute acidic waters and very low concentrations of calcium and magnesium. Frequently, but not necessarily, lakes with water like this lie on or are surrounded by ancient igneous rocks. Desmid-rich net-phytoplankton does occur in each of the Australasian dilute-water regions specified above. It is stressed that in this discussion 'desmid-rich' or 'desmid-dominated' is used in the sense of taxonomic variety, not quantity or biomass; thus Cheng (1968), for example, pointed out that in Lake

Sorell, Tasmania, desmids constitute only about 10 per cent of the total cell count even though they constitute half the total number of species recorded.

West (1909), in the first significant ecological investigation of Australian freshwater phytoplankton, made a detailed study of Yan Yean Reservoir near Melbourne. Although this is an artificial body of water (it had been in existence for forty-eight years before the commencement of West's study) it is reasonable to suppose that it was colonized from and representative of small natural bodies of water that formerly existed or still exist in the same area. West stated that the whole of the drainage basin of Yan Yean Reservoir and the three higher reservoirs that feed into it consists of granite of Silurian age. However, inspection of a recent (1967) geological map of the area shows that only the upper part of the drainage basin near Mt Disappointment in the Great Dividing Range is granite, and this is Devonian in age. Devonian and Silurian sedimentary rocks occupy an appreciable area and there is even a small amount of Recent sedimentary rock. However, although much of the catchment area is non-igneous, it is still true that the water is dilute and calcium and magnesium values are low. In 1905 the T.D.S. value for Yan Yean Reservoir was only about 60–70 p.p.m., and only slightly higher values (70–80 p.p.m.) are obtained today. Recent determinations show that the amount of both calcium and magnesium is less than 2 p.p.m. West found that the most striking feature of the phytoplankton was the large variety of desmids, and he attributed this to the nature of the geological formations upon which the reservoir is located and from which it derives its water supply. Negative features were the great scarcity of blue-green algae and a comparative scarcity of diatoms. It may be noted in passing that some desmids, including *Micrasterias hardyi* (Plate 6:1), first described by West from Yan Yean Reservoir, seem to be endemic to southeastern Australia (Tyler 1970). *M. hardyi* seems to occur mainly in dilute waters with low concentrations of calcium and magnesium (Table 6:1).

Plate 6:1 *Micrasterias hardyi:* a desmid endemic to the Australasian region and characteristic of dilute waters. *Photograph by P. A. Tyler*

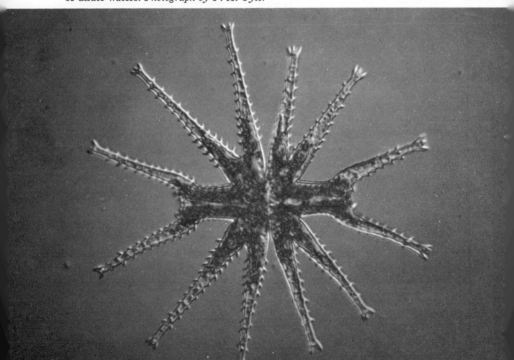

ECOLOGY OF THE MAJOR BIOLOGICAL COMMUNITIES

TABLE 6:1
CHEMICAL DATA FOR SOME HABITATS OCCUPIED BY THE
AUSTRALASIAN ENDEMIC DESMID *MICRASTERIAS HARDYI*
All Victorian records and data kindly supplied by Miss I. J. Powling of the Victorian
State Rivers and Water Supply Commission

Locality	Mean T.D.S. (p.p.m.)	Mean Ca^{++} content (p.p.m.)	Mean Mg^{++} content (p.p.m.)
Victoria			
Yan Yean Reservoir	65	<2	<2
Devilbend Reservoir	166	7	6
Glenmaggie Reservoir	42	4	2
Hume Reservoir	42	4	2
Malmsbury Reservoir	96	5	5
Newlyn Reservoir	102	5	6
Wurdee Boluc Reservoir	138	8	5
Tasmania			
Woods Lake[a]	65	3	1
Lake Sorell[b]	ca. 60	4	–
Lake Crescent[b]	ca. 60	4	–

[a] Bauld and Tyler (unpublished data).
[b] Cheng (1968).

Bayly (1964a) showed that lakes resting on organic accumulations in well-leached siliceous sand of coastal dunes in southeast Queensland and northern New South Wales have low values for salinity, calcium, and magnesium. The average salinity for this series of lakes was 39 p.p.m., all had a calcium content of less than 1 p.p.m., and with one exception the magnesium content was less than 2 p.p.m. Although it was not stressed in the above paper, subsequent examination has shown that desmids are the most significant group in net-phytoplankton collections from these lakes. Furthermore, lakes in comparable sandy regions near Cranbourne and Stratford in Victoria, and in parts of King Island, are dilute, poor in calcium, and rich in desmids— including species of *Staurastrum*, *Triploceras*, *Euastrum*, and *Micrasterias* (Brand 1967). These Australian examples may be added to that of the acidic lakes on leached sand in Florida (Hutchinson 1967:331) to emphasize further that ancient igneous rocks are not a necessary prerequisite for the development of rich desmid floras; recent aeolian accumulations of siliceous sand when leached produce the same geochemical determination. Rich desmid floras are also frequently found in acidic waters isolated from underlying rock by thick beds of peat.

Turning to the South Island of New Zealand, the catchment areas of the large lakes in Canterbury and, except for those on the plains, the smaller ones as well consist mainly of old sedimentary rocks in the form of greywackes[1] and argillites thought to be mainly of Triassic age. The large lakes in Otago, however, are surrounded by old

---

1. Hard grey sandstones or siltstones in which the mineral grains have been formed by the rapid reduction of granite to sand and then redeposited before it has been subject to chemical decay. Those with finer grain may be termed argillites because of a more clay-like texture, but they do not differ much in chemical composition from the more sandy of the series.

metamorphic rocks in the form of chlorite schists and gneisses representing a transformation of what were originally greywackes and argillites. In Southland the large Fiordland lakes (Te Anau and Manapouri) receive water mainly from old metamorphic rocks (schists, gneisses, diorites, and granites), although the eastern side of Lake Te Anau and the eastern end of Lake Manapouri touch Tertiary rocks occupying the Waiau Syncline.

Greywackes as such or in altered form seem to be poor sources of nutrients for lacustrine organisms. Thus Stout (1969a) mentioned that the chemical composition of the greywacke is such that few nutrients enter most Canterbury lakes, and further stated that the few analyses that have been made of the larger lakes indicate low values for salinity, calcium, magnesium, and phosphates. Similarly, Percival (1949) commented that Lakes Lyndon and Pearson (two smaller highland lakes in Canterbury) are very low in phosphate and nitrate. According to Stout's (1969a) analyses these two lakes are low in salinity (Lyndon ca. 36 mg/l, Pearson ca. 37 mg/l), calcium (Lyndon ca. 3 mg/l, Pearson 4–6 mg/l), and magnesium (Lyndon 0·7 mg/l, Pearson 0·5 mg/l).

In addition, Lake Pearson has a higher silicate content (2–6 mg/l) than Lake Lyndon (0·01 mg/l). The phytoplankton of both lakes is sparse but that of Lake Lyndon is rich in desmid species whilst diatoms are rare or absent (possibly because of the small amount of silicate) (Stout 1969a). Lake Pearson has fewer desmids but diatoms, including blooms of *Asterionella*, are common. Lake Sarah, also in the Canterbury highlands, was studied by Flint (1938), who found what might be described as an oligotrophic diatom-desmid plankton; *Asterionella gracillima* was the most important dominant but there was also a fairly rich desmid fauna consisting of more than twenty species. If these three lakes can be taken as representative of Canterbury highland lakes, then they are characterized by having dilute water and a desmid or diatom-desmid phytoplankton.

Lemmerman-Bremem (1899) recorded a number of planktonic algae from Lake Wakatipu, and the fact that the most abundant included *Staurastrum*, *Dinobryon*, and *Botryococcus braunii*, together with the fact that diatoms were scarce and no blue-green algae were present, suggests a dilute oligotrophic water. This of course is the case.

Despite the emphasis placed above on the chemical determination of phytoplankton populations, we hasten to point out that differences between populations are often difficult, if not impossible, to explain in chemical terms alone. Immediately adjacent lakes that are virtually identical from the viewpoint of physics and inorganic chemistry may have very different phytoplankton populations. Thus West (1909) pointed out that although Toorourrong Reservoir is just above Yan Yean Reservoir and supplied[2] most of the water that it received, the former had an abundance of *Ceratium* and the latter none. Similarly, Cheng (1968) pointed out that although the Tasmanian lakes Sorell and Crescent have many physico-chemical similarities, are less than 1 km apart, and are connected to each other by a small creek, there are many differences in their phytoplankton populations. In such cases it must be assumed that differences in morphology or age and thus the nature of bottom materials are significant. Young lakes are more likely to have a sandy or rocky bottom and little organic sediment or

---

2. Water from Toorourrong Reservoir did not flow into Yan Yean until 1885. Yan Yean Reservoir itself was completed in 1857.

reduced areas of peripheral swamps, and the reverse is true of older lakes. Increased amounts of organic matter may facilitate chelation and make inorganic materials more easily available to algae. Likewise, a large surface area permits greater wave action, which in turn may increase turbulence and assist normally benthic species to maintain a planktonic existence.

PHYTOPLANKTON TYPES AND INDICES Hutchinson (1967) proposed a provisional classification of phytoplankton associations into thirteen different types. The extent to which this classification is applicable to the Australasian scene is not well known. Furthermore, the usefulness of any proposed classification is debatable in that the composition of these associations seems almost infinitely variable. Nevertheless, if for no other reason than to stimulate further investigation, Hutchinson's classification may be given, together with some Australasian examples that apparently can be accommodated in it, as follows:
1 *Oligotrophic desmid plankton* in which the taxonomic dominants are usually species of *Staurastrum* and *Staurodesmus*. Yan Yean Reservoir and coastal lakes in leached siliceous dunes may be cited as examples of Australian lakes known to have oligotrophic desmid plankton. So too may the Great Lake and Arthurs Lakes in Tasmania (P. A. Tyler, pers. comm.).
2 *Oligotrophic diatom plankton* with *Cyclotella* as the characteristic dominant sometimes associated with species of *Fragilaria, Synedra, Rhizosolenia, Melosira,* or *Dinobryon*. Lakes Sorell, Crescent, and Leake, Woods Lake, and Tooms Lake are Tasmanian examples with this type of phytoplankton (P. A. Tyler, pers. comm.).
  1 and 2 seem to be capable also of combination as oligotrophic *desmid-diatom* or *diatom-desmid plankton* in which *Staurastrum* is associated with *Melosira*, or *Asterionella* is associated with *Staurastrum*. It seems clear that, in some parts of the world at least, *Asterionella* may be a dominant or important in the plankton of oligotrophic lakes (Flint 1938, Olive 1955, Rawson 1956, Cheng 1968).
3 *Botryococcus* plankton in which *B. brauni* is often dominant, sometimes associated with *Dinobryon, Staurodesmus,* or *Peridinium*. *Botryococcus* is often the dominant in dilute desmid-rich waters (i.e. numerically dominant in waters that contain more species of desmid than any other algal group in the net-plankton). Bayly (1964a) recorded the presence of this genus in several dilute dune lakes on Fraser Island, Queensland. *B. brauni* is very common in many Tasmanian lakes (P. A. Tyler, pers. comm.).
4 *Chrysophycean plankton* with *Dinobryon* completely dominant or associated with *Tabellaria* or other diatom genera. *Dinobryon* is often a seasonal dominant in the epilimnion of eutrophic lakes at times of phosphorus depletion, either following the spring phytoplankton peak, or towards the end of the summer stagnation period.
5 *Oligotrophic chloroccal plankton* with *Oocystis* as the dominant.
6 *Oligotrophic dinoflagellate plankton* with *Peridinium* or *Peridinium* and *Ceratium* dominant.
7 *Mesotrophic* or *eutrophic dinoflagellate plankton* with *Peridinium* plus *Ceratium* or *Glenodinium* or both.
8 *Eutrophic diatom plankton* in which one or more of the following are dominant for at least certain times of the year: *Asterionella, Fragillaria, Synedra, Stephanodiscus,* and *Melosira* (especially *M. granulata*). The phytoplankton of Lake Rotorua in New Zealand is of this type (Cassie 1969).

9 *Mesotrophic* or *eutrophic desmid plankton* in which certain species of *Staurastrum* (not those characteristic of type *1* above) and of *Cosmarium* are dominant.
10 *Eutrophic chlorococcal plankton* with *Pediastrum* and *Scenedesmus* as the most common dominants.
11 *Myxophycean plankton* with *Anacystis, Aphanizomenon, Anabaena, Oscillatoria* (particularly *O. rubescens*), *Nodularia,* or *Spirulina* dominant and forming temporary or even permanent blooms. The importance of such plankton in saline waters or high-conductivity fresh waters throughout the greater part of the Australian mainland has already been stressed.
12 *Euglenophyte plankton* characterized by dense blooms of *Euglena* and usually found in small bodies of water that are polluted with organic matter.
13 *Bacterial plankton* dominated by coloured sulphur bacteria and containing a considerable amount of hydrogen sulphide. Such plankton occurs in some meromictic lakes and also in some holomictic saline lakes.

Certain difficulties seem to arise with some parts of the above classification. Thus, for example, *Asterionella* is listed as a dominant for both oligotrophic and eutrophic diatom plankton. Flint (1938), Olive (1955), Rawson (1956), and Cheng (1968) found this genus associated with oligotrophic conditions in New Zealand, Colorado, Canada, and Australia respectively. In Europe, however, *Cyclotella* seems to be the diatom genus most usually dominant in oligotrophic waters, and *Asterionella* is a characteristic diatom dominant in eutrophic waters. Cheng (1968) reports that both *Asterionella* and *Cyclotella* occur in Lake Sorell, a fairly oligotrophic lake in Tasmania.

As an alternative to attempting to assign a particular phytoplankton assemblage to one of the above types, some sort of phytoplankton index may be calculated. The following two indices have proved useful:
a The chlorophycean index, which is the ratio of the number of species of Chlorococcales to the number of species of Desmideae. As a rough guide it has been suggested that values less than 1·0 indicate oligotrophy and more than 1·0, eutrophy.
b The compound index, which is the ratio of the number of species of Myxophyceae, Chlorococcales, centric diatoms, and Euglenophyta to the number of species of Desmideae. According to Nygaard (1949) an index of less than 1·0 diagnoses oligotrophic water, and less than 0·3 indicates a distinctly humified water. Values of more than 1·0 probably indicate mesotrophy or eutrophy and those above 5·0 very definitely eutrophy with possible faecal contamination.

Applying these indices to some oligotrophic lakes in Australasia, we obtain from the data of Flint (1938) a chlorophycean index of $\frac{9}{23}$ or 0·39 and a compound index of $\frac{12^3 + 9 + 0 + 0}{23}$ or 0·91 for Lake Sarah. The corresponding values for the phytoplankton of Yan Yean Reservoir from the data of West (1909) are $\frac{15}{55}$ or 0·27 and $\frac{4 + 15 + 3 + 0}{55}$ or 0·40; for Lake Sorell (Cheng 1968) they are $\frac{9}{14}$ or 0·64 and $\frac{1 + 9 + 2 + 0}{14}$ or 0·86.

Turning to a more productive lake, a chlorophycean index of $\frac{9}{13}$ or $0 \cdot 69$ and a compound index of $\frac{10+9+6+4}{13}$ or $2 \cdot 23$ is obtained for Lake Rotorua from data given by Cassie (1969).

Brook (1965) analyzed phytoplankton data for about 300 diverse lakes scattered throughout the British Isles and calculated the average compound quotient for the 46 most commonly occurring forms of desmid. Although his analysis confirmed the widely held belief that desmids are best represented (about 60 per cent of common forms) in oligotrophic waters, he found that there is a surprisingly high number (about 25 per cent of common forms) that are usually associated with eutrophic waters. It follows that the inclusion of *all* desmid species present in a sample when determining compound indices *decreases* the reliability of this method of assessing trophic status. Brook made a plea for the formulation of an index based on a limited number of species whose nutritional requirements and status in the plankton have been properly investigated.

Before leaving this topic it is worthwhile pointing out that although some workers have supposed that eutrophic lakes contain fewer species than oligotrophic ones, this is not always the case. An examination of the findings of Round and Brook (1959) shows that some distinctly eutrophic Irish loughs contain more phytoplankton species than some that are markedly oligotrophic. Thus, neglecting 'tychoplanktonic' species, two eutrophic localities, Rea and Arrow, contained 38 and 33 species respectively, but two oligotrophic localities, Derryclare and Kylemore, contained only 20 and 22 species. Nevertheless it is clear that the phytoplankton diversity index for saline eutrophic lakes is less than that for freshwater lakes and becomes very low in highly saline waters (see Chapter 10).

PHYTOPLANKTON DIVERSITY AND ITS BASIS It is clear from the above discussions that phytoplankton assemblages are typically complex and highly diversified. Furthermore, the Gause principle of competitive exclusion[4] seems not to hold. Several possible explanations have been advanced for this situation. One is that, despite appearances to the contrary, the limnetic habitat is not really homogeneous and actually provides more than a single niche. In other words, it presents a situation to which the Gause Principle is inapplicable. A second explanation is that the fluctuation of general conditions is too rapid to allow the establishment of any sort of permanent competitive situation, which is a prerequisite for the application of the exclusion principle. In other words, short-term, possibly day-to-day, changes

---

3. Assuming two for each of '*Oscillatoria* spp.' and '*Gomphosphaeria* spp.'; otherwise identifications were to species or it was clear that a genus contained only one unidentified species.
4. It seems advisable to give at this point a definition of both competition and the Gause Principle. The following definition of competition is taken, with a minor alteration, from Milne (1961): Competition is the endeavour of two (or more) organisms to gain the same particular thing, or to gain the measure each wants from the supply of a thing when that supply is not sufficient for both (or all). Gause's Principle may be stated as either that two (or more) species with similar ecology cannot live together in the same place or that if two (or more) species coexist it implies the existence of two (or more) different niches.

may be the basis of phytoplankton diversity and the community is largely opportunistic in nature. It is difficult to conceive how a physico-chemically uniform and mixed body of water could develop enough differentiation to provide enough distinct niches for the normally large number of algae present. Consequently the non-equilibrium hypothesis seems the more reasonable of these two theories. Short-term fluctuations may be expected to bring about the occasional chance extinction and thus the continuous reduction of diversity seems a possibility. However, truly closed systems rarely if ever exist—in most cases streams or migrating birds or waterspouts would be capable of bringing about re-introductions.

One factor that probably makes phytoplankton diversity look greater than it really is, is the occurrence of a considerable number of meroplanktonic species; that is, species which are merely 'passing through' the spatial niche in which they are usually collected.

SEASONAL ASPECTS The characteristic pattern of fluctuation in phytoplankton density in reasonably large lakes is a winter minimum, a strong rise in spring, a single maximum or two or more maxima (often with much irregularity) in summer, and a continuous fall in autumn. Except for the Laurentian Great Lakes and some highly productive ones, the 'classical' spring and autumn maxima seem not to occur in lakes and probably represent an invalid extrapolation from oceanography. In small lakes there is often a complex succession of maxima and minima at various times throughout the year.

TABLE 6:2
PHYTOPLANKTON SEASONS AND THEIR RELATIONSHIP TO
TEMPERATURE AND LIGHT LEVELS

Temperature	Light	
	'Low'	'High'
'Cold'	Winter	Spring
'Warm'	Autumn	Summer

Temperature, illumination, nutrient levels, and the interaction of these three factors are all important in determining this pattern of total fluctuation and the seasonal succession of different species. Temperature is easy to measure and definitely of some importance, illumination is very important but less easily measured, and nutrient levels are important in competitive phenomena. A basic point is that the thermal mid-summer is later than the optical mid-summer; in the southern hemisphere the maximum for illumination falls in December but that for temperature falls in February. Even at quite low temperatures increased light results in increased photosynthesis. Thus the rapid rise in phytoplankton populations in early spring is explained by the fact that at this time light is increasing at a much more rapid rate than temperature. This presupposes that in early spring the amount of photosynthesis is inadequate for rapid reproduction. This is probably a reasonable assumption, but photoperiod, chemical conditions and other factors could also play an important part in regulating the onset of rapid reproduction. Likewise, the same temperature may be found in late spring and late autumn but light and phytoplankton levels will be higher during the former than the latter. The year may be divided into four 'phytoplankton seasons', not corresponding exactly with the usual four seasons, according to the scheme shown in Table 6:2.

The fact that different phytoplankton species and genera have different optimal temperatures (e.g. *Melosira* < *Fragilaria* < *Scenedesmus*) and light intensities helps to explain some aspects of seasonal succession. However, temperature-light interaction will certainly not explain all of the observed facts of succession; chemical aspects must also be considered.

On the basis of work carried out in the English Lake District, and especially on Windermere (Lund 1965), it is clear that in early spring the water has relatively high levels of the nutrients silica, nitrate, and phosphate. The abundance of these, especially silica, together with the increase in illumination, is basic to the early rise in numbers of the diatom *Asterionella*. Eventually the amount of silica is reduced to less than $0 \cdot 5$ mg/l, the apparent lower limit of tolerance, and the *Asterionella* population declines (Lund 1950). It thus seemed that silica alone became limiting. However, Hughes and Lund (1962) discovered that the addition of small amounts of phosphate to Windermere water allowed the growth of *Asterionella* to continue until the amount of silica in the water became negligible. The relationship between the spring maximum of *Asterionella* and silica concentration, which has been repeatedly observed in Windermere, thus seems to depend on the concentration of phosphate in the water. At the same time, the minimum amount of phosphorus required for the proper functioning of an *Asterionella* cell is extremely low, and this low level is not reached in Windermere at the time of the rapid decline in population density. Following the decline of *Asterionella* there may be a slight increase in the amount of silica, but the genus does not usually occur in large numbers again until autumn. *Tabellaria* sometimes takes over from *Asterionella*, using up more of the original nutrient store as it does so. This in turn may give way to *Dinobryon*, which not only tolerates low nutrient levels, but apparently finds high levels of phosphate lethal! During the summer, when levels of organic matter and nitrate are high, blue-green algae start to flourish. However, the large amount of nitrate might be produced by the increased growth of blue-greens, rather than the converse, and it is difficult to say what stimulates the summer upsurge in myxophycean numbers.

It should be pointed out that some English lakes never have large populations of *Asterionella* even though their nutrient levels are above those in Windermere. The picture presented above is evidently not yet complete.

In the case of *Melosira* some degree of turbulence seems to be essential for the continued growth of the population; long periods of calm allow much of the population to fall out of the epilimnion.

It is likely that, eventually, seasonal succession of phytoplankton populations will be largely explicable in terms of the interaction of light, temperature, chemistry, and turbulence. It is also possible that mutual antagonism between or stimulation of two or more algal species may prove to be of importance. Finally it should be pointed out that parasitism may be of some significance in influencing algal succession; chytrid fungi can produce drastic infections of *Asterionella* in more productive waters and effectively limit the population.

TABLE 6:3
CALANOID COPEPOD ASSOCIATIONS AND THEIR SIZE DIFFERENTIATIONS IN AUSTRALIA AND NEW ZEALAND

Region	Size Description[a]	Observed Associations[b]					
Australia							
New South Wales	L	*B. fluvialis*	*B. triarticulata*	*B. triarticulata*	*B. triarticulata*	*B. triarticulata*	*B. fluvialis*
	M	[*B. minuta*]	[*B. minuta*]	—	—	—	—
	S	*C. lucasi*	*C. lucasi*	*B. minuta*	*C. canberra*	*C. canberra*	*B. minuta*
	L	*B. fluvialis*	*B. major*	*B. major*	*B. pseudochelae*		
	S	*C. ampulla*	*B. montana*	*B. pseudochelae*	*B. montana*		
Northern Territory	L	*B. triarticulata*	*B. triarticulata*	*D. lumholtzi*			
	S	*C. lucasi*	*C. canberra*	*C. lucasi*			
Queensland	L	*B. fluvialis*	*B. fluvialis*	*B. minuta*	*B. minuta*	*B. triarticulata*	*D. lumholtzi*
	M	[*B. minuta*]	[*B. minuta*]	—	—	—	—
	S	*C. lucasi*	*C. trifida*	*C. trifida*	*C. lucasi*	*C. canberra*	*C. trifida*
	L	*B. triarticulata*	*B. symmetrica*	*B. symmetrica*			
	S	*C. ampulla*	*C. ampulla*	*C. lucasi*			
Tasmania	L	*B. symmetrica*	*B. triarticulata*	*B. symmetrica*	*B. propinqua*	*B. major*	
	S	*B. rubra*	*B. rubra*	*C. gibbosa*	*C. gibbosa*	*B. pseudochelae*	
Victoria	L	*B. triarticulata*	*B. triarticulata*	*B. symmetrica*	*B. symmetrica*	*B. major*	
	S	*C. ampulla*	*C. lucasi*	*C. ampulla*	*C. tasmanica*	*B. pseudochelae*	
Western Australia	L	*B. triarticulata*	*B. triarticulata*	*C. attenuata*	*B. robusta*		
	S	*C. ampulla*	*C. tasmanica*	*C. tasmanica*	*B. geniculata*		
New Zealand							
North Island	L	*B. delicata*	*B. minuta*	*B. hamata*	*B. propinqua*		
	S	*C. lucasi*	*C. lucasi*	*C. lucasi*	*C. lucasi*		
South Island		There appear to be no records of 2 or 3 calanoid species co-existing in this region					

[a] L = large; M = medium; S = small.
[b] B = *Boeckella*; C = *Calamoecia*; D = *Diaptomus*; those in square brackets may be omitted from the association.

Zooplankton

SPECIES COMPOSITION AND SIZE DIFFERENTIATION IN CALANOIDS In contrast to phytoplankton communities, zooplankton communities tend to be quite simple in terms of species composition. This is probably because rates of reproduction are much lower than in phytoplankton and short-term fluctuations are correspondingly of much less significance. Only rarely are more than three species of calanoid copepod, two cyclopoid copepods, three cladocerans, and perhaps four or five rotifers present at one time in the limnetic region of Australasian lakes. Typically the numbers are lower than these; Timms (1968, 1970a) reported the following mean numbers for the limnetic region of a series of lentic habitats in southeastern Queensland and northeastern New South Wales respectively: Copepoda (2·3 and 2·2) and Cladocera (2·0 and 1·1). These figures may be compared with the world averages of 2·7 for Copepoda and 2·8 for Cladocera quoted by Pennak (1957).

It is a striking fact, first pointed out by Hutchinson (1951), that when two or more closely related copepod species coexist in lake plankton they nearly always differ markedly in size. Hutchinson suggested that this represented a means whereby competition is avoided by the consumption of food of different sizes. The coexistence of two differently sized calanoids in Australia and New Zealand is quite common (Bayly 1964b). Typically the larger member is a species of *Boeckella* and the smaller member is a species of *Calamoecia* or *B. minuta*. In northern parts of Australia *Diaptomus lumholtzi* sometimes constitutes the larger member. Occasionally three calanoids have been found coexisting. Table 6:3 summarizes, on a regional basis, the different associations that have been found. The distribution of many of the calanoids is also shown in Figs 6:1–6:4. Two similar-sized calanoids, *B. triarticulata* and *B. fluvialis*, have occasionally been found coexisting (Bayly 1964b) in contradiction to the typical picture. Also *C. lucasi* and *C. ampulla*, which are very similar in size, are sometimes found together in Warragamba Dam (Jolly 1966). Possibly a temporary superabundance of food, or a marked non-equilibrium condition with one species 'on the way in' and the other 'on the way out', provides the explanation of this situation. It should also be pointed out that sometimes, although the two calanoids are not identical in size, size differentiation is minimal. Thus the following mean lengths (inclusive of furcal setae) were obtained for a plankton collection taken by Dr Ann Chapman from Lake Ohakuri on the Waikato River, New Zealand: *C. lucasi* ♀ 1·13 mm, ♂ 0·97 mm; *B. minuta* ♀ 1·44 mm, ♂ 1·10 mm. Here there is little difference in size between both sexes of *C. lucasi* and the male of *B. minuta*.

FOOD AND EFFECTS OF PREDATION The various planktonic herbivores of the limnetic region of a lake, such as calanoid copepods and cladocerans,[5] feed on and frequently compete for particulate matter. Much of this lies in the approximate size range of 1–15μ. It is clear that phytoplankton constitutes the major part of this matter under normal circumstances. However, it is also clear, especially from the work of Nauwerck (1963), that detritus and bacteria may also constitute an important and sometimes dominating fraction of zooplanktonic food. Nauwerck, working on Lake Erken in Sweden, calculated that *Diaptomus (Eudiaptomus) graciloides* constituted almost two-thirds of the annual herbivore production and pointed out that

---

5. The predacious groups Haplopoda and Polyphemoidea do not occur in Australia or New Zealand.

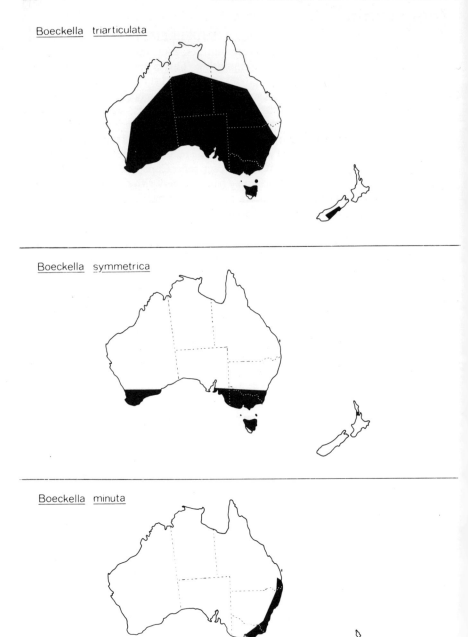

Fig. 6:1   The distribution of three species of *Boeckella* common to Australia and New Zealand

phytoplankton could be ruled out as a major source of zooplanktonic food in this particular lake. He concluded that other sources of food such as detritus and bacteria had to be considered. Saunders (1969) also draws attention to detritus and bacteria as alternative foods for phytoplankton. In one experiment he found that up to 36 per cent of the total carbon assimilated by *Daphnia* feeding under fairly natural conditions was from bacteria.

There is some evidence that adult calanoid copepods are generally much less efficient at feeding on bacteria than are cladocerans of comparable size. However, *Calamoecia tasmanica*, a small Australian calanoid, seems specialized for highly humified waters, which are often poor in phytoplankton, and probably feeds on bacteria. There is definite evidence (G. W. Brand, unpublished) that this species is capable of ingesting bacteria or at least clumps of bacteria. It may also be significant that *C. tasmanica* usually occurs alone or unaccompanied by other larger calanoids.[6] It is probable that most species of this genus are capable of some bacterial feeding and are somewhat more efficient at it than most species of *Boeckella*.

Irrespective of whether phytoplankton, detritus, or bacteria are involved, it seems clear from the work of Brooks and Dodson (1965) and especially Burns and Rigler (1967) that large zooplankters filter much more efficiently than do small ones. In addition, large zooplankters can eat large phytoplankters ($>15\mu$ long) that are too large for the small species. Because of this, small planktonic herbivores or detritus-feeders tend to be competitively eliminated by the larger species in the absence of significant fish predation. This situation normally leads to a dominance of large cladocerans and calanoid copepods. Although it would perhaps be more properly pointed out in Chapter 9, it may be noted here that the large Australian calanoids *Boeckella robusta* and *B. major*, both of which may measure up to 4 mm in length, occur only in ponds and pools and usually temporary ones at that. Such habitats of course lack fish.

Hrbáček (1962), Brooks and Dodson (1965), and Brooks (1969) have emphasized the profound effect that fish predation may have on the size composition of zooplankton communities. When fish predation is intense, as for example when there is a large population of obligate planktivores,[7] the larger zooplankters are eliminated, allowing the smaller forms (rotifers and small cladocerans) to become dominant. When there is a moderate degree of predation, it again falls mainly on the larger zooplankters, thus keeping the population density of these more efficient feeders sufficiently low to prevent elimination of the smaller competitors.

Australasian studies on the effect of fish predation on the structure of zooplankton populations are noticeably lacking. However, some observations of Jolly (1966),

---

6. The two associations of this species with *B. symmetrica* shown in Table 6:3 are each known from only one locality—Lake Purrumbete in Victoria and a farm dam in Western Australia.
7. The North American 'alewife' (*Alosa pseudoharengus*) is said (Brooks and Dodson 1965) to be an obligate planktivore at maturity, for which purpose it possesses very fine gill-rakers. The feeding habits of the smaller Australasian native fishes seem to have been insufficiently studied to enable a definite statement as to whether or not Australasian counterparts exist. Clupeid bony bream or hairback herring, *Fluvialosa richardsoni*, which occur in the Murray–Darling drainage system, feed mainly on plankton, but algae, higher aquatic plants, and small insects are occasionally found in the gut as well, and they are also partly iliophagus (J. S. Lake, pers. comm.). It is possible that Australasian obligate planktivores exist within the following genera or families: *Retropinna*, *Galaxias*, Atherinidae, and Centropomidae.

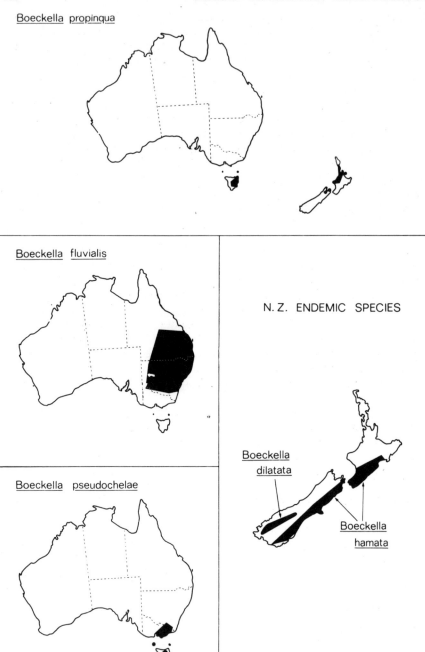

Fig. 6:2 The distribution of *Boeckella propinqua* (common to Australia and New Zealand), two species of the genus endemic to Australia, and two endemic to New Zealand

although not directly concerned with fish, may be significant in this respect. Jolly observed that the occurrence of *Daphnia carinata* and *D. lumholtzii* in Warragamba Dam is related *inter alia* to turbidity. *D. lumholtzii* which is much smaller than *D. carinata* (some forms of which are probably the largest in the genus) preferred clear water, whilst *D. carinata* favoured more turbid water. Assuming that there was at least a moderate degree of fish predation in this reservoir, it seems possible that *D. lumholtzii* had an advantage over *D. carinata* under clear conditions because of its smaller size. Conversely, fish predation could be sufficiently hindered by low transparency to give *D. carinata* the advantage because of its larger size and greater feeding efficiency.

THE FOOD OF ROTIFERS Rotifers are frequently of major importance in the zooplankton of lakes. Many are herbivorous. Thus, for example, some species of *Polyarthra* feed mainly on *Cryptomonas*, species of *Keratella*, *Notholca*, and *Anuraeopsis* will feed on *Chlorella*, and *Brachionus calyciflorus* will live on either *Chlorella* or *Scenedesmus* (Hutchinson 1967). Other rotifers are partly or wholly carnivorous. *Asplanchna*, for example, feeds on other rotifers such as *Keratella* and small microcrustaceans in addition to algae. *Ascomorpha saltans* sucks the contents out of dinoflagellates. *Synchaeta* feeds on protists and other rotifers such as *Keratella* and *Polyarthra*. A species of *Synchaeta* has also been observed to graze on *Cyclotella stelligera* (P. A. Tyler, pers. comm.).

THE FOOD OF CYCLOPOID COPEPODS All non-parasitic Cyclopoida lack a filtration mechanism like that of calanoids and are raptorial, seizing either animal or plant food. Fryer's (1957) work definitely established that both carnivorous and herbivorous species exist, the former generally being larger than the latter. It is probable that in this group the carnivorous habit is primitive.

Mesocyclops (Mesocyclops) leuckarti, which is probably the commonest cyclopoid in the plankton of Australasian lakes, is mainly carnivorous on cyclopoid and other copepods, Cladocera including *Diaphanosoma*, and rotifers. However, it is probably also partly herbivorous on large diatoms. Most species of *Eucyclops* and *Microcyclops*, genera which occasionally occur in lacustrine plankton, are mainly herbivorous. The same is probably true of *Tropocyclops*, which is a common cyclopoid in the limnetic region of Australasian lakes. Another small species, *Mesocyclops (Thermocyclops) decipiens*, occurs in the open water of some Australian lakes but nothing is known of its feeding habits. It may be noted that *Cyclops sens. str.*, which has some planktonic species in the northern hemisphere, does not occur at all in the Australasian region.

SEASONAL ASPECTS Several studies (Powell 1946, Jolly 1952, Bayly 1962, Jolly 1966, Green 1967, Timms 1967, Geddes 1968, Burrows 1968, Hussainy 1969, Timms and Midgley 1969) indicate that in most Australian and New Zealand lakes there is at least one calanoid copepod, usually *Boeckella* or *Calamoecia*, with a perennial multivoltine life history. Population densities are of course subject to a considerable amount of seasonal variation, yet this is relatively small in comparison with that for Cladocera, and the species, including at least some egg- or spermatophore-bearing females, is always present. However, in Lake Wakatipu and some other large glacial lakes in the southern half of the South Island of New Zealand, *Boeckella dilatata* has a univoltine life cycle (Jolly 1957a, 1957b), egg-bearing females occurring in numbers

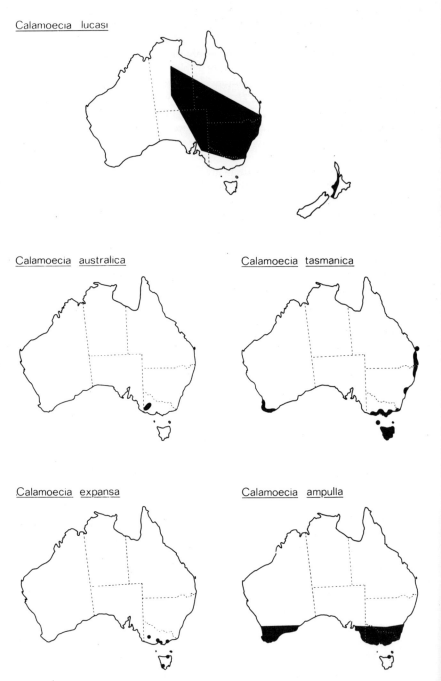

Fig. 6:3 The distribution of *Calamoecia lucasi* (the only species of the genus common to Australia and New Zealand) and four other species of the same genus

only during late spring and early summer. No seasonal study has yet been made of a calanoid population in an Australasian warm thereimictic lake, but it might be expected that the active population would disappear during the period of ice-cover, and that over-wintering would be achieved by benthic diapausing eggs or immature copepodites.

There are seasonal morphological changes in the individuals comprising a multivoltine calanoid population. Thus there are seasonal pulsations in body size which correlate inversely with water temperature, although the maximal body size is usually reached some time after minimal water temperature. The posterior lobes that arise from the last metasomal segment of females are also subject to cyclomorphosis (Bayly 1962, Timms 1967).

Mesocyclops leuckarti is a perennial species in all of the six Sydney reservoirs studied by Jolly (1966), in Borumba Dam, Queensland (Timms and Midgley 1969), in Lakes Purrumbete and Elingamite, Victoria (Hussainy 1969), and in Lake Sorell, Tasmania (Burrows 1968). The same species was present in all of a seasonally incomplete series of collections taken from a New Zealand lake by Bayly (1962). Several of these studies show that its population density is subject to wide seasonal variation, but there seems to be no consistency in its maxima.

The most seasonally variable of the zooplankton components is undoubtedly the Cladocera, species of which either are not perennial, or, if they are, usually undergo great seasonal fluctuations in density. It is clear from the several relevant studies already cited that the same cladoceran species may have its maximum abundance at different seasons in different lakes and even in the same lake in different years. However, some general trends are recognizable for some species. With respect to the occurrence of *Daphnia carinata* in lakes only, there is evidence that maximum numbers usually occur in spring from September to November. However, Jolly (1952) found that in Lake Hayes (S.I., N.Z.) *Daphnia* (sub. *D. carinata*) reached peak numbers in April and May. The seasonal occurrence of the Australian species *D. lumholtzi* has not been adequately studied. All that is established is that in two lakes at least it is not perennially present; Timms and Midgley (1969) found it present in a Queensland reservoir in all months except January, February, and March, with a maximum in August; Hussainy (1969) found it present in Lake Purrumbete, Victoria, from October to May with a peak in April; and Jolly (1966) reported that it occasionally occurred in two Sydney reservoirs and was periodically common in another. *Bosmina meridionalis* is often continuously present in Australasian lakes and the peak in population numbers is usually in autumn and winter. Burrows (1968), however, found summer maxima for this species in Lakes Sorell and Crescent in Tasmania. *Ceriodaphnia* is sometimes perennially present, but there is a tendency for numbers to be highest in spring or summer, although Burrows (1968) found high densities in Lake Sorell, Tasmania, in winter. Timms (1968, 1970a) found *Diaphanosoma excisum* common in summer in several Queensland and New South Wales localities. This is interesting in view of Hutchinson's (1967) statement that *Diaphanosoma brachyurum* is a most characteristic summer species throughout the temperate Holarctic. Another Australian cladoceran that seems to be present mainly in summer is *Moina micrura* (Timms 1970a).

DIURNAL VERTICAL MIGRATION Accounts of vertical migration of planktonic animals date from the time of Cuvier, who in 1817 wrote that in the morning and

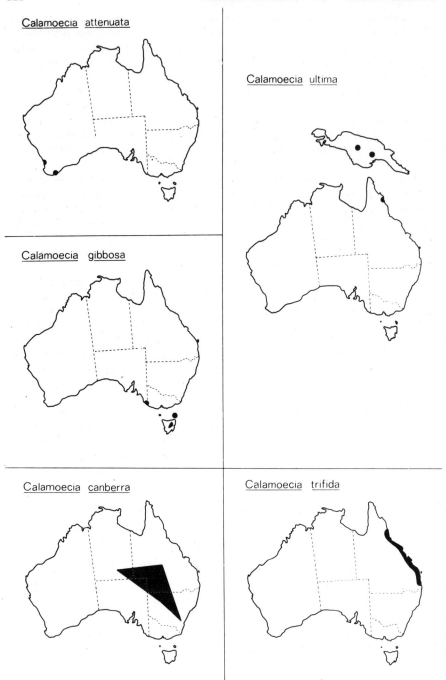

Fig. 6:4 The distribution of *Calamoecia ultima* (the only species of the genus common to Australia and New Guinea) and four other species of the same genus

ECOLOGY OF THE MAJOR BIOLOGICAL COMMUNITIES

evening and even on cloudy days *Daphnia* generally stationed themselves on the surface, but that in the heat of the day they sought the depths of the water. If he had said in the *light* rather than the *heat* of day, he would have had the distinction of not only making the first written observation on migration, but also of singling out from the very beginning what we now know to be the most important controlling factor. One must then hasten to add that a large number of other factors, many of them still ill understood, also influence migration, and sometimes exert a gross modifying effect on the main controlling factor, light. Diurnal vertical migration is a complex variable phenomenon. The basic pattern, however, may be outlined as follows. In the late afternoon upward movement commences from the day depth. This upward movement is a directional taxis involving active swimming towards a source of light that is rapidly decreasing in intensity. This upward movement continues until dusk or some later time in the night. In very deep lakes (and also in mid-oceanic regions) some species may never reach the surface, but merely rise to a higher level, or if they do it may be appreciably after dusk. The main return trip towards the day depth often commences at dawn but it may start before midnight immediately after completion of the upward migration or some time between this and dawn. If there is a considerable time-lag between completion of the upward and commencement of the downward movement, then the organisms usually remain close to the uppermost level reached by engaging in station-keeping behaviour or commence sinking as a result of inhibition of activity. The latter may result in some falling away from the surface at or slightly before midnight—a phenomenon referred to as *midnight sinking*. In species that hold station, or passively sink, for some considerable time after completion of upward movement, the increase of light at dawn may initially produce a slight upward movement (*dawn rise*), which may be kinetic in nature, before there is active downward movement away from light of rapidly increasing intensity towards the day depth. The day depth, although distinctly lower than the strata occupied at night, is variable, depending amongst other things on the degree of cloud-cover. This form of migration is common to all classes of planktonic Crustacea and many other animal taxa with planktonic species, and is found in marine as well as inland waters. It should be stressed that in many cases only part, and sometimes only a small part, of the total population participate in the sequence of events described.

The first investigation of diurnal vertical migration in an Australasian lake seems to be that of Powell (1946), carried out on Lake St Clair, Tasmania (Plate 3:6), during the years 1937–38. On the basis of three-hourly sampling at the surface and at a depth of 4·6 m once a month for a year, he showed that *Boeckella longisetosa* (sub. *B. propinqua longisetosa*) usually performed a distinct migration. From 9 p.m.* to 9 a.m. numbers at the surface and 4·6 m were about the same, but at other times they were quite different. In surface waters a minimum usually occurred at 12 noon or 3 p.m. and a maximum usually at 6 p.m. With respect to the migration of this species, Powell thought that the year was divisible into two periods: April to September and October to March. He considered that migration was less marked in spring and summer, during which time he supposed the population preferred higher light intensities. Although his two periods are somewhat arbitrary in that neither is completely homogeneous from the viewpoint of behavioural pattern, it does seem from his data that migration, especially the 6 p.m. maximum, was more definite in autumn and winter. It should be pointed out, however, that the more definite 6 p.m. maximum during autumn and winter could be largely an artefact produced by this standard sampling time falling closer to true dusk during these two seasons than during spring and summer. It is noteworthy that Powell observed on at least two different occasions that the more darkly coloured reddish specimens of *B. propinqua longisetosa* avoided strong surface light more than the lightly coloured greenish ones. This may be cited as evidence against the theory that carotenoids have a protective function against excessive ultra-violet radiation. Another significant observation is that the males and females of this calanoid react differentially to changes in light intensity, males being the more sensitive. For the Cyclopoida and Cladocera (neither further differentiated) of Lake St Clair, Powell found a pattern of migration that was fairly similar to that of *B. propinqua longisetosa*, except that the Cyclopoida avoided strong light more than the other two groups and appeared to have a distinctly deeper optimum day depth. Also the surface maximum for Cladocera was not uncommonly at midnight rather than 6 p.m.

Jolly (1952) showed from diurnal surface sampling that in Lake Hayes 'Copepoda' (mainly *Boeckella* subsequently known to be *B. dilatata*), *Bosmina* (sub. *B. meridionalis*), *Ceriodaphnia* (sub. *C. dubia*), and *Daphnia* (sub. *D. carinata*) all underwent vertical migration during at least one of the five different periods of investigation. The pattern of migration differed from the typical one in that the migrants did not occupy the uppermost levels all night but did so at dawn.

Bayly (1962) found from horizontal towing of s Clarke-Bumpus sampler at various depths in Lake Aroarotamahine (Mayor Island, N.I., N.Z.) that the males but not the females of *Boeckella propinqua* (cf. *B. prop. longisetosa* above), and also *Daphnia carinata*, migrated in reverse to normal manner at dusk in January when there was strong thermal stratification. The same unusual phenomenon was shown by *D. carinata* at dusk in May when the lake was isothermal. It was suggested (Bayly 1963) that high pH was a causative factor in this and another case of reversed migration, that of *D. lumholtzi* in Lake Rudolf in Africa (Worthington and Ricardo 1936). Timms (1967) investigated vertical movements in a *Boeckella minuta* population by making simultaneous top and bottom tows in a shallow Queensland pond. First there was marked crowding of the surface water at dusk, but in later night collections the numbers at top and bottom became progressively more equal (cf. above pattern found by Jolly). With the onset of dawn there was a distinct migration into the lower layers which remained crowded for the rest of the day. It may be noted, in contrast to Bayly's findings, that in this case the whole of the *Boeckella* copepodid population (adult males and females and the immature copepodites) participated in migration. Jolly (1965) reported the results obtained from two-hourly surface sampling of Lake Taupo on four different occasions—three when the lake was stratified and one when it was in complete circulation. On all occasions the numbers of *B. propinqua* were maximal about two hours after sunset, after which a rapid descent occurred, and minimal from 10 a.m. to 2 p.m. The cladocerans *Ceriodaphnia dubia* and *Bosmina meridionalis* tended to have maximal surface numbers at about the same time as *B. propinqua*, but except for one occasion (not the holomictic period) they tended to remain in the surface waters for a longer period than the calanoid. This difference between calanoids and cladocerans is in agreement with the findings of Powell (1946). It is possible that the distinct lag in Lake Taupo between dusk and maximal surface numbers is merely a reflection of its considerable depth ($z_m = 163$ m, $\bar{z} > 100$ m).

ECOLOGY OF THE MAJOR BIOLOGICAL COMMUNITIES

Green (1967), working on Lake Pupuke near Auckland, showed that a significant proportion of the total copepodid population of *Calamoecia lucasi*, including egg-bearing females, underwent a distinct migration; maximum numbers in surface waters were found at 9 p.m. and midnight, and at depths greater than 25 m densities were greatest at 9 a.m., noon, and 3 p.m. Green found that cyclopoid copepods and *Bosmina meridionalis* also migrated in Lake Pupuke, and that in the case of egg-bearing *B. meridionalis* females midnight sinking occurred as well as a marked dawn rise.

Geddes (1968) investigated a population of *Boeckella triarticulata* in a shallow Melbourne pond by taking twelve two-hourly surface tows in mid-April only. The adults commenced upward migration after 4 p.m. but before 6 p.m. and their numbers in surface waters increased steadily throughout the night to reach a maximum at about 4 a.m. Minima in surface numbers occurred at midday and 4 p.m. The surface concentrations of immature copepodites doubled between 4 p.m. and 6 p.m., fell considerably by 8 p.m., and then rose again to a second lesser maximum at midnight. Numbers then remained fairly high until 4 a.m. when a decline set in. The migration pattern for the nauplii was fairly similar to that of the immature copepodites except that the midnight surface maximum was greater than that at 6 p.m. Geddes also produced evidence, from surface sex ratios, of adult females migrating differentially; relatively more males than females came to the surface during the night and moved out of the surface waters during the day.

If nothing else, the above investigations show what a variable phenomenon diurnal vertical migration is. It is well established that not only do different species behave differently, but the behaviour of the same species may be quite different in one locality from that in another, or even in the same locality from one season to another. It has been known for some time, and also clearly emerges from the Australasian studies, that responses to light may differ markedly between different stages and sexes of the same species. Some planktonic animals become more sensitive to light as they grow older and others less so. Despite variability in the results of Australian and New Zealand studies, some general points emerge. It is clear that in *Boeckella* the migration patterns of adult males and females are usually different, males being apparently more sensitive to light and certainly having more extensive and definite migrations. It may be noted that in the case of the much-studied marine *Calanus* the reverse is true; the females are more strongly migratory than the males (Nicholls 1933). Also in lacustrine populations of *Boeckella*, there is often a striking maximum in surface densities at dusk or within the 2–3 hours following dusk. After this maximum there is a rapid decline in numbers which is much more marked than that occurring at dawn. In other words, unlike many forms, *Boeckella* seems not to have large surface numbers for most the whole period of darkness, but only for 2–3 hours somewhere in the period dusk to midnight. Evening surfacing with a rapid decline before or about midnight also commonly occurs in the marine *Calanus* (Esterly 1912, Nicholls 1933).

THE SIGNIFICANCE OF DIURNAL VERTICAL MIGRATION The widespread occurrence of vertical migration, together with the fact that migrations are often very extensive (some marine species daily migrate through a total distance of 800 m [2×400 m]) and energy-consuming, strongly suggest that the habit must be of fundamental importance in the lives of migratory planktonic organisms. However, if there is one fundamental and universal significance attached to the habit, we seem still not to know what it is. In discussing this question it seems desirable to mention some facts obtained from marine and estuarine studies rather than restrict consideration to inland waters. Hardy (1956) put forward the view that vertical migration has been evolved because it gives the animal a continual change of environment, with improved feeding possibilities, which would otherwise be impossible for a more or less passively drifting organism. He points out that there are differences in speeds of surface and deeper currents, and that dropping out of surface waters for an interval each day is a means of accomplishing movement relative to the surface. Against Hardy's theory, it seems fairly clear that in some cases the effect of vertical migration is quite the opposite of bringing about the continual sampling of new environments; it maintains the organism in the broad environment of a particular geographical region (Packard 1955). Thus in southern Antarctic seas the copepods *Metridia* and *Calanus*, and the euphausiaceans *Euphausia frigida* and *E. triacantha*, make extensive vertical migrations of the order of 200 m daily. Because of the existence of two opposed currents—a cold surface current about 100 m thick away from the South Pole and a warmer deep current towards the Pole—the effect of these migrations is to keep these populations in a particular geographical region between the edge of the pack-ice and the Antarctic convergence. Another very interesting fact is that these genera vary the extent of their migration seasonally in accordance with variations in the thickness of the currents, apparently so that the restrictive effect will always be achieved. Ekman (1953) reported the collection of dead specimens only of *Metridia gerlachei* in an 'extrapatriation area' south of New Zealand. This indicates a very strong selection pressure against too northerly a distribution for this species in surface waters.[8] It is probable that this mortality in the *Metridium* population arose from its failure to return to the deeper south-moving current after transport northwards in surface currents. Mackintosh (1937) stated that unless a planktonic organism in the Antarctic surface water was 'caught up in a returning current' it would continue to move northwards 'finally reaching a foreign environment in which it could not possibly survive'. In estuarine planktonic populations too, it seems that vertical migration serves the important function of helping to offset the ever-present tendency to be transported seawards. This is made possible by the fact that in many estuaries there is a deep landward current beneath the surface seaward one. The inhibition of vertical migration by haloclines (Grindley 1964) almost certainly represents a mechanism that prevents or helps to prevent the destruction of estuarine populations by sudden flooding. As such it emphasizes the close relationship between station-holding and migration. Finally, whilst on this theme, it may be noted that the number of truly cosmopolitan marine planktonic species is relatively very small; most are adapted to life in a particular geographical region. So too with estuarine plankton; it seems that there are no cosmopolitan estuarine copepods—certainly those in Australian and New Zealand estuaries are almost entirely endemic.

Harris (1953) suggested that the evolutionary advantage of diurnal vertical migration is the maintenance of animal plankton at suitable depths for feeding in the euphotic zone. He supposed that the vertical movements themselves are the accidental consequence of using light as the only consistent means of depth control available.

---

8. Ekman (1953) pointed out that *M. gerlachei* has occasionally been found as far north as 17°N in the Atlantic in deep water.

When this theory was advanced it was thought that pressure reception by invertebrates with the same compressibility as water was probably impossible. However, it has since been shown (Knight-Jones and Qasim 1955, 1959) that some invertebrate planktonic organisms are sensitive to changes in pressure. The exact mechanism of this pressure perception is still not clear but has been the subject of some investigation (e.g. Digby 1961).

McLaren (1963) explained vertical migration on the basis that it is more efficient to feed at high temperatures and to grow at low temperatures. After considering the data available for the temperature dependency of physiological processes in crustaceans, he concluded that as soon as a fairly marked thermal stratification develops vertical migration is metabolically advantageous. At temperatures around 20°C a difference of as little as 1°C or even 0·5°C between surface temperature and that at the day depth may be enough to confer some advantage over an animal that continuously remains in the warm epilimnion close to the phytoplankton. However, the McLaren theory offers no explanation for the regular occurrence in a small minority of populations of a reverse-to-normal pattern of migration, and the undoubted occurrence of this phenomenon challenges it. The existence of reversed migration suggests that the most important thing about migration is that daily movement occurs, its direction and relationship to vertical temperature differences being of secondary importance. Another difficulty with the McLaren theory is that although a vertical temperature difference of 0·5°C may seem small, it is quite large compared with actual vertical differences found throughout most of the day in some polymictic lakes and dimictic or cheimomictic lakes during periods of holomixis. Yet in such lakes or periods normal vertical migration continues. But for the occasional occurrence of reversed migration and the extremely variable nature of vertical migration even within the same species, the continuation of migration under these circumstances would be explicable on the basis that once the habit had evolved it was genetically fixed and occurred irrespective of anomalous environmental situations. However, all the available evidence indicates that the environment has a very direct modifying effect on the apparent basic necessity for the whole population not to remain continuously at the same depth; the particular pattern of migration fits the particular local circumstance or conditions.

Hutchinson's (1967) suggestion that vertical migration may be significant in niche diversification warrants consideration. However, in inland waters this necessity is obviously less for the migratory fraction of the plankton, that is the specifically simple zooplankton, than the diverse phytoplankton for which generally there has been no demonstration of well-marked diurnal vertical migration (but see Berman and Rodhe 1971). Hutchinson also suggested that vertical migration may be interpreted as protective nocturnalism, that is exploitation of the cover of darkness for protection from predation whilst feeding. Wynne-Edwards (1962) seemed to suggest that vertical migration evolved so that males and females can be concentrated near the surface at night for the purpose of displaying themselves to each other and assessing, in view of the existing population density, whether copulation, re-dispersion, or adjustment of mortality is the best course of action. His theory can hardly be taken seriously. The possibilities that vertical migration is significant as a means of preventing local accumulation of toxic metabolites or the local depletion of oxygen seem not to have been previously investigated or suggested. If, however, a high population density of animals were to build up at a particular level which was continually occupied, then these seem real possibilities with potentially serious consequences.

According to Rudjakov (1970) the downward movement of zooplankton is not an active one, but is entirely due to passive sinking. He pointed out that the observed rates of sinking in anaesthetized animals are quite great enough to account for observed rates of descent in natural populations. In anaesthetized *Daphnia* '*magna*', *D.* '*longispina*', and *Diaptomus coeruleus*, for example, the sinking velocities are 16–36 m/h, *ca*. 18 m/h, and 4–8 m/h respectively. In the case of some planktonic animals the passive-sinking rate may be as high as 70 m/h. The Rudjakov theory of vertical migration of zooplankton is that the life of an individual consists of alternating phases of high and low activity. During the phase of low locomotory activity a passive motion of the organism takes place, directed upwards or downwards depending on the level of this activity, the direction and velocity of vertical transport, changes in the density and viscosity of the water, and daily variations in the specific gravity of the organism. During the phase of high activity, the organism actively travels upwards or downwards, depending on the direction of its motion during the passive phase. Some of the peculiar or apparently anomalous aspects of vertical migration seem to be explicable by the Rudjakov theory. Thus midnight sinking and dawn rise may be a consequence of there being two peaks of activity—a main one in the evening and a smaller one in the morning, with something of a lull around midnight. Although illumination may act in a general sort of way as a synchronizer in this circadian rhythm of activity, its influence is not simple or direct; both peaks (in animals with bimodal activity) usually persist under conditions of constant or rapidly changing illumination and are not, therefore, specifically geared to changes in illumination. Reversed migrations may be explained in two ways. First as cases of daytime rather than nocturnal activity. This, however, is unlikely to be the explanation of reversal when typical migration also exists or is the norm in the same species. There is, however, a second explanation: reversed migrants may still have a nocturnal-activity maximum, but as their buoyance is positive they ascend passively during the daytime and move downwards actively during the evening. The Rudjakov theory seems to be more comprehensive and flexible than many other theories that have been advanced.

It is quite possible that the search for a universal significance of vertical migration will be endless for the reason that there isn't one; the habit may well have a different significance in different circumstances. It seems difficult to believe, for example, that a daily migration of about 1 m by *Daphnia* in a small pond lacking unidirectional currents has the same significance as a euphausiacean daily migrating through 200 m or more in southern Antarctic seas. Rudjakov (1970) argues that vertical migration is not the result of adaptation to a planktonic mode of life but is still performed as a 'carry-over' of activity rhythms that were inherent in the ancestors of present-day plankton. He even suggests that in some cases daily vertical migrations may be simply the result of unfavourable changes in the living conditions of the animals (e.g. loss of stored fat).

CYCLOMORPHOSIS The term cyclomorphosis refers to the seasonal polymorphism that is often observed in quite a variety of planktonic organisms. The phenomenon has been extensively studied in the cladoceran genus *Daphnia* in which the usual pattern of seasonal variation involves the gradual expansion, in successive generations during summer, of the anterior part of the head to form a helmet. By

ECOLOGY OF THE MAJOR BIOLOGICAL COMMUNITIES

mid-summer this helmet, which is extremely variable in shape, often becomes disproportionately large. In the most common species in the southern half of Australia and New Zealand, *D. carinata*, cyclomorphosis seems to influence not merely the form of the front of the head, but the size of the keel or carina which sweeps right round from the ventral rostrum to the postero-dorsal aspect of the head (Fig. 6:5). The existence of extreme cyclomorphic forms in *D. carinata* makes this species a remarkably variable one, which fact was fully appreciated by Sars (1914).

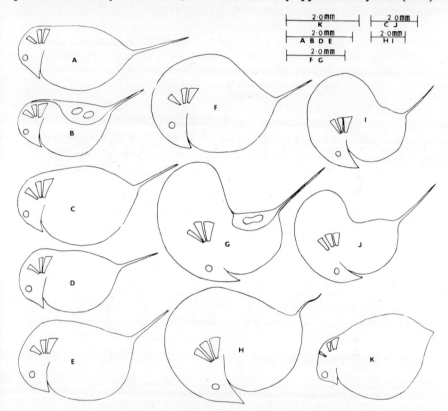

Fig. 6:5 Variation in the keel or carina of *Daphnia carinata*. This figure does not depict a cyclomorphic series from the one population or locality, but illustrates the extent of polymorphism in specimens from different localities. Nevertheless true cyclomorphosis involving the shape of the keel does occur (A–H after Sars, I–K original). K was drawn from a specimen collected from Lake Cootapatamba near Mt Kosciusko (see text)

Since that time, however, no one seems to have sampled a population of *D. carinata* at regular intervals throughout the year and documented in published form the full cycle of morphological changes. It is the experience of one of us, however, that forms in which the carina is exceptionally large (Fig. 6:5F–J) nearly always come from shallow ponds rather than lakes. Not only is the head-keel of *D. carinata* subject to great variation, but so too is the relative size of the postero-dorsal carapace spine. The latter variation, however, may not be truly cyclomorphic in nature; it seems that this spine may become very small or virtually absent (Fig. 6:5K) in populations that are subjected to very low temperatures (Bayly 1970b). In *D. lumholtzi* cyclomorphosis seems to involve the variable production of a forward-directed spine from the front of the head and also variations in the lengths of the spines on the fornices covering the bases of the antennae.

Copepods are generally regarded (Hutchinson 1967) as showing only small seasonal changes in form, yet some of the Australasian boeckellids show, in addition to seasonal pulsation in body size, a well-marked cyclomorphosis with respect to the length of the lobes on the last metasomal segment of the female (Bayly 1962, Timms 1967) (Fig. 6:6). The ratio (length of lobe):(body length or metasome length) is greatest in late winter or early spring when body length is maximal, and least in summer when the body size is minimal (Table 6:4). It seems that cyclomorphosis is much more striking in boeckellids than in diaptomids.

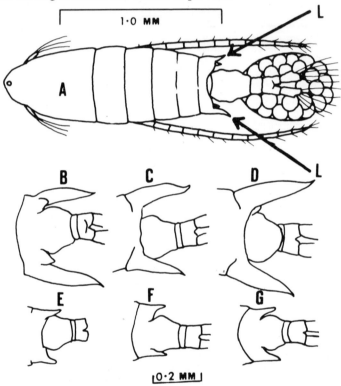

Fig. 6:6 Cyclomorphosis in the lobes (L) on the last metasomal segment of the female of *Boeckella propinqua* (A). B–C: dorsal and ventral aspects of the lobes of late winter or early spring forms; E–G: ventral and dorsal aspects of the lobes of mid-spring or summer forms. *After Bayly (1962)*

Cyclomorphosis is especially well shown in loricate rotifers such as *Brachionus* or *Keratella*. In these animals it usually involves the production of reduced spines on small summer forms or the addition of posterolateral spines or processes. Food and other (non-food) organisms seem to have an influence on the latter phenomenon.

ECOLOGY OF THE MAJOR BIOLOGICAL COMMUNITIES   123

A well-developed cyclomorphosis is exhibited in the dinoflagellate *Ceratium hirundinella*. This species sometimes has only two spines or horns but usually there are three. In summer, however, it is not uncommon for a fourth horn to be produced, and there is some evidence that this retards the rate of sinking.

Many cyclomorphic changes can be described by the following relationship:

$$y = bx^k$$

in which y is the size of the part of the animal which is mainly involved in the change of shape and x is some standard linear dimension. When relative growth occurs in accordance with this equation the process is termed allometry or, perhaps better, *heterauxesis* (Hutchinson 1967). When k is less than unity, the part is relatively smaller in large than in small stages of growth and the process may be termed *bradyauxesis*. When k is greater than unity, the part is relatively larger in large than small stages and the term *tachyauxesis* is used. In practical investigations of heterauxesis it is best to convert the above equation to logarithmic form as follows:

$$\log y = \log b + k \log x.$$

It is also clear that many cyclomorphic changes, such as the length of the helmet in cladocerans, the length of posterior metasomal lobes in boeckellids, and the number of horns on *Ceratium hirundinella*, correlate with temperature. However, despite our ability to describe many cyclomorphic changes in mathematical terms and to correlate them with temperature, their adaptive significance, if any, still seems to be obscure. Where it produced extra spines or projections on summer forms, it was once supposed to represent an adaptation to retard the rate of sinking in the face of the lowered viscosity of warmer waters. This now seems an unlikely explanation. Readers are referred to Hutchinson (1967) for a detailed discussion of the phenomenon.

Non-planktonic biota

The non-planktonic biota falls fairly naturally into five categories: *benthos*, *macrophytes*, *Aufwuchs*, *nekton*, *psammon*, and *pleuston* (see Chapter 3 for definitions). These will be considered separately.

BENTHOS

Figure 3:2 presents a gross division of the benthos into two strata, a littoral one in the trophogenic zone and a profundal one in the tropholytic. This division was appropriate for the initial discussion of the major ecological regions and biological communities of lakes, but of course limnologists have attempted to be more precise about the vertical zonation of the benthos. Unfortunately, however, this attempted precision has itself caused considerable confusion, and a variety of terms and definitions of the same term arose (cf. Welch 1935, table 32). Recently Hutchinson (1967) has discussed the terminology briefly and proposed a compromise one; his ideas are expressed in diagrammatic form in Fig. 6:7. It is unlikely that any single scheme will obtain universal acceptance, or indeed is universally applicable, but Hutchinson's seems to provide a reasonable working basis.

Figure 6:7 indicates that the vertical profile of the lake edge is not a smooth curve; as a result of wave erosion *typically* it tends to conform to the shape shown, with the development in the lake of a littoral shelf (= shore terrace) in which the coarser particles are nearer the lake edge and the finer ones furthest from the shore.

The composition and abundance of the littoral benthic fauna are closely determined by the degree of shelter that exists, in that in exposed situations subject to significant wave action there is little or no development of vegetation and the movement of loose substratum particles frequently creates difficult abrasive conditions for animals.

TABLE 6:4
CYCLOMORPHOSIS IN THE POSTERIOR METASOMAL LOBES OF FEMALES OF BOECKELLA MINUTA
Data after Timms (1967) for a population in a Queensland pond

Date	20.ii.1964	9.iv.1964	8.v.1964	4.vi.1964	2.vii.1964
(mean length of lobes) / (mean body length)	0·125	0·134	0·150	0·154	0·151

Date	30.vii.1964	27.viii.1964	5.ix.1965	8.x.1964
(mean length of lobes) / (mean body length)	0·161	0·157	0·168	0·149

Conversely, in sheltered positions there is often a rich development of vegetation and correspondingly a diverse and abundant fauna. As will be indicated later in the consideration of pond biota (Chapter 9), certain similarities of composition exist between the biota of permanent ponds and of the sheltered littoral regions of lakes. Certain similarities (not to be stretched too far) also exist between the faunas of exposed lake shores and stony streams.

With increasing depth the benthic fauna in general becomes less diverse and also less abundant in terms of biomass. In profundal regions the commonest forms include bivalves, oligochaetes, and chaoborid and chironomid larvae, but the profundal fauna in composition reflects *sensu lato* the general level of productivity of the lake via the expression of this upon chemical conditions operating in the tropholytic zone and especially at the lake bottom-water interface. Thus, as discussed in more detail in Chapter 5, in highly productive (eutrophic) lakes contributing large amounts of organic material to the hypolimnion, oxidative processes result in varying degrees of hypolimnetic oxygen depletion. Oxygen may disappear completely from the hypolimnion in very productive lakes and, additionally, waters near the bottom may contain large concentrations of reducing substances such as hydrogen sulphide. On the other hand, in lakes of only moderate or low productivity, the amount of organic material contributed to the hypolimnion is insufficient to cause significant oxygen depletion there.

Needless to say, the effects of such differences are far-going. In strongly reducing situations normal respiration is not possible and the fauna consists only of some few protistan types which respire anaerobically. In progressively less severe situations there is the development of a specialized fauna of higher types (metazoans), and in this connection the association of particular genera and species of the Chironomidae

ECOLOGY OF THE MAJOR BIOLOGICAL COMMUNITIES

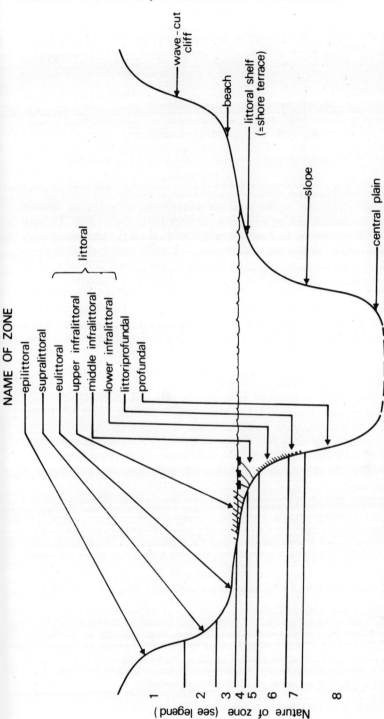

Fig. 6:7 Vertical zonation of lake regions, *following Hutchinson (1967)*. 1: zone uninfluenced by spray; 2: zone subject to wave action and sometimes inundated; 4: zone of emergent rooted vegetation; 5: zone of floating-leaved rooted vegetation; 6: zone of submerged rooted or adnate macrophytes; 7: zone of scattered autotrophs if these are present, sometimes zone of abundant blue-green algae, always a transitional zone; 8: zone of bare mud, lacking vegetation (if profundal extends below 600 m, then it is sometimes referred to as the abyssal)

with various degrees of oxygen depletion at the mud-water interface is of considerable interest. The association was first noticed by Thienemann (1920) and refined soon afterwards by Lenz (1925, 1927) and Lundbeck (1926, 1936) in particular and later by Brundin (1949, 1956). Initially the association was of oligotrophic lakes and the genus *Tanytarsus*, and of eutrophic lakes and *Chironomus*, but the more detailed association became:

    oligotrophic lakes: a *Orthocladius*
                             b *Tanytarsus*
    mesotrophic lakes: a *Stictochironomus*
                             b *Sergentia*
    eutrophic lakes: a *C. bathophilus* (= *anthracinus*)
                             b *C. plumosus*

The scheme was proposed as valid for central Europe, but was later shown to be partly applicable elsewhere in the northern hemisphere, although its detailed application has been criticized by many (e.g. Deevey 1941, Stahl 1959). To what extent such a scheme applies to Australia and New Zealand is unknown, despite Brundin's (1958) comment that the scheme does have world-wide applicability.[9]

It would be appropriate to round out this very brief and general discussion of the benthos by referring to some Australian and New Zealand work. Unfortunately this is not possible to any great extent. No published works relating comprehensively to the benthos of Australian freshwater lakes exist, and even for New Zealand, where freshwater lakes are much more common, the only accounts available are general ones (Cunningham *et al.* 1953, Stout 1969b). There does exist, however, some unpublished information on an Australian freshwater lake, Lake Purrumbete, a lake 45 m deep in western Victoria, and this has kindly been placed at our disposal for summary by Mr B. V. Timms who is currently investigating the lake.

The fauna recorded thus far from Lake Purrumbete and an indication of the vertical range of its components and their abundance is given in Table 6:5. In this particular lake, the deeper strata, predictably, are dominated by chironomids and oligochaetes (molluscs are rare), and chaoborid larvae, so typical of the profundal benthos of many northern hemisphere freshwater lakes, are absent. The table does not indicate clearly the decreased species diversity with increasing depth; this is shown more clearly in Fig. 6:8. With increasing depth there is also a rapid decrease in benthic biomass, and this relationship, based on twelve monthly samples over a period of a year, is shown in Fig. 6:9. The average annual (1969–70) weighted[10] benthic biomass (wet weight) for the whole lake was $5 \cdot 47$ gm/m², a value slightly more than that recorded for 20 Swedish and 54 Finnish lakes (Lundbeck 1936), slightly less than that for 30 lakes in Connecticut, U.S.A. (Deevey 1941), and much less than that recorded for some temperate eutrophic lakes (e.g. $39 \cdot 3$ gm/m² for Lake Esrom, Denmark, Berg 1938). However, as Ökland (1964) has stressed, such

---

9. Although the ideas of Brundin as an authoritative and meticulous worker on the family Chironomidae should be accorded full weight, it seems that the idea of a southern-hemisphere applicability of a lake typology founded upon northern profundal chironomids was based primarily upon an examination of southern Andean oligotrophic lakes where Brundin found a *Tanytarsus* community. According to Freeman (1961) at least some of the components in the typology are not so far recorded from Australia.
10. A factor obtained by determining the area of the bottom at each depth was used. The weight of mollusc shells is not included in this value.

ECOLOGY OF THE MAJOR BIOLOGICAL COMMUNITIES

comparisons need to be judged with reservations. For a further, more detailed consideration of regional comparisons of benthic standing crops, and of the relationship between benthic standing crops and depth, readers are referred to Chapter 3.

TABLE 6:5
BENTHIC FAUNA OF LAKE PURRUMBETE, VICTORIA
All data from Mr B. V. Timms (pers. comm.)

Taxon[a]		Approximate abundance[b]	Vertical range (m)
Porifera:	Spongillidae	+	2·5– 5·0
Turbellaria:	Dugesia sp.	+	0·5– 5·0
	Phaenocora sp.	++	0·0–45·0
Nematoda:	Dorylaimus stagnalis	++	0·5–14·0
Oligochaeta:	Branchiura sowerbyi	+++	0·5– 1·0
	Lumbriculus variegatus	+++	1·0–22·5
	Limnodrilus hoffmeisteri		0·5–45·0
	Potamothrix bavaricus }	++++	1·0–22·5
	Tubifex tubifex		0·5–45·0
Crustacea:	Candonocypris sp.	+++	
	Cypridopsis sp.	+++	
	Gomphocythere australica	++	
	Newnhamia fenestrata	++	
	Pseudochydorus sp.	+	mainly littoral
	Ilyocryptus sp.	+	
	Heterias sp.	+	
	Austrochiltonia subtenuis	++	
Insecta:	Ephemeroptera, nymph	+	littoral
	Zygoptera, nymph	+	
	Chironomus occidentalis	++	5·0–35·0
	Cryptochironomus sp.	++	0·5– 1·0
	Unident. tanypodine	+++	0·5–45·0
	Unident. chironomid A	++	0·5–22·5
	Unident. chironomid B	+	7·0–22·5
	Unident. chironomid C	+	0·5
	ceratopogonid	+	1·0–22·5
	Ecnomus sp.	+	7·0
	Xuthotricha sp.	+	1·0
	Unident. trichopteran	+	1·0
	Unident. pyralid	+	2·5
Hydracarina		++	0·5–22·5
Mollusca:	Unident. gastropod A	++++	0.5–35·0
	Unident. gastropod B	++	0·5–14·0 } but
	Unident. gastropod C	++	0·5–22·5 } mainly
	Unident. planorbid	+	1·0–14·0 } littoral
	Unident. sphaerid	+++	0·5–22·5

[a] Identification of some taxa incomplete. Definitive identifications by: Mrs Allison, University of Canterbury (Turbellaria); Mr M. Saur, C.S.I.R.O. (Nematoda); Prof. R. Brinkhurst, University of Toronto (Oligochaeta); Dr S. U. Hussainy, M.M.B.W., Melbourne (Ostracoda); Dr N. N. Smirnov, Moscow (Cladocera); Dr J. Martin, University of Melbourne (Chironomidae); Mr E. F. Riek, C.S.I.R.O. (Trichoptera); others by Mr B. V. Timms. Unident. = unidentified.
[b] + (rare), ++ (not common), +++ (common), ++++ (abundant).

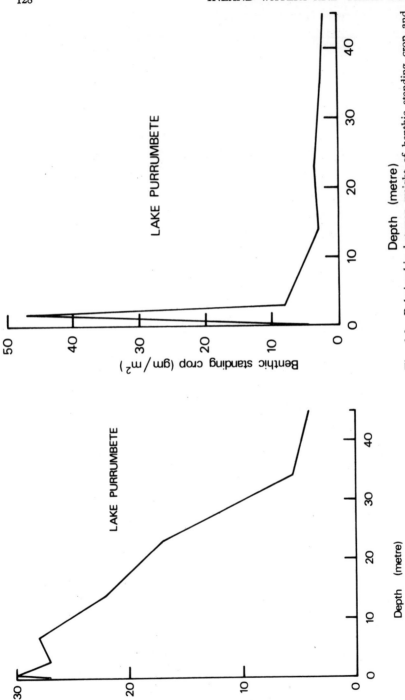

Fig. 6:8 Relationship between number of benthic species recorded and depth in Lake Purrumbete, Victoria. Based on unpublished information provided by B. V. Timms

Fig. 6:9 Relationship between weight of benthic standing crop and depth in Lake Purrumbete, Victoria. Wet biomass only is considered and the relationship based on twelve monthly samples (1969-70). Based on unpublished information provided by B. V. Timms

ECOLOGY OF THE MAJOR BIOLOGICAL COMMUNITIES

No comparable quantitative data on the benthos of New Zealand lakes are available. We may note, nevertheless, Stout's (1969b:461) general comments on the profundal benthos of Lake Lyndon, Canterbury: 'Below the lower limit of rooted vegetation, the fauna shows an increase in the number of midge larvae or blood worms (*Chironomus zealandicus*) and the presence of bivalve molluscs such as the pea mussels (*Pisidium novaezealandiae*) and the large dark freshwater mussel (*Hyridella menziesi*), while the snails, worms, and mites decrease in numbers or are absent.'

MACROPHYTES

From Fig. 6:7 we can see that the three zones of the infralittoral correspond to three major structural groups of macrophytes: emergent rooted species, rooted species with floating leaves, and rooted species with completely submerged leaves. Additionally, we should note the occurrence of some macrophytes which have no permanent connection with the substratum and float freely—usually on or near the surface. There is strong evidence (e.g. the presence of a well-developed internal system of aeration) that all the angiosperm macrophytes, that is most aquatic macrophytes, evolved from terrestrial ancestors and are merely secondary invaders of the aquatic environment. The composition and extent to which each zone develops, if at all, depends upon several factors including especially the degree of shelter from wave action, nature of the substratum, stability of water level, morphology of lake edge, and clarity of the water. Geographical factors are also implicated. In general macrophytes do not extend deeper than about 11 m.

Examples of some of the principal genera involved with at least some species in each of the structural groups are (there are many more):

1 Emergent rooted macrophytes (down to about 150 cm water depth): *Eleocharis, Glyceria, Phragmites, Scirpus, Juncus.*
2 Floating-leafed, rooted macrophytes (from about 0·5 to 3·5 m water depth): *Apogonetum, Nymphaea, Nuphar, Potamogeton, Nymphoides.*
3 Completely submerged, rooted macrophytes (down to about 10 or 11 m): *Elodea, Lagarosiphon, Potamogeton, Vallisneria, Chara, Nitella.*
4 Free-floating macrophytes (relatively sheltered situations only): *Eichhornia, Lemna, Wolffia, Salvinia, Ceratophyllum, Azolla.*

The genera listed occur either in Australia or New Zealand or both countries.

According to Sculthorpe (1967), many of the Australian indigenous aquatic macrophytes are pan-tropical, Asiatic, or Malaysian species which extend into Australia. Amongst the few endemic forms are *Myriophyllum verrucosum* and some species of *Cycnogeton, Haloragis,* and *Maundia.* New Zealand, too, has few endemic forms. Moreover, there is a particular shortage of indigenous submerged species with only five forms recorded, four of which occur mainly in saline waters (*Lepilaena bilocularis, Potamogeton pectinatus, Ruppia spiralis, Zannichellia palustris*). Sculthorpe goes on to say that it is not surprising therefore that several alien species have been able to spread throughout much of New Zealand. Much information on the spread of adventive aquatic macrophytes in Australia and New Zealand is summarized by Sculthorpe, and we have abstracted from his data (table 11.5) and added to them to produce Table 6:6 which lists the alien species in Australia and New Zealand. Many of these are of potential or actual economic importance in that they may or do create nuisance conditions as a result of heavy infestation (but see also Chapters 12 and 2).

TABLE 6:6
ALIEN AQUATIC MACROPHYTES OF AUSTRALIA AND NEW ZEALAND
Most data from Sculthorpe (1967), remainder from a variety of sources

Species	Australia[a]	New Zealand[a]
Callitriche intermedia	+	+
C. stagnalis	+	+
C. heterophylla	—	+
Glyceria maxima	+	+
Iris pseudacorus	—	+
Potamogeton crispus	—	+
Ranunculus aquatilis (? R. fluitans)	—	+
Rorippa nasturtium-aquaticum	+	+
Elodea canadensis	+	+
Eichhornia crassipes	+	—
Salvinia auriculata	+	—
Myriophyllum brasiliense	+	+
Trapa natans	+	—
Vallisneria spiralis	+	+
Egeria densa	—	+
Hydrocleys nymphoides	—	+
Aponogeton distachyos	+	+
Lagarosiphon major	—	+
Alisma plantago-aquatica	+	+
Sagittaria trifolia	+	—
Ludwigia palustris	+	+

[a] + present, — not recorded.

In addition to permanently aquatic macrophytes, there are of course many macrophytes which are characteristically associated with the eulittoral zone of lakes. Special mention may be made of the monocotyledonous family Centrolepidaceae which has some endemic species in Australasia. *Centrolepis* (=*Gaimardia*) species are known from three large sub-alpine lakes (Lakes Brunner and Te Anau in New Zealand, and Pedder in Tasmania [Cheeseman 1925, Bayly, Peterson, Tyler, and Williams 1966]). At Lake Pedder it is apparently submerged in winter. *Hydatella*, another centrolepid genus, occupies a similar habitat, but in warmer regions—two species occur in Western Australia, and one, *H. inconspicua*, in Northland, New Zealand (see Cheeseman 1907, Edgar 1966). Other Australian endemic eulittoral plants are known (e.g. *Milligania* nov. sp. on the shores of Lake Pedder [W. Jackson pers. comm.]).

AUFWUCHS
Considerable semantic confusion surrounds the use of those terms that apply to microscopic plants and animals which live attached to or closely associated with submerged materials. And the situation has not been helped, in our view, by any of the recent texts in which definitions are provided (e.g. Dussart 1966, Hutchinson 1967). Without considering in detail here the variety of definitions for terms proposed, we adopt as a working basis the following terminology derived from current popular usage, so far as we can determine this.

ECOLOGY OF THE MAJOR BIOLOGICAL COMMUNITIES

The *Aufwuchs* community comprises all plants and animals (for the most part microscopic) attached to or sessile on submerged material. It also includes microscopic free-living forms inhabiting or closely associated with the assemblage of attached or sessile forms. The plant component of the Aufwuchs is collectively termed the *periphyton* (irrespective of the nature of the substratum and the etymological derivation of the word). Both plant and animal components of the Aufwuchs may be divided into those that are attached to or associated with mud or sand surfaces (*epipelic* species), with plant surfaces (*epiphytic* species), with rock or stone surfaces (*epilithic* species), and with animal surfaces (*epizooic* species). It is customary, however, for these terms to be used only in connection with discussions of the periphyton.

To a significant degree the nature of the surface determines the composition of the Aufwuchs community, and a further factor of importance for the periphyton is light intensity and hence depth. Nevertheless on all types of surface the animals involved are predominantly protozoans and rotifers, and diatoms, filamentous green algae, and blue-green algae are important plant groups (Round 1965).

NEKTON

A full consideration of all those organisms that are nektonic in Australian and New Zealand lakes is beyond the scope of this book. We mention merely that nektonic organisms world-wide comprise mainly fish, and some of the larger crustaceans and certain insects (especially some Hemiptera, Diptera, and Coleoptera). A summary of the principal nektonic invertebrates of Australian and New Zealand freshwater lakes is presented in Table 6:7.

The fish faunas of Australia and New Zealand are depauperate, and in Australia at least, because of the relative paucity of freshwater lakes, most species are associated with river systems. The two faunas are considered further in Chapter 8.

Most nektonic organisms are carnivores.

PSAMMON

A variety of organisms has been found inhabiting sand and the interstices between sand grains at lake margins. Predictably, most are microscopic, but some macroscopic forms are also recorded. The fauna commonly occurring includes protozoans, rotifers, turbellarians, nematodes, tardigrades, copepods, and various insect larvae, although other forms occur, as was found by Bayly *et al.* (1966) in their investigation of the macroscopic fauna of the quartzite beach of Lake Pedder, Tasmania (Table 6:8). Many species of algae have also been recorded from the upper layers of sand beaches. Marked stratification of controlling factors (e.g. oxygen, carbon dioxide, pH, temperature) is an important feature in the ecology of this community.

PLEUSTON

Several sub-communities of the pleuston may be distinguished. The microscopic sub-community, the neuston, is one, and includes bacteria, many algae (chrysophyceans, diatoms, xanthophyceans, and others), and protozoans (e.g. *Arcella*) living either on the upper level of the surface film and termed the epineuston, or the lower level and termed the hyponeuston. Another sub-community, the epipleuston, comprises a variety of scavenging hemipteran bugs which really live above the surface film and use it only for support and a source of entrapped food. Such forms for

TABLE 6:7
PRINCIPAL NEKTONIC INVERTEBRATES RECORDED IN SOME
AUSTRALIAN AND NEW ZEALAND FRESHWATER LAKES

Taxon		Australia[a]	New Zealand[a]
CRUSTACEA			
Notostraca:	*Lepidurus*[b]	+	+
Anostraca:	*Branchinella*[b]	+	—
Mysidacea:	*Tenagomysis*	—	+
Amphipoda:	*Austrochiltonia*	+	—
	Paracalliope	—	+
Decapoda:	*Paratya*	+	+
	Caridina	+	—
	Macrobrachium	+	—
INSECTA			
Hemiptera:	*Micronecta*	+	—
	Diaprepocoris	+	+
	Sigara	+	+
	Agraptocorixa	+	—
	Anisops	+	+
	Enithares	+	—
Diptera:	*Chaoborus* larvae	+	—
	culicid larvae and pupae	+	+
Coleoptera:	several families including especially Dytiscidae	+	+

[a] + present, — not recorded.
[b] Of sporadic occurrence only.

TABLE 6:8
MACROSCOPIC FAUNA OF PSAMMON IN QUARTZITE BEACH AT LAKE
PEDDER, TASMANIA
Data rearranged from Bayly *et al.* (1966)

Taxon		Fine sand[a]	Gravelly sand[a]
Nematoda		++	—
Oligochaeta		++++	++++
Bivalvia		++	—
Crustacea			
Ostracoda:	*Ilyodromus* sp. a.	+	—
	Ilyodromus sp. b.	—	+
	Eucypris sp.	—	+
Isopoda:	Phreatoicidea	+++	+++
	Asellota	—	+
Insecta			
Diptera Chironomidae:	Orthocladiinae sp. a.	—	+
	Orthocladiinae sp. b.	—	+
Non-Chironomidae		++	++

[a] ++++ dominant, +++ abundant, ++ present but not rare, + rare, — absent.

Australian lakes include *Naeogeus* and *Merragata* (Hebridae), *Hydrometra* (Hydrometridae), *Mesovelia* (Mesoveliidae), and several genera of the Gerridae and Veliidae. *Hydrometra*, *Mesovelia*, and some Gerridae also occur in New Zealand. Those submerged members of the pleuston which do not appear at the surface film continuously, the meropleuston, include in Australia *Scapholeberis* (Cladocera), *Macrogyrus* (Gyrinidae), and several genera of the Culicidae and Ceratopogonidae.

Finally we should stress that the development of a pleuston is dependent upon the occurrence of sheltered conditions. Clearly, such a community would not develop if water turbulence at the surface were excessive, and for this reason it is mostly restricted to the littoral region. Indeed, it would perhaps have been more valid to have discussed this community in our consideration of pond biota (Chapter 9). Pleuston also occurs along the sheltered margins of streams, even some that are quite swiftly flowing.

part three
RUNNING WATERS

Freshwater stream in northeastern Victoria. *Photograph by David Deakin*

chapter seven
NON-BIOLOGICAL FEATURES

As remarked elsewhere (Chapter 10), Australia is the most arid of the world's inhabited continents. However, for the amount of work that has been published on the ecology of Australian rivers and streams, limnologists outside Australia might well be forgiven for thinking that no running waters exist in Australia at all! In some ways, therefore, the task of writing this chapter and the next with respect to Australian running waters is unfortunately a rather simple one. We need provide merely a summary of the pertinent physical, chemical, and biological features of running waters sparsely illustrated with the few Australian data available. The task is made all the easier by the recent publication of Hynes's (1970b) admirable and comprehensive world-wide account of the ecology of running waters. Rather more information is available for New Zealand streams and rivers, but even that country can hardly be said to be well served with accounts of its running-water environments.

Before our main discussion in this and the following chapter, it is as well to summarize briefly the salient features which distinguish running waters (*lotic* environments) from those which are standing (*lentic* environments). Running waters are characterized by their:

1 unidirectional flow,
2 often, considerable fluctuation in flow rates,
3 relatively unstable bottom and shoreline areas,
4 linear morphology,
5 relative shallowness,
6 biota which generally has adaptations to the unidirectional flow,
7 (as a rule) greater turbidity, oxygen concentrations, and terrestrial/aquatic nutrient interchange.

PHYSIOGRAPHICAL BACKGROUND
In spite of their frequent topographical prominence, rivers and streams contain only a small portion ($\ll 1$ per cent) of the total volume of water that occurs on the surface of the world's land areas, and of this portion Australia, in particular with respect to its area, has an extremely small fraction. Thus the total annual run-off (i.e. water which re-enters the sea from the land surface) in Australia, *ca.* 346 km^3, is only slightly in excess of the annual discharge from one European river, the Danube, with *ca.* 282 km^3.

There is indeed only one major river system in Australia, the Murray-Darling,[1] and this, though (combined) the fourth longest river system in the world (5,270 km, exceeded only by the Nile, Amazon, and Mississippi-Missouri) and with the world's sixth largest drainage basin, has an annual average discharge of only ca. 15 km^3, a value far below that of other large rivers in the world (see Table 7:1). The total number of separate Australian rivers is of the order of 400, with a total length of some 133,000 km and with each km on average draining 60 km^2 of land. In proportion to its area many more rivers exist in New Zealand, each draining on average a much smaller area. Some values for annual discharge (flow) and drainage areas of the principal Australian rivers are given in Table 7:2.

TABLE 7:1
ANNUAL AVERAGE DISCHARGE OF SOME OF THE WORLD'S PRINCIPAL RIVERS
From various sources

River	Average annual discharge (km^3)
Yangtze, China	895
Danube, Europe	282
Ganges, India	180
Columbia, U.S.A.	177
Indus, India	109
Nile, Africa	89
Sacramento-San Joaquin, U.S.A.	40
Murray, Australia	21

TABLE 7:2
MEAN ANNUAL DISCHARGE AND DRAINAGE AREAS OF SOME IMPORTANT AUSTRALIAN RIVERS
Basic data abstracted from the Resources Information and Development Branch of the Department of National Development, Canberra (1962), and converted to metric units

River[a]	Mean annual discharge (km^3)	Total drainage area (km^2)
Murray-Darling	21·08	1,060,000
Mitchell, Alice, and associated rivers, Qld	10·25	69,400
Burdekin, Qld	8·98	130,800
Fitzroy, W.A.	4·79	82,100
Derwent, Tas.	4·30	9,800
Clarence, N.S.W.	3·78	22,800
Murrumbidgee, N.S.W.	3·69	91,500
Daly, N.T.	2·63	53,100
Snowy, Vic.	2·34	15,800
Yarra, Vic.	1·21	4,070
Fortescue, W.A.	0·79	46,400

[a] Qld, Queensland; W.A., Western Australia; Tas., Tasmania; N.S.W., New South Wales; N.T., Northern Territory; Vic., Victoria.

---

1. An exceedingly readable account of the physiography, history, and some biological features of this system has been given by Gill (1970).

NON-BIOLOGICAL FEATURES

The broad nature of drainage systems or hydrological regions is discussed more fully later (Chapter 10; see also Figs 10:1, 10:2), and we need reiterate here only that four types can be recognized: *exorheic* (coastal drainage), *endorheic* (internal drainage), *arheic* (no superficial drainage), and *cryptorheic* (underground drainage). Within each drainage system a variety of local drainage types may be distinguished according to present pattern (cf. Fig. 7:1) or historical events (i.e. superimposed, inherited, antecedent, or defeated drainage). Twidale (1968) provides various Australian examples and further explanation. Referring to the actual rivers or streams involved, physiographers also distinguish between those which are *consequent*, that is follow the initial slope of the land, *subsequent*, that is develop along zones of geological weakness, and *obsequent* and *resequent*, that is develop in association with subsequent streams and flow, respectively, in a direction opposed to or the same as the consequent stream. *Insequent* streams fall into none of these categories. In addition, hydrologists (e.g. Leopold, Wolman, and Miller 1964) may analyze drainage patterns according to the relationship between stream length and stream order, analyses which, it has been suggested (Abell 1961, Hynes 1970b), may be of considerable value to biologists.

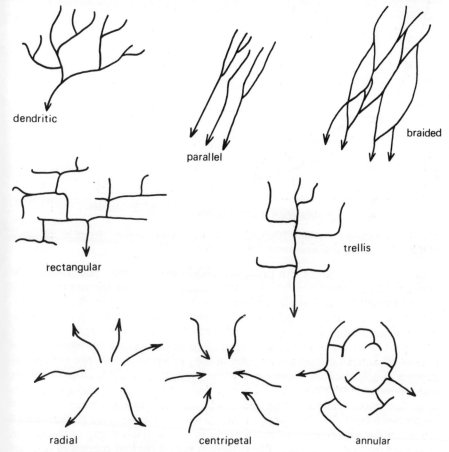

Fig. 7:1 Local drainage patterns

Stream order is an indication of the hierarchical status of the stream in terms of its tributaries: first-order streams have no tributaries, second-order streams are formed by the confluence of two first-order streams, third-order streams from the confluence of two second-order streams, and so on.

Streams and rivers are geomorphological agents of considerable importance, as is well recognized by physical geographers. The extent of their importance is related *inter alia* to the volume of their discharge, the duration of flow, and the flow velocity. In general the upper reaches of a river or stream are erosional (or degrading) in that they remove more material than they deposit in the area they drain or pass over, whereas the lower reaches are depositional (or aggrading) in that more material is deposited than is eroded. As a consequence there is a trend for the longitudinal profile (*thalweg*) of all rivers and streams to assume a shallow concave shape (Fig. 7:2), and this, given time (which only rarely happens), has the tendency gradually to coincide with the base-level profile when no further erosion and consequently deposition occur, and the longitudinal profile is said to be graded.

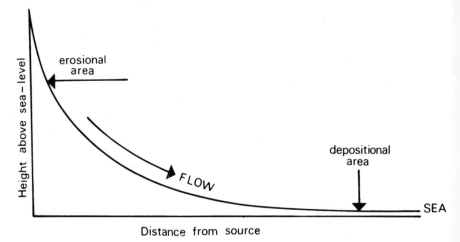

Fig. 7:2 Longitudinal profile (thalweg) of a hypothetical stream. Steepness of curve exaggerated

Erosion is the result of two processes, corrosion or solution, and corrasion or abrasion. The former process, as the terms indicate, results in the addition to the river or stream of soluble material which once dissolved is not again deposited. The latter process contributes largely insoluble material, the *load*, to the river or stream. The amount, size of particles, and the distance the load is transported depend principally upon the discharge and rate of flow of the water body involved. The amount of material carried is a measure of the *capacity* of the river or stream and depends largely upon the volume of the discharge; the size of the largest particles that can be carried is a measure of the *competence* of the river or stream and depends upon flow rates. That part of the load which is so heavy that it is never actually in suspension is termed the bed-load; it is moved only by traction. Load particles which are sometimes suspended and move partly by traction and partly in suspension constitute the load of saltation. The remaining part of the load consists of particles more or less permanently in suspension.

NON-BIOLOGICAL FEATURES

Some important physiographical features associated with the erosional properties of rivers and streams include stream piracy, extension of valleys headwards and seawards, the alteration of valley transverse profiles, lateral immigration of streams and rivers, river meanderings, and slumping of and tunnel formation in banks. Details on these can be gained from a number of more appropriate texts (e.g. Hills 1959, Drury 1966, Twidale 1968, Leopold *et al.* 1964), and here no further comment is necessary. Figure 7:3 illustrates some of the major features associated with a river meander; such meanders and their ultimate offsprings, ox-bow lakes (see Chapter 3), are particularly characteristic of many lowland rivers in Australia. Examples of the various erosional processes are easily provided by many Australian and New Zealand rivers and streams.

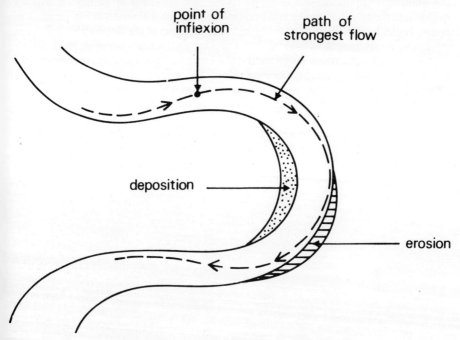

Fig. 7:3   Major physiographical features associated with a river meander

PHYSICAL FEATURES

The main morphological parameters for streams and rivers are maximum and mean depth, width at actual water surface, width at bank-full discharge, surface area, cross-sectional area, water volume, extent of shoreline, mean (transverse) slope, shoal area, longitudinal gradient, and discharge. All of these terms are self-explanatory.

The salient physical feature is the unidirectional flow of water. The flow rate may undergo considerable fluctuation according to season or local meteorological conditions and flow may even cease altogether at certain times. It is convenient to recognize those streams and rivers which always contain flowing water, *permanent* streams or rivers, those which flow only seasonally and are otherwise dry, *intermittent* streams or rivers, and those—mostly in arid regions—which flow only when unpredictable rain has fallen, *episodic* streams and rivers (Plates 7:1–7:4). Of course it is not always possible rigidly to distinguish these categories of running-water locality from each other, and many rivers in semi-arid parts of Australia, for example, consist of a series of pools during the dry season but with extensive subsurface flow between pools.

Seasonal and secular variations in flow rates or discharge are most pronounced in the continental region of Australia, and less so in Tasmania and in New Zealand, although even in these regions they may be considerable. Because of the construction of impoundments (themselves a reflection of the unreliability of Australian rainfall) on most of the larger Australian rivers, as well as other human activities, it is now not possible to ascertain the exact pattern of natural fluctuations in the discharge of most Australian rivers. However, some actual data relating to the River Murray (which has numerous impoundments along its course) are provided in Tables 7:3 and 7:4 to illustrate the extent of secular and seasonal fluctuations that now occur in permanent rivers. It can be seen that, despite the moderating effect of the impoundments, large variations in discharges still occur. Even greater fluctuations occur in many intermittent and episodic rivers and streams; the latter in Australian desert regions may be dry for many years but become flooding torrents following sudden rain (Plates 7:1–7:4). A similar situation, on a seasonal basis, frequently prevails in intermittent streams and rivers in monsoonal regions of Australia.

Plate 7:1 Cooper's Creek, Queensland, looking north from Arninga Waterhole (25° 40′ S, 142° 40′ E) towards Windorah. Photograph taken 8 April 1949 from a height of 4,600 m. *By courtesy of the R.A.A.F.*

TABLE 7:3
SECULAR VARIATION IN DISCHARGE OF RIVER MURRAY NEAR MILDURA (AT EUSTON GAUGING STATION) OVER A TEN-YEAR PERIOD (1956–66)
Basic data abstracted from Bibra and Mason (1967) and converted to metric units

Year	Monthly minimum (km^3)	Monthly maximum (km^3)	Annual total (km^3)	Annual total as percentage of annual mean[a]
1956–57	0·13	8·53	41·31	441
1957–58	0·09	0·58	2·87	31
1958–59	0·16	2·32	11·11	119
1959–60	0·10	0·63	3·36	36
1960–61	0·12	2·71	13·49	144
1961–62	0·09	0·72	4·30	46
1962–63	0·07	0·61	3·72	40
1963–64	0·09	0·97	5·53	59
1964–65	0·06	3·36	12·60	135
1965–66	0·11	0·82	3·12	33

[a] Based on mean for period 1930–66, i.e. 9·36 km^3. Original.

TABLE 7:4
SEASONAL VARIATION IN DISCHARGE OF RIVER MURRAY NEAR MILDURA (AT EUSTON GAUGING STATION) FOR PERIOD 1930–66
Basic data abstracted from Bibra and Mason (1967) and converted to metric units. All values as km^3

Month	Minimum	Maximum	Mean
January	0·026	2·071	0·425
February	0·011	0·698	0·197
March	0·048	0·450	0·192
April	0·021	0·993	0·248
May	0·032	1·859	0·408
June	0·054	3·774	0·537
July	0·136	6·392	0·857
August	0·138	8·531	1·259
September	0·112	6·008	1·434
October	0·076	5·316	1·531
November	0·049	4·367	1·298
December	0·025	3·649	0·973

The discharge values of hydrologists are less interesting to ecologists than the flow rates operating where the biota actually lives. However, such rates are usually more difficult to determine than discharge values. Hynes (1970b) summarized the various sorts of methods that have been used by biologists to gather information on this subject. Hydrologists have provided formulae for the determination of flow rates in channels, but since natural water courses rarely correspond in morphology to these, such formulae have limited usefulness to biologists. Let it suffice for us to say that flow in rivers and streams is a complex process and the determination of actual rates of flow involves parameters of length, width, depth, discharge, nature of the bottom, and longitudinal gradient. For Australian rivers and streams that flow only intermittently, an indirect way of estimating flow conditions during floods has been proposed by Williams (1969), based on the physical nature of the dry bed.

Within the water body itself flow rates are different at different points; in general they are greatest near the surface and middle of the river or stream and least near the banks and bottom. This pattern is the result of course of frictional forces, and these have such an effect that for a distance of about 1–3 mm above stones or other parts of the substratum flow rates are very small. Small or non-existent flow rates, so-called zones of dead water, also occur downstream of protruding obstructions on the beds of rivers and streams. Contrary to popular belief, the intuitive assessment of both authors, and the assumptions of several early authors (e.g. Tansley 1939), hydrologists claim

Plate 7:2 Dendritic drainage patterns in the flooded 'Channel Country' (Cooper's Creek) of Queensland. Photograph taken 8 April 1949 from a height of 4,600 m. *By courtesy of the R.A.A.F.*

that *mean* flow rates are not greatest in headwaters and decline seawards; downstream there is an overall *increase* in mean flow rate, although the increase itself becomes progressively less as the longitudinal gradient decreases (Leopold 1953). This, however, may be merely a reflection of the manner in which *mean* flow rates are computed by hydrologists.

Because of irregularities on the banks and beds of rivers and streams, water in these exhibits turbulent flow; that is, it is subject to eddying and local circulation. Only rarely in nature do surface water masses flow in parallel lines, that is exhibit laminar flow. Occasionally, when extremely high natural flow rates occur as for example in waterfalls, sudden spurts of water may result and flow is then said to shoot or jet.

The range of flow rates in rivers and streams is from zero to > 800 cm/sec. (Hynes 1970b). However, rates over 600 cm/sec. are rare, and rates usually do not exceed 300 cm/sec. General correlations have been drawn between flow rates and the nature of the substratum of rivers and streams (cf. Table 7:5), but these correlations are by no means as simple and straightforward as their proposers thought. Thus the mean particle size of river substrata generally decreases seawards despite the concomitant increase in the *mean* flow rate. The explanation partly involves the gradual comminution of particles, stones, and so on as they progress downstream, but other phenomena are undoubtedly involved. A correlation between flow rates and the type of bottom deposit that is moved at a given flow rate is a somewhat better way of viewing the matter, although even then water turbidity can complicate the issue. Table 7:6 illustrates this point and gives one approximate correlation of this sort that has been published.

TABLE 7:5
RELATION OF FLOW RATES AND NATURE OF STREAM AND RIVER BEDS
After Minnikin (1920)

Flow rate (cm/sec.)	Nature of bed	Habitat
> 121	rock	torrential
121–91	heavy shingle	torrential
91–60	light shingle	non-silted
60–30	gravel	partly silted
30–20	sand	partly silted
20–12	silt	silted
< 12	mud	pond-like

Temperatures of running waters,[2] it can be said broadly, are more variable than are those of standing bodies of water, but their absolute range is smaller. Because a large number of local factors operate in determining the temperatures of rivers and streams, it is rather difficult to generalize on this subject, but it is nearly always possible to discern a relatively smooth seasonal pattern superimposed upon which

---

2. We are grateful to numerous State Fisheries officers for making available to us much pertinent but unpublished information which has been useful in the compilation of this part of our account.

may be a daily pattern of variation. Additionally, temperatures are lower in localities at higher altitudes than in those at lower ones. Factors which are important in determining temperatures include the presence of lakes or impoundments along the course of the river or stream, the nature of the source of the water (e.g. from springs, run-off, lakes), the presence or absence of shading vegetation on banks, and, of increasing significance in this modern world, the occurrence or not of thermal pollution.

In Australia and New Zealand seasonal maxima usually occur in January or February, and minima in July or August. Absolute seasonal values correlate of course with geographical position. In eastern Australia rivers of the semi-arid inland have a range of about 11–29°C (the Darling River at Bourke, N.S.W.), but lower and higher values have been reported on occasion, the latter particularly in backwaters where 36°C may be attained (Lake 1967c). The rivers of the southeastern uplands, on the other hand, have maximal summer temperatures generally not much greater than 20°C but winter temperatures can fall to 0°C. In those rivers which flow coastally from the southeastern uplands the seasonal range is about 7–27°C. Upland stream temperatures both in southeastern Australia and elsewhere in Australia as well as New Zealand occupy somewhat narrower seasonal ranges, but are known to have broader diurnal ones than rivers, particularly in summer.

Plate 7:3 The Warburton River, South Australia, looking westward from a point (27° 45′ S, 137° 38′ E) close to its entry into Lake Eyre North, which is in the background on the left. Photograph taken 11 May 1950 from a height of 4,600 m. *By courtesy of the R.A.A.F.*

NON-BIOLOGICAL FEATURES

TABLE 7:6
MEAN FLOW RATES OF CLEAN AND MUDDY WATER NECESSARY TO MOVE VARIOUS TYPES OF BOTTOM DEPOSIT ON RIVER AND STREAM BEDS
After Schmitz (1961) and Hynes (1970b)

Type of deposit	Critical mean flow rate cm/sec.	
	Clean water	Muddy water
Fine grained clay	30	50
Sandy clay	30	50
Hard clay	60	100
Fine sand	20	30
Coarse sand	30–50	45–70
Fine gravel	60	80
Medium gravel	60–80	80–100
Coarse gravel	100–140	140–190
Angular stones	170	180
Grass sods	180	180

Plate 7:4 The Warburton River, South Australia, looking eastward from its point of entry into Lake Eyre North (27° 50′ S, 137° 13′ E). Photograph taken 11 May 1950 from a height of 4,600 m. *By courtesy of the R.A.A.F.*

Thermal stratification is of minor importance in running waters and only develops in stream and river pools with minimum flow. Nevertheless it is a phenomenon that has been noted in pools in some Australian rivers, where temperature differences of almost 9°C may develop in pools only slightly deeper than 1 m (Morrissy 1967). Despite contra-indications from overseas work, such vertical temperature differences may have important biological repercussions; it is suggested, for example, that it is only the occurrence of such thermally stratified pools in South Australia that permits the persistence there in many rivers of the introduced brown trout (Morrissy 1967).

Characteristically, most lowland rivers in Australia—and to a less extent in New Zealand—are slightly to highly turbid. In the major inland rivers of New South Wales, for example, that is the Darling, Lachlan, and Murray-Murrumbidgee systems, Secchi-disc transparencies are rarely over 2 m and are very frequently less than 1 m (Lake 1967c). In part, especially in the more temperate areas of Australia, the high turbidity results from the greater land erosion now taking place following clearance of the natural vegetation to make way for pastoral activities, but natural erosional phenomena are also partly involved. Upland streams, like those world-wide, are usually very much clearer.

CHEMICAL FEATURES

Stream and river salinities do not reach the high values of certain inland lakes (see Chapter 10), and although exceptionally some Australian saline streams are known with salinities as high as 6,000 p.p.m., the vast majority of running waters world-wide contain less than 3,000 p.p.m. total salts, and outside endorheically-drained areas less than 1,000 p.p.m. There is not as close a correlation between climate and stream and river salinities as might at first be assumed. Some exceedingly arid parts of Australia may have rivers which are fresh (e.g. parts of northwestern Western Australia) when flowing or when consisting of permanent pools interconnected by subsurface flows, and some markedly saline rivers occur in comparatively well-watered areas (e.g. western Victoria). The nature of drainage basins, local geology, land use, nearness to the coast, as well as the seasonal pattern of rainfall and the climate in general, are some of the more important determinative factors.

In countries so geographically diverse as Australia and New Zealand little point is served in attempting to survey absolute salinity values—even assuming that this were possible comprehensively, which it is not. However, some actual values are given in Table 7:7 to provide an indication of the range of values encountered in Australia. Most (not all) upland streams and rivers predictably contain less salts than lowland localities and, indeed, within reasonably circumscribed areas or for one river system a fairly close correlation between altitude and salinity may emerge, as, for instance, in Tasmania (Williams 1964b) or in the River Murray (cf. Table 7:7).

The sources of contained salt are mainly twofold: one is corrosion in the drainage basin; the other is rain which in turn derives its salts from either the sea ('cyclic salts'), terrestrial dust, or nowadays atmospheric pollutants added by man. The relative importance of each derivation depends upon location, local geology, and climate; rain near the coast, for example, contains much more salt than rain inland (cf. Hutton and Leslie 1958). The comparative importance of salts from rain and from corrosion for any one river or stream is still a matter for some argument. Thus Currey (1970) has argued that most of the salt in the Wannon River, western Victoria, is derived from local geological sources, whereas others argue that cyclic salts in rain are more

TABLE 7:7
SALINITIES OF SOME AUSTRALIAN RIVERS AND STREAMS
Basic data from various sources

River or stream	Type[a]	Position[b]	Salinity (p.p.m.)	Sample details	Data source
Shannon	A	Tas.	21[c]	—	(1)
Snowy	A	Vic.	50–100[c]	3 readings, 1947	(2)
Murray[d]	A	N.S.W./Vic.	41	mean, Jingellic	(3)
Murray[d]	B	N.S.W./Vic.	79	mean, Torrumbarry	(3)
Murray[d]	B	N.S.W./Vic.	152	mean, Euston	(3)
Murray[d]	B	S.A.	277	—	(4)
Darling	B	N.S.W.	225–371	3 stations, January 1969	(5)
Paroo	B	N.S.W.	248	1 station, January 1969	(5)
Collie	B	W.A.	159–374	8 samples, 1940–44	(6)
Little River	B	Vic.	350–6,000[c]	20 samples	(7)
Tennant Creek	C	N.T.	156	1 sample, May 1965	(8)
Sixth Creek	D	S.A.	287[c]	1 sample, September 1966	(9)
Drysdale	D	W.A.	16·5	1 sample, June 1965	(8)

[a] A, permanent upland river; B, permanent lowland river; C, intermittent lowland river; D, permanent upland stream.
[b] Tas, Tasmania; Vic., Victoria; N.S.W., New South Wales; S.A., South Australia; W.A., Western Australia; N.T., Northern Territory.
[c] Strictly, total dissolved solids.
[d] Stations are progressively downstream.

Data sources: (1), Williams (1964b); (2), State Rivers and Water Supply Commission (1954); (3), Bibra and Mason (1967); (4), Livingstone (1963); (5), Williams, Walker, and Brand (1970); (6), Samuel (1951); (7), Webster (undated); (8), original; (9), Morrissy (1967).

important as the original source. The subject of ion supply to inland waters is discussed in more detail by Gorham (1961) and in Chapter 1.

Fluctuations in salinity are significantly related to discharge values, so that highest salinities usually (but not invariably) occur during times of low flow and *vice versa*. The phenomenon is one well known to hydrologists. Some data for a number of Victorian rivers (Table 7:8) illustrate the point. Since discharge values frequently display a seasonal pattern of variation (cf. Table 7:4), it follows that salinities do so also. Moreover, like temperature, vertical salinity stratification may develop in pools along the course of rivers at certain seasons. River pools which display such stratification are said to be important in various Western Australian rivers where during the rainy season surface waters in pools may be quite fresh and lower ones quite saline (Simpson 1928).

TABLE 7:8
RELATIONSHIP BETWEEN SALINITY AND DISCHARGE IN TEN VICTORIAN RIVERS
Basic data abstracted from Bibra and Mason (1967) and converted to metric units

River	Maximum salinity recorded (p.p.m.)	Contemporaneous discharge value (m^3/sec.)	Minimum salinity recorded (p.p.m.)	Contemporaneous discharge value (m^3/sec.)
Avoca	2,900	0·008	110	0·168
Fiery	2,800	0·420	250	11·98
Barwon	1,650	2·128	280	42·34
Campaspe	1,260	0·168	40	5·60
Goulburn	260	3·780	50	259·0
Murray (Euston)	260	52·08	100	71·71
Agnes	180	0·084	30	0·476
Yarra (Warrandyte)[a]	120	9·69	30	4·93
Acheron	60	5·32	20	30·91
Mitta Mitta	50	5·71	19	45·28

[a] Note unusual *direct* rather than inverse relationship.

The major inorganic ions of running waters in Australia display a variety of combinations in terms of relative contributions to total salinity (cf. Table 7:9), and it is certainly not valid to assume that ion combinations are predominantly of the divalent cations plus bicarbonate ion type, as is said to apply to most running waters elsewhere (Hynes 1970b). In Tasmania, for example, sodium is more likely to be the dominant cation than calcium or magnesium, and even in the most dilute waters (salinity <100 p.p.m.) chloride may be the dominant anion (Williams 1964b). Most rivers of southwestern Western Australia, the temperate part of that State, and of Victoria likewise are chemically not 'bicarbonate' waters; most such rivers are dominated by sodium and chloride ions (Samuel 1951; Webster undated). The large inland rivers of southeastern Australia, on the other hand, may contain as many or more bicarbonate anions and divalent cations as chloride and sodium ions (Williams, Walker, and Brand 1970), and similar ion proportions prevail in several rivers of northern Australia (unpublished). Except in the more saline rivers and streams which, reflecting a similar situation to that in standing waters, apparently are always dominated by

TABLE 7:9
CHEMICAL COMPOSITION OF SOME AUSTRALIAN RIVERS
Basic data from various sources

River or stream	Type[a]	Position[b]	Type of data[c]	Na^+	K^+	Ca^{2+}	Mg^{2+}	HCO_3^- + CO_3^{2-}	SO_4^{2-}	Cl^-	Source of data
Kiewa	A	Vic.	X	8	—	7	2	26	20	4	(1)
			Y	40.0	—	41.2	18.2	49.4	48.2	12.9	
Maribyrnong	A	Vic.	X	68	—	14	20	72	24	122	(2)
			Y	55.6	—	13.2	31.3	22.2	9.4	64.6	
Wannon	A	Vic.	X	680	—	78	117	200	183	1,250	(2)
			Y	68.4	—	9.0	22.6	7.6	8.8	81.5	
Paroo	B	N.S.W.	X	50.0	9.8	12.1	4.6	111.0	0.0	60.1	(3)
			Y	63.8	7.3	17.8	11.1	53.4	0.0	49.7	
Darling	B	N.S.W.	X	56.5	7.4	21.2	14.9	221.4	0.0	49.7	(3)
			Y	49.9	3.8	21.5	24.8	73.7	0.0	28.4	
Murrumbidgee	B	N.S.W.	X	6.9	5.5	9.0	5.4	39.6	12.0	9.9	(4)
			Y	22.4	10.4	33.6	33.6	48.5	18.6	20.9	
Murray	B	N.S.W./Vic.	X	16.1	3.0	5.3	3.9	36.8	9.5	23.8	(5)
			Y	51.2	5.6	19.4	23.8	44.1	14.5	49.0	
Tennant Creek	C	N.T.	X	23.0	6.0	4.0	10.7	70.2	12.8	29.4	(6)
			Y	44.6	6.8	8.9	39.7	51.2	11.9	36.9	
Sixth Creek	D	S.A.	X	46	—	26	25	165	21	70	(7)
			Y	37.3	—	24.6	38.0	54.1	8.2	36.7	

[a] A, permanent upland river; B, permanent lowland river; C, intermittent lowland river; D, permanent upland stream.
[b] Vic., Victoria; N.S.W., New South Wales; N.T., Northern Territory; S.A., South Australia.
[c] X = p.p.m.; Y = equivalent % of total cations.

Data sources: (1), State Rivers and Water Supply Commission (1954); (2), Webster (undated); (3), Williams, Walker, and Brand (1970); (4), Cassidy (1949); (5) Anderson (1945); (6), original; (7), Morrissy (1967).

sodium and chloride ions, exact correlations between salinity and ion proportions are difficult to discern, although they possibly do exist within localized areas. So few data are available to us concerning the chemistry of New Zealand running waters that we cannot discuss them in the present context.

Finally in this chapter it is appropriate to consider briefly the subject of dissolved gases in running waters. In general these gases appear to be more or less in equilibrium with atmospheric gases, although of course local variations frequently occur due to such causes as organic pollution (see Chapter 12), the presence of springs, plant growth, and so on. The two main gases of interest are oxygen and carbon dioxide and, as in standing waters, the concentrations of these are often inversely related to each other. In highly turbulent rivers and streams oxygen concentrations are usually near or in excess of saturation values; in slowly flowing waters, however, oxygen concentrations may be far from saturation values. Diurnal changes in oxygen concentrations in rivers have been used as a basis to compute primary production (see Chapter 2 for details).

chapter eight
BIOLOGICAL FEATURES

COMPOSITION OF THE BIOTA

Understandably, an autochthonously developed phytoplankton is universally absent from quickly flowing streams, but where running waters originate from a large body of standing water they frequently carry over significant amounts of phytoplankton from the lentic environment. This so-called potamophytoplankton, under many conditions, can continue active metabolism until it is finally washed into the sea. When rivers are large, relatively slowly flowing, and have perhaps numerous peripheral quiet regions and some deep pools or anabranches wherein flow is negligible, appreciable phytoplanktonic biomass may also develop *in situ* and contribute significantly to the potamophytoplankton. Diatoms are generally the most common plants of the potamoplankton (Hynes 1970b).

As a rule, most of the algae of running waters[1] are microscopic and attached. Collectively such algae are conveniently referred to as the periphyton community, as are those in standing waters (see Chapter 6). The epilithic and epiphytic forms are frequently the most important and include a wide variety of genera in many different algal groupings. The epipelic and epizoic algae are less diverse, the former consisting mainly of diatoms, coccoid Chlorophyceae, euglenoids, and some blue-green algae (Round 1965). Several authors (e.g. Round 1965) have listed genera which are common components of each of the four sub-communities, but it should be emphasized that considerable overlaps occur, and that many if not most of the microscopic algae involved are opportunistic.

Since most of the algae of running waters are cosmopolitan or almost so (Blum 1956), no useful purpose is served here by attempting to list the river and stream algae recorded from Australia and New Zealand. Suffice it to say that in those few and for the most part non-intensive studies of Australian and New Zealand running waters that include accounts of the microscopic algae (e.g. Playfair 1914, Phillips 1931, O'Farrell 1949, McMichael 1952, Jolly and Chapman 1966), no significant differences emerge to distinguish them from studies on running waters elsewhere.

---

1. A general adjectival term for use in referring to stream and river inhabitants (plant or animal) is rheophilous (from the Greek *reos*, current, and *philos*, loving).

Not all rheophilous algae are microscopic—there are many genera which are quite visible to the naked eye. Several rhodophytes (e.g. *Batrachospermum*), charophyceans (e.g. *Chara*), filamentous chlorophyceans (e.g. *Cladophora, Ulothrix, Spirogyra*), chrysophytes (e.g. *Vaucheria*), and cyanophytes (e.g. *Oscillatoria, Phormidium*) are of this sort; all of the examples quoted here have been recorded from Australia or New Zealand. The remaining macrophytes comprise bryophytes (mosses and liverworts), lichens, and angiosperms (flowering plants). Of these the bryophytes and most of the restricted number of lichens known to lead a submerged existence are characteristic of swiftly flowing rivers and streams where they occur attached to partially or completely immersed solid substrata. Many genera of bryophytes are known from such habitats, including *Campylopus, Aneura, Lepidozia,* and *Calypogeia*, all of which were found by McMichael (1952) in Warrah Creek, N.S.W. Some few angiosperms also characteristically grow attached to solid substrata in rapidly flowing waters; they belong to the families Hydrostychaceae and Podostemaceae. The former occurs only in Africa and Madagascar, but the latter is pantropical and one species has been recorded from Australia. This is *Torrenticola queenslandica*, recorded thus far from three localities in northeast Queensland, about 15-30 km apart and on two different river systems: Johnstone River (near Innisfail in 1873), Babinda Creek, a tributary of the Russell River (north of Innisfail in 1949), and Coolamon Creek, a tributary of the North Johnstone River (in 1970). The species has also been recorded from southeast New Guinea. Descriptions and illustrations of it are provided by Van Steenis (1949, 1952).

Most other angiosperms of running waters develop roots and are therefore restricted to localities which have substrata soft enough or sufficiently comminuted to be penetrable by roots. Hence these angiosperms are generally confined to comparatively slowly flowing regions of rivers or streams where depositional phenomena are more in evidence than erosional ones. Finally, a few angiosperms occur as free-floating forms. They are largely tropical or semi-tropical species of slowly flowing waters. One, *Eichhornia crassipes*, the water hyacinth, has been introduced into Australia where it has already become pestiferous in certain regions (see Chapter 12) although it is not yet the nuisance that it is in some other tropical and semi-tropical parts of the world.

The composition and density of the zooplankton of running waters, as for the phytoplankton, depend significantly upon the number and sort of associated standing waters and the flow regime. Paralleling the composition of the phytoplankton of rivers, there are no animals which are obligatorily planktonic in *modus vivendi*; the zooplankton of rivers must in general be considered as consisting mainly of opportunistic or stray species from standing waters. This being so, one would expect river zooplankton to be similar in composition to that of standing waters. Such seems not the case, however, and in large rivers, at least, planktonic crustaceans—which feature so prominently in lake zooplankton—are mostly unimportant whereas planktonic rotifers are important. The zooplankton of rivers is, moreover, often very much less abundant as compared with lakes than the phytoplankton (Hynes 1970b).

No investigations of any consequence have been conducted on the plankton of the larger Australian rivers, so that unfortunately we cannot reinforce our general statements on river plankton with more precise local information. It is conceivable in view of the length, small discharge, and low velocity of the Murray-Darling river system that the zooplankton—if not the phytoplankton—of this system is much closer to the composition of Australian lacustrine zooplankton than our general statements predict. However, if this were the case, one notable seasonal exception would be the occurrence in the river plankton of eggs of the golden perch, *Plectroplites ambiguus*, which is a planktonic spawner living typically in the warmer, sluggish rivers of southeastern Australia (Lake 1967c).

The benthic invertebrate fauna of running waters displays considerable uniformity of *general* composition on a world-wide basis (except in some especially isolated regions such as volcanic islands) (Hynes 1970a, 1970b). It also includes representatives of nearly all major taxa known from fresh waters and, additionally, many taxa are either restricted to running waters or reach their maximum abundance and density there. There are, however, regional faunistic features and these are particularly obvious in running waters in Australia and New Zealand which have been isolated by marine barriers for an extremely long period. Some of the more obvious (there are many others) of these features are:

1 The dominance of the freshwater mussel family Hyriidae (cf. Unionidae in the northern hemisphere).
2 The occurrence in Tasmanian streams of the extremely primitive and unmodified syncarid crustacean, *Anaspides tasmaniae* (absent elsewhere).
3 The almost complete absence of isopods in the family Asellidae (common in the northern hemisphere).
4 The occurrence (particularly in Tasmania) of phreatoicid isopods (absent in the northern hemisphere except India).
5 The presence of the plecopteran families Eustheniidae, Gripopterygidae, and Austroperlidae (absent in the northern hemisphere).
6 The absence of several important northern-hemisphere families of Ephemeroptera, and the dominant position of the Leptophlebiidae. In Australia and New Zealand this family has undergone adaptive radiation to produce species which occupy niches elsewhere characteristically occupied by species from other families.
7 The occurrence of aquatic mecopteran larvae of the family Nannochoristidae (no aquatic mecopterans are known from the northern hemisphere).
8 The absence of some common northern-hemisphere families of the Trichoptera (cf. Phryganeidae), and the adaptive radiation of some trichopteran families that do occur, especially the Leptoceridae.

A number of New Zealand investigators have produced reasonably comprehensive lists of the macroscopic invertebrate fauna in a variety of New Zealand rivers and streams (e.g. Phillips 1929, Allen 1951, Hirsch 1958, McLean 1966, McLay 1968, Stout *et al.* 1969, Burnet 1969). The fauna of Australian running waters, in sad contrast, is much less known (cf. O'Farrell 1949, McMichael 1952, Jolly and Chapman 1966), a discrepancy which we hope will constructively embarrass Australian freshwater biologists! Merely to illustrate in a general way the nature of the diversity of bottom invertebrates in Australian and New Zealand running waters, Table 8:1 lists this fauna in two small and physically rather similar rivers, one in the South Island of New Zealand, the Kakanui River (and its tributary, Island Stream), the other in New South Wales, Cox's River. The dominance of insects in these localities suggested by Table 8:1 parallels the dominance of insects in running waters throughout the world, although absent from the table are several taxa which frequently occur in other streams or rivers in temperate parts of Australia and in New Zealand. Such taxa include particularly several genera of amphipods and larvae of the dipterous family Blepharoceridae.

TABLE 8:1
COMPOSITION OF THE BENTHIC INVERTEBRATE FAUNA
(MACROSCOPIC ONLY) IN THE KAKANUI RIVER (AND TRIBUTARY
ISLAND STREAM) IN THE SOUTH ISLAND OF NEW ZEALAND, AND IN
COX'S RIVER, NEW SOUTH WALES, AUSTRALIA
Original data rearranged and partly reclassified from McLay (1968) and Jolly and Chapman (1966). The data for Cox's River relate to a downstream unpolluted station on the river ('CD')

Taxon	Kakanui River and tributary[a]	Cox's River[a]
Porifera: Spongillidae	—	+
Cnidaria: Hydra	—	+
Turbellaria	+	+
Nematoda	—	+
Polyzoa	—	+
Oligochaeta	+	+
Hirudinea: Glossiphoniidae	—	+
Mollusca: Sphaeriidae	+	—
Ferrissiidae (= Ancylidae)	—	+
Planorbidae	+	+
Hydrobiidae (= Potamopyrgidae)	+	—
Crustacea: Atyidae	—	+
Ephemeroptera: Baetidae	—	+
Leptophlebiidae	+	+
Siphlonuridae	+	—
Caenidae	—	+
Plecoptera: Eustheniidae	+	—
Gripopterygidae	+	+
Odonata: Zygoptera	—	+
Anisoptera	—	+
Hemiptera: Veliidae	—	+
Megaloptera: *Archichauliodes* sp.	+	+
Coleoptera: Parnidae	+	—
Hydrophilidae	+	—
Psephenidae	—	+
Trichoptera: Leptoceridae	+	+
Sericostomatidae	+	—
Hydroptilidae	+	+
Hydropsychidae	+	+
Rhyacophilidae	+	+
Polycentropidae	+	+
Odontoceridae	—	+
Diptera: Simuliidae	+	+
Tipulidae	+	+
Chironomidae	+	+
Lepidoptera: Nymphulinae	—	+

[a] + present, — not recorded.

The exact composition of the benthic invertebrate fauna of running waters obviously differs from point to point within any given body of running water, depending upon the duration and intensity of the various controlling factors. Thus, in only slowly flowing regions and peripheral pools, the benthic fauna approximates in composition to that of ponds and other small standing waters. Similarly, differences in composition also occur according to the nature of the substratum; quite a different fauna develops in rocky stretches than develops in silting stretches, for example.

Finally, in this brief discussion of faunal composition, we need to refer to the vertebrate components. World-wide these are mainly fish, but also inhabiting running waters in Australia are larvae (tadpoles) of several genera of frogs, about a dozen species of tortoise, two crocodile species, and a few mammals. None of these latter components occurs in New Zealand, which has no native aquatic mammals at all and where even the native frogs (*Leiopelma* spp.) do not have aquatic tadpoles.

The fish fauna of both Australia and New Zealand is depauperate, and it seems that Australia has only two and New Zealand no species of fish that are truly freshwater in the sense of being primarily freshwater (Norman 1963). The two Australian species in question are the Queensland lungfish, *Neoceratodus forsteri*, in certain rivers of southeastern Queensland, and the spotted barramundi, *Scleropages leichhardti*, in rivers of the Gulf of Carpentaria and Queensland. Both are of considerable zoological interest. All other Australian and New Zealand native fish, ca. 130 species in Australia and 30 in New Zealand (Lake 1971, Stokell 1955), are to be regarded as secondarily freshwater in habitat. In any event the fish faunas of both countries display little diversity compared with other regions of the world, and even the vast area drained by the Murray-Darling river system in Australia supports only about 26 native species. Of particular interest and some importance in rivers and streams of both countries are species of galaxiids, southern 'trout'; Australia has about 25 species (Frankenberg 1969), New Zealand about half that number (McDowall 1964a, 1964b).

With regard to vertebrates other than fish in Australian running waters, special mention may be made of the reptiles and mammals. All the tortoises (order Chelonia) belong to the group that retract their heads sidewards into their shells (suborder Pleurodira) and are almost entirely aquatic throughout life. Four genera occur, *Chelodina, Emydura, Elseya*, and *Pseudemydura* (Goode 1967). The commonest species, *C. longicollis*, occurs in the River Murray and elsewhere, and the River Murray and its tributaries also contain the largest species, *C. expansa*, as well as *Emydura macquari*. Our knowledge of the tortoise fauna of northern rivers is still incomplete. Of the crocodiles, one, *Crocodilus johnstoni*, is confined to rivers in northern Australia, whereas the other, *C. porosus*, typically occurs in northern Australian estuaries but can penetrate upstream long distances and well into the freshwater reaches of rivers. The most interesting mammal present is without doubt the duck-billed platypus, *Ornithorhynchus anatinus*, almost to be considered a zoological aberration and belonging to the extremely primitive mammalian order, the Monotremata. It inhabits lakes and rivers in eastern Australia from Cape York to Tasmania and is still common in the Murray and Murrumbidgee River areas. No marsupial mammals are characteristically associated with Australian rivers in the manner of the platypus, but some native placental mammals are (*Hydromys* spp.). *Hydromys chrysogaster* is the common eastern water-rat of Australia.

CONTROLLING FACTORS

A variety of factors has been shown to control the distribution and abundance of the rheophilous biota. Some are much more important in a general sense than others, and the effect of some is more or less confined to or especially significant for only certain parts of the biota. There may be, moreover, a considerable degree of correlation between factors (e.g. current speed, substratum type, and oxygen concentration) so that it is frequently not easy to distinguish precisely the effect of one from that of others.

Those factors of greatest importance are current speed, temperature, nature of substratum, and the concentration of various dissolved substances (especially oxygen, carbon dioxide, and salts). In addition, the following factors may be important or significant either seasonally or at other times, or for specific parts of the biota:

1 frequency of desiccation (and availability of interstitial water) and flooding;
2 intensity of scour;
3 presence or absence of rapid stretches, waterfalls, anabranches, peripheral and on-course pools;
4 depth;
5 turbidity;
6 light (and shade);
7 biotic factors (e.g. interspecific competition, grazing effects);
8 nature of catchment area;
9 zoogeography;
10 extent of human influence (e.g. pollutional loading, number of impoundments, type of introduced biota.)

Hynes (1970b) has extensively reviewed and discussed the large number of studies concerned with factors controlling the lotic biota. One group of studies of particular relevance and to which we may make separate reference was presented at an international conference in 1961 and later published (*Symposium über den Einfluss der Strömungsgeschwindigkeit auf die Organismen des Wassers* 1962).

TABLE 8:2
EFFECT OF FLOODING UPON THE NUMBER AND BIOMASS OF BENTHIC INVERTEBRATES IN THE HOROKIWI STREAM, NEW ZEALAND. SEVERE FLOODING OCCURRED ON 12–14 FEBRUARY 1941
Basic data from Allen (1951)

Zone of stream[a]	Percentage of December 1940 numbers and biomass (wt)			
	22, 24 February 1941		9, 10, 21 April 1941	
	No.	Wt	No.	Wt
I	22·5	23·5	68·4	75·8
II	14·5	11·7	38·4	31·7
III	48·3	27·7	109·5	48·3
IV	34·9	14·9	75·7	68·3
V	39·2	17·5	90·6	29·4

[a] Arbitrary zonation after Allen (1951). In zones I, III–V, the stream bed was stable; in zone II it was unstable and aggrading.

Because of the paucity of Australian and New Zealand investigations on this subject, it is not possible for us to illustrate comprehensively the effect of each of the above factors by reference to local examples. However, since Australian and New

Zealand rivers and streams characteristically display wide fluctuations in flow rates (see Chapter 7), it is pertinent to mention briefly one local study which clearly illustrates the importance of flooding as a controlling factor. We refer to Allen's (1951) work on the Horokiwi Stream on the North Island of New Zealand. This author, as part of a classical study of fish production (see Chapter 2), investigated the density of the bottom fauna during 1940 and 1941. In 1940 relatively dry weather prevailed and no significant floods occurred. Then, on 12–14 February 1941, there was a major flood, and the collections which followed (22, 24 February) showed a pronounced reduction both in the numbers and biomass of the bottom fauna. A slight flood occurred shortly after 24 February, but a series of collections made on 9, 10, and 21 April 1941 indicated that the fauna had recovered to a considerable degree its former density. These changes are shown on a quantitative basis in Table 8:2. Allen's data also clearly showed the differential effect of the flooding upon the fauna; in general those species inhabiting rapid regions of the stream were better able to withstand flood-damage than those in quieter regions. Table 8:3, in which some of Allen's data are reproduced, quantifies this differential aspect. Another New Zealand study of the effect of flooding upon the benthic fauna of streams is that of McLay (1968). No details of this need be given since the results parallel Allen's.

TABLE 8:3
DIFFERENTIAL EFFECT OF FLOODING UPON PART OF THE BENTHIC INVERTEBRATE FAUNA OF THE HOROKIWI STREAM, NEW ZEALAND
Effect shown as a percentage decrease in density after flooding (mean density of taxon in February 1941 following flood × 100/combined mean density of taxon in April and December 1940, i.e. before flood). Modified from Allen (1951)

Taxon[a]	%
Trichoptera: *Hydropsyche* sp.	0·84
Rhyacophilidae	0·25
Helicopsyche sp.	0·51
Olinga sp.	0·37
Sericostomatidae, type C	1·05
Pycnocentrodes sp.	0·09
Ephemeroptera: *Coloburiscus* sp.	0·07
Deleatidium sp.	0·21
Plecoptera: Gripopterygidae	0·29
Coleoptera: Parnidae	0·19
Parnidae adults	0·20
Diptera: Chironomidae	0·06
Simuliidae	0·05
Mollusca: *Potamopyrgus* sp.	0·10
Oligochaetes	0·21

[a] Insects represented by immature stages unless otherwise indicated.

ADAPTATIONS

Since the planktonic biota of running waters is only facultatively rheophilous, consisting largely of opportunistic and stray species, it is not surprising that none of its components shows any special adaptive modifications for this environment. By and large the same also seems to be true of the rooted macrophytes, and those morphological characteristics which appear to distinguish the macrophytes of running waters from those of still waters are frequently phenotypically induced, or are general characteristics—such as flexible stems and creeping habit—that in a sense pre-adapt plants possessing them for life in rivers and streams. A few specific adaptations, however, are displayed by some attached higher plants (mosses, liverworts, a few angiosperms) in that they may develop rhizoid-like holdfasts, a flattened thallus, or have sticky seeds. Epilithic and epiphytic algae, too, may have flattened bodies and rhizoid-like structures, and many are stalked and use adhesive substances to fix themselves to submerged surfaces.

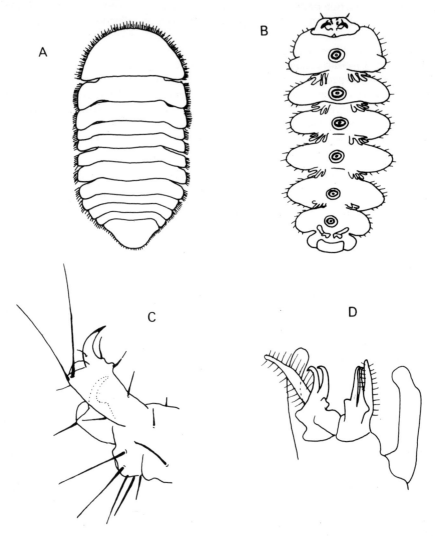

Fig. 8:1 Some morphological adaptations to flowing water. A: dorso-ventral flattening of psephenid larva; B: ventral suckers of blepharocerid larva; C, D: respectively, grappling apparatus at hind end of abdomen of campodeiform caddis larva and melagopteran larva. All drawings based on specimens collected from various Victorian streams

BIOLOGICAL FEATURES

By far the widest array of adaptations to unidirectional flow is displayed by animals, particularly the insects. These adaptations may be morphological, behavioural, or physiological, and many animals have adaptations spanning all three categories. The most general characteristic displayed is a behavioural one; with some exceptions, notably amongst the fish, the animals of rivers and streams *avoid* as far as possible exposure to the direct effect of the current. Thus most occupy positions where the speed of the current is small or negligible, that is in the boundary layer about 1 to 3 mm above submerged objects or in the dead space behind (see Chapter 7), in spaces below and between stones, within interstitial spaces of the substratum, and inside plant remains.

Many morphological adaptations in effect are structural modifications which make such a behavioural pattern more efficacious. Thus the extreme dorso-ventral flattening exhibited by psephenid beetle larvae (Fig. 8:1a) enables the larvae to live on stones in rapidly flowing parts of streams and rivers because the larvae live within the boundary layer on the upper surface of the stones. Other stream animals which are flattened—for example, in Australia, larvae of *Aldia* (a gripopterygid stonefly) and several leptophlebiid larvae—occupy spaces beneath stones or other regions within the substratum and beyond possible exposure to the force of the current. Here, also, flattening is suitably adaptive. Yet other animals are typically flattened irrespective of whether they live in lentic or lotic environments, and they are pre-adapted for life in the latter sorts of environments. Triclads provide an example. Few invertebrates are streamlined, because streamlining conveys few benefits in animals so small. Many fish, on the other hand, being much larger, are streamlined, and general correlations have been drawn for fishes between the roundness of their bodies in cross-section and their ability to resist currents.

Several types of structures used for adhering firmly to the substratum have evolved in invertebrates as well as in some fish and frog tadpoles. True hydraulic suckers, however, are rather rare amongst stream-dwelling invertebrates (amongst free-living forms only in the Blephaceridae, Fig. 8:1b), although several fish families and frog genera of running waters include species with them. In the frogs of course the suckers occur in the aquatic tadpole stage. Australian examples are provided by Cox's gudgeon (*Gobiomorphus coxii*), a small goby in which the ventral fins are modified to form a suction cup so efficient that even vertical surfaces can be climbed (Lake 1959), and by *Hyla lesueri*, a hylid frog whose rheophilous tadpoles have a large mouth with numerous labial papillae forming a suction cup (Fig. 8:2) (Martin 1967a).

Friction devices occur more frequently than suckers in invertebrates and usually consist of spines located either marginally or arranged on definite pads. Thus species of *Deleatidium* (a leptophlebiid genus recorded from New Zealand; it was also thought to occur in Australia but the species there are now regarded as members of the genus *Atalophlebioides*) have nymphs with seven pairs of single gills which can be so disposed as to form a ventral adhesive friction pad enabling the nymph to live in the most rapidly flowing parts of streams. Basically similar structures also exist in many rheophilous fish.

Hooks and grappling structures for adhesion to submerged objects are possessed by many river and stream insect larvae, and several sorts have evolved. Most campodeiform (caseless) caddis as well as megalopteran larvae develop strong claws and hooks at the hind end of their abdomen (Fig. 8:1c, d); others, for instance certain dipterous larvae, develop circlets of small hooks on their prolegs. Silk and allied secretions are used by representatives of several insect groups to aid adhesion, for example by some chironomid and simuliid species, and most pupating caddis. Some caddis, we note in passing, may also aid adhesion by selecting particularly large stones for incorporation in their cases, and such stones, the suggestion is, act as ballast.

A final direct morphological adaptation that should be mentioned is the genetically determined tendency to lose or reduce wings, displayed by many adult insects whose immature stages live in streams. This tendency is understandable in that lotic environments are continuous and not, like lentic ones, discrete. Dispersal of the biota is therefore less dependent upon its powers of overland transport. Other reasons are undoubtedly involved, including the obvious disadvantages of flight in the exposed conditions where many upland streams typically occur. A number of species of stonefly have adults which are apterous or have reduced wings and so provide examples of this phenomenon. The apterous stoneflies of New Zealand (Wisely 1953, Illies 1963) or Australia (Illies 1968), however, are not particularly appropriate examples, for most of them have semi-terrestrial or terrestrial immature stages; the only definite examples so far of an apterous species with an aquatic immature stage is the New Zealand species *Apteryoperla* (= *Aucklandobius*) *longicauda* (Illies 1963).[2]

Fig. 8:2 Mouth disc of *Hyla lesueuri* tadpole. Note numerous labial papillae. *After Martin (1967a)*

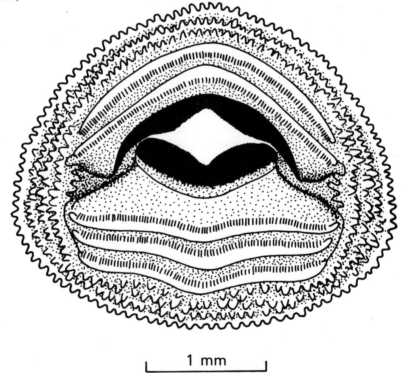

1 mm

[2]. Note added in proof. Dr H. B. N. Hynes (pers. comm.) has recently discovered the nymph of the only known Australian apterous stonefly, *Leptoperla darlingtoni*. It is fully aquatic in small streams near the summit of Mount Donna Buang, Victoria.

BIOLOGICAL FEATURES

A number of indirect morphological adaptations to life in rivers and streams also exist. Those animals which avoid the current by burrowing into the substratum, for example, are frequently worm-like (oligochaetes, various dipterous larvae) or have numerous hairs which keep their respiratory surfaces clear of clogging particles (certain mayfly and stonefly nymphs). Whilst such adaptations are primarily directed more towards a burrowing existence, they may also be thought of as indirectly induced by the above-surface effects of current. At this point we may note, too, that considerable densities of animals found perhaps more 'typically' at the surface may occur even at depths down to about half a metre in stony stream substrata (Schwoerbel 1961, Coleman and Hynes 1970), densities much greater indeed than were previously thought to exist at such depths.

Apart from the general behavioural pattern common to most rheophilous animals of avoiding the current (see above), there are a number of more specialized behavioural responses that are clearly of adaptive value in lotic environments. Many of these are important during only a short time in the life of the animal possessing them. Of those examples that spring to mind we mention the habits of many ephemeropteran and odonatan nymphs, and all plecopteran nymphs, of emerging (releasing the imago) only after the nymph has crawled to a suitable surface *above* water level. Ovipositing aerial insects show several specialized responses. Some actually crawl beneath the water surface to lay eggs, others lay eggs at the water's edge or on overhanging vegetation from where the immature stages may crawl or drop into the water (e.g. most dipterous species and all Megaloptera), and others 'dive-bomb' the water surface. An associated behavioural pattern is the tendency for many aerial insects to fly upstream before ovipositing, and this presumably serves as a compensation movement offsetting any long-term trend for the species to be washed downstream. In fact it has been suggested that perhaps all rheophilous invertebrates have an innate tendency to migrate upstream and that this is for the same reason.

Insectan behavioural patterns that operate over longer periods than the time of adult emergence and oviposition include several complex ones associated with feeding. Many of these are linked with the development of specialized anatomical structures. Thus many campodeiform caddis larvae (e.g. Polycentropidae, Hydropsychidae) have salivary glands secreting silk and this is spun by the larva into a net located in an appropriate position with respect to the current. Nets may assume a variety of shapes, but for all the basic aim is to entrap organisms and other food particles suspended in the current. Carnivorous larvae may directly attack ensnared prey, whereas other larvae may have mouthparts modified to form a brush which is used to sweep the net clear of particles. Nets may be so fine that mesh apertures are less than $4 \times 32\mu$ in size, as noted by Sattler (1963) for some South American species. Feeding in most simuliid larvae likewise is effected by a combination of a given behavioural pattern and specialized anatomical structures. Their mouthparts are modified to form a fan-like structure which functions as a sieve; each larva spreads its mouthparts out and orientates itself so that only its head protrudes above the boundary layer and is exposed to the current—whence, of course, the food particles are then strained.

Physiological adaptations to life in running waters include (there are many more) those associated with the process of respiration, with the need to have an 'emergency' means of escaping the current, and with the special problems posed by intermittent streams and rivers. Ordinarily, few problems of obtaining dissolved oxygen exist in natural streams, where oxygen tension is usually near or above saturation level.

However, when gaseous as distinct from dissolved oxygen is required, special problems arise. This is the case for adult Aphelocheiridae (hemipterans formerly and sometimes still regarded as part of the Naucoridae) and Elminthidae and Hydraenidae (Coleoptera), and in the pupal stages of several dipterous groups (e.g. some empidids and tipulids, all simuliids and blepharocerids). All have been recorded from Australia. The Aphelocheiridae and Elminthidae have solved the problem by the development of a *plastron*, a submerged air-bubble which functions as a 'physical' lung, and the pupae by developing spiracular gills, air-filled cuticular extensions of the body which function in a manner akin to a plastron.

With respect to emergency escape mechanisms, a particularly effective one exists in some veliid bugs and staphylinid beetles which live in close association with the water surface. They possess a gland that can release when needed a chemical which has the property of lowering the surface tension of the water and thus acting as a chemical propulsant. This mechanism can result in remarkable bursts of speed by the insect involved, at least for short intervals.

Finally, a whole series of basically physiological adaptations exists in those animals living in intermittent rivers and streams. They range from the occurrence of a diapause period in the life-history (e.g. an egg diapause in certain simuliids, a nymphal diapause in certain plecopterans) to extended hatching periods and breeding cycles closely correlated with flow regimes. *Paratya australiensis*, an atyid prawn, seems to provide a good example of the latter adaptation. This species occurs in southeastern Australia where it commonly lives in rivers and streams which essentially become interconnected pools with negligible surface flow-through during the summer. The adults can maintain their position in a current, but their planktonic larvae cannot. Breeding is so timed that the females become berried (carry external eggs) shortly before the stream or river dries up into pools. By the time the larvae are released the current is negligible and they are not therefore swept away (except in exceptional summer floods). Then, by the time the current is once again strong, the larvae have reached a size at which they can avoid being swept away.

Many Australian fish also display reproductive adaptations to the fluviatile regime of flood and drought so characteristic of Australian running waters (Chapter 7). Some of these have been discussed by Lake (1967a, 1967b). They include the physiological triggering by flood conditions of spawning, the occurrence of quick-hatching pelagic eggs with large perivitelline spaces to protect the embryo from injury by jarring, or of demersal adhesive eggs in nests, and the laying of eggs in hollow logs. Lake (1970 and pers. comm.), further, has stressed the significance for fish reproduction of decreased oxygen concentrations and increased turbidities of rivers in the Northern Territory and Queensland. Reproductive adaptations to these conditions are many, and several examples have been tabulated by Lake; his table forms the basis of our Table 8:4. It may be noted, in passing, that contrary to previous assumptions the lung of the Queensland lungfish does *not* appear to be a specific adaptation to decreased oxygen concentrations according to Grigg (1965), in the sense of being a regularly used air-breathing organ. *Neoceratodus* appears to be well adapted to lowered oxygen concentrations without its lung, although it does use this in conditions of severe hypoxia, and the lung is a useful accessory in times of activity (G. C. Grigg, pers. comm.). A few fish, for example *Madigania unicolor* of the Lake Eyre drainage basin and elsewhere, can withstand the complete drying of a river by aestivating (Llewellyn 1968).

BIOLOGICAL FEATURES

TABLE 8:4
REPRODUCTIVE ADAPTATIONS OF SOME AUSTRALIAN FISH TO
CONDITIONS OF LOW OXYGEN CONCENTRATION AND HIGH
TURBIDITY
After Lake (1970)

Fish	Mode of reproduction
Scleropages leichhardti Hexanematichthys leptaspis Glossamia aprion	Incubation in buccal cavity
Kurtus gulliveri	Eggs carried by male on hook formed from the supraoccipital[a]
Tandanus tandanus (and probably most species in same family, i.e. Plotosidae)	Nest built in shallow water, and parent fans eggs
Neoceratodus forsteri Nematocentrus maculata Melanotaenia nigrans Hypseleotris compressus	Adhesive eggs are scattered on to plants in shallow water. In some species, adults fan and protect eggs
Plectroplites ambiguus	Eggs planktonic and hatch within 48 hours

[a] Not by a projection of first dorsal fin ray as stated in Lake (1970) (Lake, pers. comm.).

DRIFT

We have already referred to the presence of suspended organisms in running waters in our discussion of the potamoplankton, and noted that significant amounts may be derived as outflow material from standing bodies of water. Associated with this feature, many workers have been able to demonstrate increased densities of the bottom invertebrate fauna commencing immediately downstream of lakes and reservoirs, although by and large the increases do not extend far downstream. Such increases are clearly due to the extra amount of allochthonous food supply available.

In addition to this source of suspended organisms, considerable numbers and amounts are also derived from the *typically* benthic *in situ* fauna (and at times other than during a flood). Many investigators over the last decade and a half have shown that the occurrence of this material is not accidental, and that it occurs in running waters apparently world-wide, is inclusive of most components of the benthic fauna, and is by no means insignificant in terms of biomass. The phenomenon is now referred to as *drift*. Two studies of drift which are of especial interest here are those of Morrissy (1967) on a small stream near Adelaide and of McLay (1968) on the Kakanui River, New Zealand. Morrissy's study in particular should be noted as a thoroughgoing local evaluation of the phenomenon, and should not be overlooked because of its relative inaccessibility (it is an unpublished thesis). The extensive use of drift as a collecting technique by Brundin (1966) in his important study of some southern-hemisphere rheophilous chironomids should also be mentioned.

Although almost all members of the benthic fauna are involved, the extent to which each participates in drift is of course dependent upon the degree of activity of the animal involved and its ease of detachment from the bottom, to mention the most obvious factors. Molluscs, hydracarines, cased caddis larvae, and beetles especially tend to be underestimated in drift collections regarding their proportion in the benthic fauna. There is nevertheless a general correlation between the composition of the drift fauna and the bottom fauna, as is shown, for example, by McLay's (1968) work, part of the results of which form our Table 8:5.

TABLE 8:5
RELATIONSHIP BETWEEN THE COMPOSITION OF DRIFT AND BOTTOM FAUNA
All data as percentages, and from McLay (1968, Table 2)

Taxon	Kakanui River[a]		Island Stream[b]	
	Bottom fauna	Drift	Bottom fauna	Drift
Oligochaetes	0·8	3·2	8·0	11·0
Ephemeroptera: *Deleatidium*	25·1	27·3	20·8	4·6
Coloburiscus humeralis	0·9	1·8	0·9	0·9
Trichoptera: *Pycnocentrodes* sp.	5·9	2·4	8·4	2·8
Oxyethira albiceps	0·1	0·6	0·3	20·3
Olinga feredayi	5·2	2·2	4·1	—
Hydropsyche colonica	5·9	1·0	0·1	1·4
Rhyacophilidae	2·1	6·0	0·6	3·2
Helicopsyche iltona	—	—	1·5	—
Coleoptera: Parnidae adults	1·6	9·4	—	0·4
Parnidae larvae	24·0	4·4	0·9	1·4
Diptera: Chironomidae	22·2	33·7	28·3	38·2
Austrosimulium sp.	0·4	6·4	0·9	9·7
Ostracoda	—	—	—	0·9
Mollusca: *Potamopyrgus antipodum*	4·7	0·2	20·8	2·8
Planorbis corunna	—	—	—	1·8
Others	1·1	1·4	4·0	1·5

[a] Samples collected 3 January 1965.
[b] Samples collected 29 December 1964.

With regard to the amount of drift material, absolute recorded values are somewhat difficult to compare since a variety of units of expression has been used. Many authors for instance, have used as a unit the number or weight of individuals passing a given point in twenty-four hours; others the number or weight of individuals per volume of water filtered; others merely the number or weight of individuals caught by a given net during a specified time interval. At all events, it is abundantly clear that large amounts of drift material may be involved on occasion. Berner (1951) calculated that 64×10^6 individuals, weighing 20,000 g, passed a given point on the Missouri River during one interval of twenty-four hours. Values from smaller localities are naturally much less impressive, but are nevertheless sizeable. To provide a local example, we note that McLay (1968) collected 376 individuals during twenty-four hours (2 February 1965) from only 8,160 l of water filtered in the Kakanui River.

The amount of drift material shows marked fluctuations according to season, time of day, discharge rates, temperature, nature of substratum, position of collection, population densities of the stream fauna, and other factors. Perhaps the most interesting aspect is that of diurnal fluctuation which, in terms of both species composition and total amount, is certainly an established feature noticed by all investigators. The general rule is for the amount to increase at night and display a peak shortly after sunset (with sometimes a smaller secondary peak shortly before dawn). Data selected from the account of Morrissy (1967) will indicate this general pattern in an Australian stream (Table 8:6). Light intensity has been shown to be the main controlling factor.

TABLE 8:6
DIURNAL FLUCTUATIONS IN DRIFT IN SIXTH CREEK,
SOUTH AUSTRALIA, 20–21 FEBRUARY 1964
Data modified from Morrissy (1967, Table 4.3). Boxed data = data obtained during hours of darkness

Time interval[a] (hr)	Flow rate through net (m^3/hr)	No. of individuals in net	Wt of individuals in net (mg)	Drift rate (no. of individuals /m^3/hr)
1100–1200	7·8	43	24·7	5·5
1300–1400	7·5	44	26·0	5·9
1500–1600	7·2	21	8·1	2·9
1700–1800	6·3	11	4·9	1·7
1900–2000	5·9	52	37·5	8·8
2000–2100	5·8	142	123·7	24·5
2100–2200	5·8	113	89·7	19·5
2300–2400	6·0	120	82·8	20·0
0100–0200	6·0	96	68·4	16·0
0300–0400	6·1	76	46·3	12·5
0400–0500	6·2	57	34·3	9·2
0500–0600	6·4	24	17·5	3·7
0700–0800	6·7	20	10·3	3·0
0900–1000	7·3	22	9·1	3·0

[a] Sunset, 1907 hr; sunrise, 0553 hr.

The significance of drift is still largely unresolved. At first, following the work in Sweden of Roos (1957) who showed that more adult insects (with aquatic immature stages) flew upstream than downstream, the significance seemed to be clear; there was a 'colonization cycle' with adults flying upstream to oviposit, and juveniles drifting downstream. However, by no means all insects behave like this (and besides many aquatic invertebrates cannot), so some other explanation must be involved. It has been argued that drift represents excess secondary production and results from competition for space in overcrowded populations (e.g. by Waters 1966), an argument, however, denied by some workers (e.g. Elliot and Minshall 1968). Hynes (1970a) has suggested that perhaps the importance of drift has been overestimated since, despite the large amount of material that may occur in nets, this actually represents only a small loss to the fauna. Upstream movements of stream faunas, we note, though demonstrated (Bishop and Hynes 1969), do not appear to be sufficiently large to counteract downstream losses.

As for the actual distances travelled by drifting organisms, very little work has been performed on this aspect of drift. One investigator (Waters 1965) suggested that about 50 to 60 m was the normal distance travelled in the particular stream he studied. Some investigators have suggested much longer distances—hundreds of metres. A few (Elliot 1967) suggest only short distances are travelled. Whatever the distance, the occurrence of drift does mean that colonization of denuded downstream areas rapidly occurs.

LONGITUDINAL ZONATION

In Chapter 7 we discussed the physical changes in lotic environments as they occurred from the upper reaches down to the river or stream mouth, noting that the upper reaches are in general erosional or degrading and the lower ones depositional or aggrading. Correlated with these physical changes of course are numerous biological ones, and these, predictably in view of man's innate desire to categorize phenomena, have been the subject of various zonational classifications.

Some of the earliest attempts to classify longitudinal biological changes occurring in running waters were based upon the distribution of fish species. In Europe, for example, four main zones were delimited: an upper zone dominated by trout (*Salmo*), followed consecutively by a grayling (*Thymallus*) and barbel (*Barbus*) zone, and finally a bream (*Abramis*) zone. However, such fish zonations have little if any applicability outside the region on whose fish they are based and, moreover, are of rather limited conceptual help to ecologists.

Basically similar zonational patterns have also been discerned in many other groups of rheophilous biota, both plants and animals (Hynes 1970b has provided a long list of pertinent studies), but again, *separately* such patterns must be regarded as primarily of local interest and use only. On the other hand, it has been suggested that if communities *as a whole* are considered, relatively distinct zonations emerge which appear to be applicable on a more or less cosmopolitan basis. Illies, in a series of papers (e.g. 1961, Illies and Botosaneau 1963), has proposed a zonation of running waters based on this sort of premise as well as some physical conditions (profile, substratum type, monthly mean temperature). In summary, the zones proposed are as follows:[3]

1 Eucrenon Spring region.
2 Hypocrenon That part of system flowing directly from the spring.
3 Rhithron. Physical features: that part of system flowing directly from the hypocrenon to a point where the annual range of mean monthly temperature begins to exceed 20°C (i.e. the beginning of the potamon); water flow rate high; flow turbulent; discharge small; oxygen concentrations high; substratum of rock, pebbles, gravel, sand (silt and mud only in sheltered regions). *Fauna:* more or less cold stenothermal, rheophilous, requiring high concentrations of oxygen, frequently with marked adaptations to unidirectional flow. Benthic forms include Ephemerellidae, Leptophlebiidae, Gripopterygidae, Blepharoceridae, Simuliidae, Podonomidae, Psychodidae, Elmidae, Psephenidae, Helodidae, Hydraenidae, Rhyacophilidae, Odontoceridae, Glossosomatinae, Philopotaminae (except *Chimarrha*), Ancylidae, and many others. Nektonic forms include galaxiid fish. A plankton is not or only scarcely present. *Further zonation:* mainly on a faunistic basis

---

3. In the faunal lists only taxa which occur in Australia or New Zealand are included.

BIOLOGICAL FEATURES

the rhithron has been divided further into an upper *epirhithron*, a middle *metarhithron*, and a lower *hyporhithron*.

4 *Potamon. Physical features:* that part of system below rhithron to point of entrance to sea; annual range of mean monthly temperature exceeds 20°C or, in the tropics, with summer mean monthly temperatures greater than 20°C; water flow rate near bottom is slow; flow more or less laminar; deeper pools may develop oxygen deficit partly due to complete light extinction and mud formation; discharge rates show considerable annual fluctuation; substratum of sand and mud but can include gravel. *Fauna:* eurythermal or warm stenothermal, rheotolerant, mostly derived from families whose main development is in still waters. Benthic forms include Siphlonuridae, Caenidae, Chironomidae, Culicidae, Tabanidae, Stratiomyidae, Corixidae, Notonectidae, Haliplidae, Dytiscidae, Leptoceridae, Hydroptilidae, and many others. Nektonic forms include many endemic fish families. A plankton is richly developed. *Further zonation:* mainly on a faunistic basis the potamon has been divided further into an upper *epipotamon*, a middle *metapotamon*, and a lower *hypopotamon*. The hypopotamon is the brackish water region that is affected by the sea.

The position of the division between zones will vary according to several factors and Fig. 8:3 illustrates how the division between the major zones of rhithron and potamon may move according to latitude and altitude.

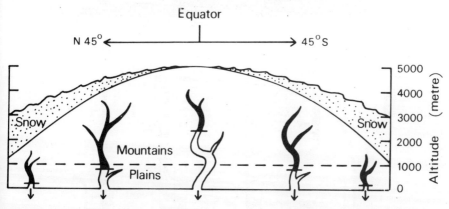

Fig. 8:3 Schematic diagram to illustrate how position of division between rhithron (black) and potamon (clear) portions of rivers may vary according to altitude and latitude. *Modified from Illies (1961)*

There have been no investigations as yet to determine the extent to which this zonational concept applies to Australian and New Zealand running waters. It has been applied (outside Europe) with some success to South American (Illies 1964) as well as southern African waters (Harrison 1965), and there seems no good reason why it should not also apply to lotic environments in the remaining major land masses of the southern hemisphere with free water, namely Australia and New Zealand. An examination of this subject by antipodean limnologists is awaited with interest! It is as well for us to reiterate, however, Hynes's (1970b:397) general conclusion on the concept of zonation: 'the idea of zonation is useful in general descriptive way but... attempts at precise defination of zones are of doubtful ecological value.'

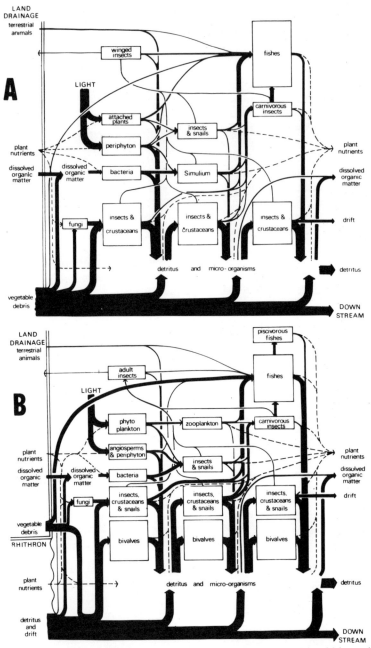

Fig. 8:4 Diagrammatic representation of trophic relationships within rhithron (A) and potamon (B) ecosystems. Relative sizes of boxes indicate biomass and arrow width is proportional to supposed relative importance of energy pathway. Dotted lines represent pathways of salts in solution. *Reproduced with permission from H. B. N. Hynes (1970b)* The Ecology of Running Waters, *published by University of Liverpool Press, Liverpool*

RUNNING WATERS AS ECOSYSTEMS

It is only in the past few years that running waters have received serious attention as examples of *whole* ecosystems, and it is certainly true that our knowledge of lotic ecosystems *in toto* contrasts sadly with the extent of our knowledge of lentic ones (although even for these a great deal remains to be discovered). At this point in time, then, no firm, overall picture can be given of trophic relationships, energy pathways, and related features in running waters. This situation promises to change in the not too distant future, since this area of limnology is currently being explored by several active groups of investigators.

One of the first points to emerge from recent studies has been that, unlike 'typical' terrestrial or lake ecosystems, streams in particular but all running waters in general appear to derive a substantial if not a major part of their energy for maintenance from allochthonous sources; that is, a good deal or most of the energy necessary to operate lotic ecosystems is derived from outside the system. For the most part lotic ecosystems are predominantly heterotrophic (Nelson and Scott 1962, Hynes 1963, 1969, 1970b, Minshall 1967). A further difference from 'typical' ecosystems that it is pertinent to mention here is that *in situ* biogeochemical cycles, so much an integral part of most ecosystems, can scarcely be so in streams and rivers, where of course there is a constant downstream displacement of material.

It has been suggested (e.g. Hynes 1963, Minshall 1967, 1968, Kaushik and Hynes 1968) that most of the allochthonous energy input for streams (and to a less extent for rivers) enters as leaves from terrestrial trees and shrubs. In the northern hemisphere there will be considerable seasonal fluctuation in the amount of such input but in the tropics, it seems, and, we can predict, also in Australia and New Zealand (where deciduous trees are not widespread), seasonal fluctuations in input will be less important. It is now known that a large fraction of the dry weight of fresh leaves is leached away relatively quickly after being placed in water; this material, together with dissolved organic material from other sources—*in situ* algal excretions and from terrestrial sources—forms one category of available energy. The manner of its utilization has still to be unravelled in detail, but it appears that bacteria (and perhaps water turbulence too) play a critical role in precipitating it and making it available as food for other heterotrophs, including, for example, simuliid larvae (Fredeen 1963).

Microbial decomposition of the leaves follows the leaching phase and during this it seems that fungi, particularly the aquatic hyphomycetes, are especially important since they have cellulase enzymes and can thus begin the degradation of leaf cellulose. Once fungal populations have developed, larger heterotrophs ('detritovores') begin comminution of the leaves.

To a large extent the nature of the trophic relationships and the relative importance of the numerous energy pathways involved will vary longitudinally. In the rhithron, for instance, the main energy source is likely to be allochthonous organic material which is many times recycled by the detritovores. In the potamon, on the other hand, organic material derived from the rhithron constitutes a major energy source. Hynes (1970b) has illustrated the principal differences of this sort between rhithron and potamon and his schematic representations are reproduced in Fig. 8:4.

part four
OTHER BODIES OF FRESH WATER

Waikerei Geothermal Project . *By courtesy of the New Zealand Government Tourist Bureau*

chapter nine
OTHER BODIES OF FRESH WATER

There is a wide variety of freshwater bodies which it is inappropriate to discuss as running waters (Chapters 7 and 8) and which are too small to be regarded as lakes (Chapters 3–6). We are referring of course to such water bodies as ponds, roadside puddles, springs, farm dams, water-filled depressions in trees, cave waters, and so on. Little useful purpose is served by any attempt to classify them together and we shall deal with these aquatic situations on the basis of a gross division into major environmental types.

UNDERGROUND WATERS

A considerable volume of water occurs beneath the land surface. Part is held in the soil, the amount depending *inter alia* particularly upon the size of the pores between the soil particles. Part may sink beneath the soil and be held in and move through permeable rock strata, where it may accumulate in caves to form underground streams and standing bodies of water of various sizes, particularly in limestone areas. The surface of the water in permeable rock strata is termed the water-table; water above this is vadose water, water below, ground water. Areas of ground water under pressure are known as artesian basins; the principal ones in Australia are mapped in Fig. 11:2. The largest, the Great Australian Artesian Basin, is one of the world's biggest.

It is possible to distinguish various sorts of aquatic environments in subsurface water. Two principal ones generally of interest to limnologists are interstitial waters, that is waters occurring in the interstices between substratum particles (the hyporheic biotope), and larger continuous water bodies occurring in caves and wells. The fauna of interstitial waters (the interstitial or phreatobious fauna) has only relatively recently been subjected to thorough investigation, but has already proven to be one rich in interesting forms. It has been scarcely examined in Australia and New Zealand, but preliminary results indicate that it is as interesting and diverse here as elsewhere. Thus Schminke and Noodt (1968) have recently reported syncarid crustaceans belonging to both the Stygiocaridacea (formerly known only from South America) and the Bathynellacea from New Zealand; and a variety of new arthropods collected by Schminke from Australian interstitial waters is in the process of being described (H. K. Schminke, pers. comm., Weigmann and Schminke 1970). Nicholls (1946) early drew attention to the important position of syncarids in the interstitial fauna, and the nature of this fauna as a whole is comprehensively discussed and reviewed by Delamere-Deboutteville (1960) and Husmann (1966). Husmann (1966, 1970) has presented a detailed classification of the various biotopes that can be distinguished in interstitial ground waters.

Water bodies in caves and wells have been more intensively investigated and for longer than interstitial waters. They are especially well studied in Europe and reference may be made to the treatises of Chappuis (1927) and Vandel (1965). The fauna is broadly referred to as cavernicolous, and divided into those forms that are obligate cavernicoles, troglobites, and those that are merely facultative ones, troglophiles. Most aquatic cavernicoles are arthropods, predominantly malacostracan crustaceans and beetles, but other groups have been recorded including the Protozoa, Turbellaria, Oligochaeta, Gastropoda, fish, and Amphibia. Common adaptations to the continuous dark and the usually constant temperatures include apterism (in the insects), attenuation of appendages, reduction of eyes, loss of pigmentation, and aperiodicity of life-cycles.

In Australia cave and well forms which are truly troglobitic recorded thus far are the atyid prawns *Stygiocaris lancifera*, *S. stylifera*, *Parisia gracilis*, and *P. unguis*, and the fish *Milyeringa veritas* (family Eleotridae) and *Anommatophasma candidum* (Synbranchidae). These records relate to underground waters in two areas, the North West Cape in Western Australia and near Katherine in the Northern Territory (Whitley 1945, Holthuis 1960, Mees 1962, Williams 1964a). The atyids are interesting in that both the endemic genus *Stygiocaris* and the genus *Parisia* appear to be most closely related to Madagascan atyids, and both genera are represented by two species living sympatrically. Cave troglophiles recorded thus far include *Anaspides tasmaniae* (Williams 1965b), *Salmo trutta* and a species of gammarid amphipod in various Tasmanian caves (Goede 1967), and *Paratya australiensis australiensis* in a Victorian cave (Dew 1963). Since this list of Australian aquatic cavernicoles relates to extremely few localities, significant extensions to it are to be expected. If encouragement to extend it be needed, may we recall that a completely new *order* of the Crustacea, the Spelaeogriphacea, was found only recently (by amateur zoologists) in a cave in South Africa (Gordon 1957).

The fauna of caves and wells in New Zealand is likewise rather ill known. The most comprehensive report concerning it is that of Chilton (1894) who investigated the fauna of wells sunk in the gravels of the Canterbury Plain. He recorded oligochaetes, turbellarians, *Potamopyrgus spelaeus* (a gastropod), *Cruregens fontanus* (an anthurid isopod), and several species of phreatoicids and amphipods. It has been suggested that the fauna of these wells is a mixture of surface and interstitial forms (Vandel 1965), but at least one of the crustaceans recorded by Chilton has now been found in a cave (Schminke and Noodt 1968). No troglobitic atyids have been recorded from New Zealand, but the epigean species *Paratya (Paratya) curvirostris* occurs facultatively in caves, as also do various fish (trout, eels, and *Gobiomorphus*) (Yaldwyn 1959).

In addition to the two main sorts of subsurface aquatic fauna discussed above, we may also briefly mention that a variety of aquatic crustaceans—frequently blind and white—has been recorded in Australia from basal pools in terrestrial crayfish burrows, in boggy ground, and associated with non-thermal springs. Such forms in general appear to be too large to be truly interstitial, yet cannot be regarded as cavernicolous. They include several species of phreatoicid isopods, amphipods, and some syncarids. The most recent record (Swain, Wilson, Hickman, and Ong 1970) is of a new syncarid genus, *Allanaspides*, obtained from crayfish burrows and small bodies of water in Tasmania.

SPRINGS

Natural springs arise where underground water re-surfaces. Three principal sorts are usually recognized: those where the water flows away down a gradient (rheocrenes), those where it, initially at least, is contained in a basin (limnocrenes), and those where it egresses into a marsh (helocrenes). These are paralleled by artificial bores which either discharge subsurface water under its own pressure or use an external power source (frequently the wind) to pump it to the surface.

The chemical composition of spring water depends largely upon that of its underground source. Thus, in springs discharging water from the Great Australian Artesian Basin, the dominant ions are calcium and bicarbonate, whereas elsewhere in Australia they are frequently sodium and chloride. There are some mildly saline springs in Australia but most are fresh ($<3\%_0$ salinity). Their composition is to a large degree constant with time, but some differences may occur in the water before and after discharge resulting from the release of free carbon dioxide and the resultant effect on pH/carbonate/bicarbonate equilibria. One consequence of this is frequently the deposition of significant amounts of calcium carbonate as tufa or travertine and its gradual build-up as a rim around the spring. This is clearly seen in many springs associated with the Great Australian Artesian Basin, and such springs ultimately form *mound springs* which may be extremely large and high (Plate 9:1). The Report of the Third Interstate Conference on Artesian Water (1921) and Gregory (1906) give good descriptions of mound springs. The chemical composition and distribution of Australian spring and bore waters has been briefly reviewed by Williams (1967a).

The temperature of springs is generally relatively constant—more so in fact than is usual in surface waters—except when the water arises from shallow strata and has not been underground for long. Mostly, temperatures are either slightly lower than those of adjacent surface waters or are considerably higher. Springs with elevated temperatures are termed *thermal springs* and in Australia most are confined to the area underlain by the Great Australian Artesian Basin (Waring 1965); their distribution, together with that of thermal bores, is indicated in Fig. 9:1. The highest temperature recorded is about 85°C (Innot Creek, Queensland), and for a bore, 100°C (Waring 1965). A temperature of about 85°C represents the thermal upper limit for the continuation of active life (bacteria).

Thermal springs in New Zealand are far more spectacular than those in Australia and are well-known tourist attractions. Those near Rotorua, North Island, include geysers, blowholes, and boiling mud pools. Many non-thermal springs also occur; the Waikoropupu Springs, at the northern tip of the South Island, are some of the world's largest.

Extremely little is known of the biology of Australian springs. Reference has already been made to the various malacostracans which have been collected from non-thermal springs, and from thermal waters we now mention *Caridina thermophila*,[1] *Phreatomerus*

---

1. Riek's (1959 : 249) suggestion that this atyid's adaptation to hot water is a recent development cannot be upheld.

Plate 9:1 Mound springs in central Australia. The foreground shows a spring near the point of egress of water. Another mound spring with well-developed rim is shown in the background. *Photograph by K. F. Walker*

latipes, and *Gambusia affinis* (Riek 1953, Chilton 1922, and unpublished). Apart from such faunal records little published information is otherwise available.

Much more is known concerning the biology of New Zealand springs. Reflecting a similar situation to that elsewhere, in thermal springs blue-green algae are the dominant plants present and the fauna mainly comprises a restricted number of species of ephydrids, chironomids, beetles, hemipterans, nematodes, rotifers, and ciliate protozoans (Stoner 1923, Winterbourn and Brown 1967, Winterbourn 1968, 1969, Dumbleton 1969). In the non-thermal springs at Waikoropupu (temperature $11 \cdot 5°C$), the commonest animals are amphipods, trichopteran larvae, *Potamopyrgus antipodum*, and *Paranephrops planifrons* (Wells and Taylor 1970).

WATERS ASSOCIATED WITH TERRESTRIAL VEGETATION

Small isolated bodies of water in tree-holes and in the axils of certain terrestrial plants have been studied by several investigators, not least because some of the insects inhabiting them may be of medical or agricultural interest. Maguire (1963) has given a relatively complete list of these studies.

Water bodies of this sort are of fairly common occurrence in the better-watered parts of Australia, and also occur in semi-arid parts. In the latter they may represent the only free water over large areas and consequently their biota is of considerable

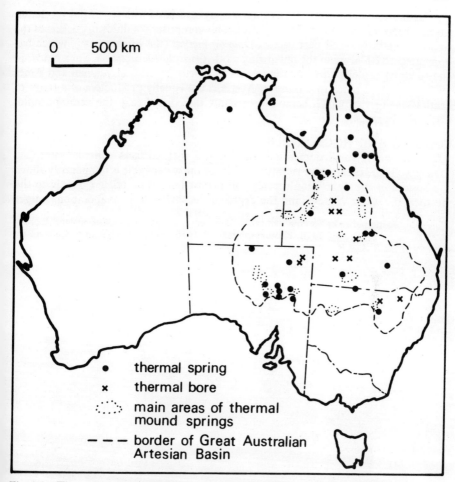

Fig. 9:1 Thermal springs and bores in Australia. *Redrawn after Waring (1965)*

interest. The biota of some in semi-arid New South Wales has been reported on briefly by Dyce (1964), who recorded a wide variety of forms: *Hyla* (Amphibia), larval culicids, ceratopogonids, psychodids, syrphids, chironomids, tabanids, and stratiomyids, as well as beetles, hydracarines, oligochaetes, ciliates, and certain crustaceans. In his discussion Dyce distinguished two main sorts of locality, those within hollow 'pipes' in the tree and containing rain water from a few days to more than a year, and those occurring in pockets at the base of the tree and originating from rain water which had seeped through the fibrous chewings of longicorn grubs. The chemical composition of contained water will of course be different from that of the initial rain water, not only because of materials leached out from the tree into the standing water, but also because the foliage, branches, and stems of trees are known to alter the composition of rain water as it passes through and along them (e.g. Voigt 1960).

INLAND WATERS AND THEIR ECOLOGY

A further and most bizarre type of water body associated with terrestrial vegetation has also been reported on by Dyce (1970). This comprises rot fluids in pockets of the stems of the introduced pest cactus, *Opuntia inermis* (the prickly pear), which has been attacked by larvae of the burrowing moth, *Cactoblastus cactorum*. Such fluids are known to be exploited by aquatic insects in North America (Ryckman and Ames 1953), and Dyce notes that those in Australia are equally productive of a range of beetles and flies of which some are certainly introduced (e.g. the ceratopogonid, *Culicoides loughnani*).

PUDDLES AND ROCK-POOLS

After rain many small depressions on soil and rock surfaces contain water (see Plate 9:2). If the substratum is impermeable (or the water-table is temporarily above the bottom of the depression), water will persist for a time related mainly to the weather, the water volume, and the degree of its shelter from evaporation. Larger

Plate 9:2 Portion of Lake Yamma Yamma and adjacent temporary pools, Queensland, looking south from 26° 27' S, 141° 30' E. Photograph taken 8 April 1949 from a height of 4,600 m. *By courtesy of the R.A.A.F.*

OTHER BODIES OF FRESH WATER

bodies of water may be more or less permanent; smaller ones will soon disappear. The fauna and various physico-chemical features of such waters have been investigated for the most part on a restricted geographical basis, or with emphasis on the biology of a particular faunal group. Rain-pools on soft substrata have been most investigated, and we draw attention to the work (on waters outside Australia and New Zealand) of Gauthier (1938), Rzóska (1961), Hall (1961), Moore (1963), Yaron (1964), Hartland-Rowe (1966), and Eriksen (1966). In summary, a variety of insects and other aquatic animals which have some stage in their life-cycle resistant to desiccation has been recorded, and especially typical are species of Anostraca, Notostraca, and Conchostraca. Considerable fluctuations in dissolved oxygen concentration, pH, and other factors may occur both diurnally and over a period of a few days to weeks. Ecological succession of both biological and environmental characteristics is frequently a feature.

Rain-pools on soft substrata have been little studied in Australia as total habitats,

Plate 9:3 *Triops australiensis australiensis* in the last remnants of a rain-pool in central Queensland. *Photograph by M. Happold*

but a large number of faunal records relate to them, particularly those in inland areas. Especially characteristic in inland arid regions is the shield shrimp *Triops australiensis australiensis* (Plate 9:3), and in less arid regions, *Lepidurus apus viridis* (Williams 1968b). Anostracan and conchostracan genera frequently present are *Branchinella* and *Lynceus, Limnodopsis, Cyzicus, Eulimnadia,* and *Limnadia*. Many smaller (and therefore less spectacular) forms also occur, and in a small, temporary rain-pool recently examined on the outskirts of Melbourne the following groups were present: protozoans, rotifers, cyclopoid and calanoid copepods, ostracods, coleopterans, and culicid and chironomid larvae. Cladocerans were not recorded in this particular pool but are known to occur in similar ones; the endemic genus *Saycia*, for example, appears to be confined to pools that are of a temporary nature.

Several frog species are associated with temporary rain-pools which they use as breeding loci (Main 1968). Many in arid regions rely entirely on such waters (in fact all Australian desert frogs lay their eggs in water and have aquatic larvae). Commonly, these species have terrestrial embryonic development, delayed hatching, and relatively short larval lives. *Adelotus brevis* is a species that occurs in the eastern coastal zone and breeds in roadside ditches. *Cyclorana* and *Neobatrachus* are two genera living in desert conditions in Western Australia which breed in claypans after rain; the former has a larval life of only 30–50 days and most species of the latter have larval lives of *ca.* 40 days (Main 1968).

Rain-filled rock-pools appear to develop similar faunas to rain-pools on soft substrata. Edward (1964, 1968) and Fairbridge (1945), discussing the fauna of pools on granite in Western Australia, noted the occurrence of larvae of *Allotrissocladius amphibius, Paraborniella tonnoiri* (both chironomids), ceratopogonids, and the copepod *Boeckella opaqua*, as well as Cladocera, Conchostraca, Anostraca, Ostracoda, rotifers, nematodes, rhabdocoels, and tadpoles of *Crinia pseudinsignifera*. The chironomid larvae are of particular interest in that they can survive the complete drying out of a rock-pool by building tubes in the bottom mud and retreating into these to form capsules which can withstand drying and temperatures up to 56°C. Occasionally, after a sudden fall of rain, the larvae may leave the capsules and feed actively until the pool dries again. Some individuals of *A. amphibius* appear to be able to survive total dehydration, but it is not known if they are truly cryptobiotic in the sense of Hinton (1960) who, referring to the larvae of *Polypedilum vanderplanki* in temporary African rock-pools, found that the larvae could have their metabolism brought reversibly to a standstill and could survive almost total dehydration. In the dehydrated condition the larvae could be immersed in liquid helium ($-270°C$) or heated to over 100°C and still recover. Neither adults nor larvae of *Limnadia stanleyana*, a conchostracan occurring in temporary rock-pools in eastern Australia, can withstand drying, but eggs are quite resistant to desiccation. The ecology of this species has been intensively studied by Bishop (1967a, 1967b, 1968) and he has been able to show how photoperiodically-induced diapause and associated low-temperature quiescence of eggs synchronizes the life-cycle with seasonal changes to ensure the continuity of the species. Associated with the conchostracan studied by Bishop were larvae of *Paraborniella tonnoiri, Allotrissocladius, Dasyhelea* (Ceratopogonidae), rotifers, and *Dorylaimis* (Nematoda) (Bishop and Dyce 1968).

PONDS (INCLUDING FARM DAMS)

No clear distinction can be drawn between rain-filled depressions on soft substrata and ponds, and this is particularly the case when ponds which do not contain water all the time are considered. Equally, at the other end of the scale, limnologists from Forel (1892) onwards have had difficulty in distinguishing a pond from a lake. We adopt (see also Chapter 3) the arbitrary position of most modern authors: a typical pond is a body of water shallow enough for rooted vegetation to be established[2] over most of the bottom. From water bodies thus defined we must then exclude those already discussed in this chapter. Recalling Charles Darwin's view of what constitutes a species, we suggest that a pond is a body of standing water so termed by a good limnologist! Fortunately, from our viewpoint, lochans, meres, sloughs, tarns, and other names given to pond-like bodies of water in Britain and North America are names not generally used in Australia and New Zealand.

Two main sorts of ponds are distinguishable: those where water is continually present (*permanent* ponds), and those where it is present only seasonally (*temporary* ponds). The latter are sometimes further classified according to the season they contain water. Farm dams and sewage ponds may be regarded as special sorts of man-made permanent ponds. Throughout most of Australia and many parts of New Zealand the former are nowadays one of the most ubiquitous forms of standing-water body. Most are shallower than 10 m, the approximate lowest depth to which rooted vegetation can generally extend. Conditions in sewage ponds are so unnatural that these localities are not further considered in this chapter.

The principal limnological feature of ponds is their physico-chemical and biological variability, and it is this aspect which is given prominence in the following discussion. The restriction of most of our discussion to a consideration of Australian and New Zealand ponds, we note, is particularly relevant, for ponds are much more a product of local geographical conditions than many other sorts of inland waters. Comments on Australian ponds are based mainly on the accounts of Weatherley (1958a, 1958b), Brand (1967), Geddes (1968), Timms (1967, 1970a, 1970b, 1970c), and Morrissy (1970), and on New Zealand ponds of Byars (1960), Stout (1964, 1969b), and Barclay (1966). Some unpublished data on a small pond near Melbourne are also briefly referred to.

The pattern of temperature fluctuations depends largely upon geographical position, pond morphometry, and the degree of shading and protection from the wind. The seasonal range of course is largely a function of geography. Generally in shallow ponds water temperatures follow air temperatures but are less variable. The data of most authors, however, suggest that surface water temperatures are lower than air temperatures. This indeed may be the situation during most of the day (when temperatures are usually measured by investigators), but we must emphasize that the opposite may apply during the night and perhaps over a period of several days at certain seasons. This is shown by Fig. 9:2 in which are reproduced two weekly thermograph records of surface water and adjacent air temperatures at a small, shaded, and shallow pond near Melbourne. The direct correlation between the absolute range of diurnal variation in air and water temperatures can also be seen from this figure.

Frequently, in the past, thermal stratification in ponds has been regarded as unimportant or non-existent. Indeed, Muttkowski (1918) defined a pond as a body of water which was not large enough to stratify thermally. It is clear nevertheless that some differences in temperature may exist with respect to depth even in shallow

---

2. This is not to imply that bottom vegetation necessarily *is* established.

ponds, and we support Eriksen's (1966) contention that thermal stratification in ponds may be more common than is recognized. The phenomenon is not, of course, nearly as permanent or pronounced as it is in deep, thermally stratified lakes, but it may have biological reverberations nonetheless, particularly in the hotter parts of Australia. Both diurnal and seasonal events are involved. The general picture is that in shallow ponds temperature differences between upper and lower layers of water are minimal at night and during the winter, and maximal in the day and during the summer. The greatest difference recorded so far in Australia is slightly in excess of 10°C, an afternoon value recorded by Morrissy (1970) in a farm dam in southwestern Western Australia (surface = 26·05°C, bottom [ca. 2 m] = 15·65°C). In New Zealand Barclay (1966) has recorded (at 2 p.m.) a difference of 8·5°C in a pond only 35 cm deep. We stress, however, the transitory nature of any large temperature differences that develop in shallow localities; these are mainly the result of marked diurnal changes in air temperature and the absence of significant wind action. In deeper localities, especially when these are protected from the wind, vertical temperature differences may be much less variable and transitory, and such localities may behave thermally as lakes (though stratification in them is apparently much less precise) (Timms 1970c).

Fig. 9:2 Thermograph records of surface water temperature (broken line) and adjacent air temperature (unbroken line) at a small, shallow, shaded pond near Melbourne. Left-hand chart is for the period 6–12 January 1962, right-hand one for the period 27 July–2 August 1962. The original charts are circular and the records are to be read clockwise. Records for a single day run between the central segments numbered 1 to 7, and each segment covers the time period 6 a.m.–6 a.m. The temperature scale is the slightly curved line at the top of each figure; values are °C

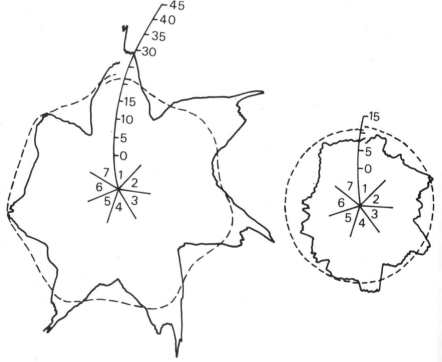

OTHER BODIES OF FRESH WATER

All grades of turbidity exist in Australian and New Zealand ponds, although it is true to say that many, perhaps most, farm dams in Australia are highly turbid. The biological repercussions of this high turbidity are obviously important, for high turbidity results in decreased light penetration with significant consequences for the development of plant and, in turn, animal communities. It may well be that most Australian farm dams are much less productive than they are potentially capable of being. Certainly, few of them have substantial amounts of submerged (or emergent) macrophytic vegetation, although this is likely to be a result of both high turbidity and astatic water levels. In ponds and dams used by sheep and cattle as watering places, part of the turbidity will be caused by the trampling of the stock. But since most Australian dams, at least, are formed by excavations in clay or deliberately lined with clay some turbidity must be caused by 'natural' semi-permanent suspension of clay particles, particularly in those dams quite exposed to wind action. Wind-deposited dust is no doubt a further source of suspended material. The high turbidity of Australian farm dams, it has been suggested, is important in producing thermal stratification (Morrissy 1970); essentially all the light energy is converted to heat at or near the surface.

Orderly seasonal patterns of fluctuation in the concentration of dissolved oxygen in shallow ponds are either lacking or, at best, are ill defined. Nevertheless the highest values, frequently supersaturation ones, usually occur in summer, the lowest in winter, and in temporary ponds initial concentrations are low; concentrations gradually increase during the time ample water is present, and then fall steadily as the pond dries out. Diurnal and vertical fluctuations are somewhat better defined. Concentrations are highest in the afternoon and lowest after dawn, and more oxygen is usually present in surface than bottom waters during most of the day and for most of the year. Part of the explanation for the vertical differences is undoubtedly the greater amounts of oxygen gained by the surface waters from the atmosphere by diffusion and, especially in turbid dams, the greater amount of photosynthesis at the surface. It has also been suggested for dams, however, that there is a rapid consumption of oxygen by bottom deposits (Weatherley 1958a). Deeper, sheltered ponds (e.g. large farm dams) where thermal stratification persists, may develop severe deoxygenation in the lower layers of water during the period of stratification.

Seasonal patterns of fluctuation in pH, like those of dissolved oxygen, are absent or ill defined. Most authors have found that by and large the highest values occur in summer, but some record no consistent seasonal variation and others a trend towards increased acidity in summer. Local and non-seasonal variation may also be important, as, for example, that caused by flood waters (B. V. Timms, pers. comm.). Timms (1967), working on a small pond in Brisbane, found that values were higher on the whole in the littoral region and in surface waters than in the limnetic region and bottom waters, were higher on warmer days than colder ones, and were highest ($>9 \cdot 0$) where weedbeds were active. Absolute values for a majority of inland localities studied thus far are in excess of $7 \cdot 0$ and may occasionally be as high as $9 \cdot 5$ (cf. Stout 1969b). Coastal (but athalassic) ponds and lagoons in sandy regions of eastern Victoria and of King Island, on the other hand, commonly have values between $5 \cdot 0$ and $7 \cdot 0$ (Brand 1967).

The nature of those fluctuations which occur in the concentrations of total dissolved solids and major ions has been best studied by Timms (1967, 1970b, 1970c), who has investigated, in addition to the Brisbane pond already mentioned, several small dams and a natural shallow lagoon in northeastern New South Wales. Apart from one not considered here, all the dams are <7 m deep and though not farm dams are said to be similar to them. In summary, for these localities he has shown that:

1 Considerable fluctuations in salinity may occur.
2 The extent of salinity fluctuations is related to water renewal rates (i.e. catchment area: surface area of pond).
3 Inorganic ions may concentrate at different rates as water levels fall (data for Ellalong Lagoon only).
4 Although no distinct seasonal pattern is obvious, there is in general an inverse correlation between rainfall and the concentration of total dissolved solids.

Some of Timms's data on Ellalong Lagoon are reproduced in Fig. 9:3 to illustrate some of these points.

One parameter not closely studied by Timms but obviously related to his findings is the extent of fluctuation in water volume and area. Such fluctuations are probably substantial over most of Australia in ponds not replenished from underground sources, and we refer to Weatherley's (1958a) data on eight Tasmanian farm dams which show that, even in Tasmania, the most equable and best-watered Australian State, great fluctuations in area and volume occur (Table 9:1). It should be noted, however, that significant numbers of farm dams and other ponds—particularly in inland Australia—are largely filled from underground sources (cf. Williams and Siebert 1963) and also occur in regions with a much better-defined seasonal periodicity of rainfall. The chemical nature of these is likely to be quite different from that of the localities studied by Timms.

TABLE 9:1
EXTENT OF SEASONAL FLUCTUATIONS IN AREA AND VOLUME IN EIGHT SMALL FARM DAMS IN TASMANIA
After Weatherly (1958a)

Dam	Area (ha)		Minimum as % of maximum	Volume (m^3)		Minimum as % of maximum
	Maximum	Minimum		Maximum	Minimum	
1	0·11	0·04	39	—	—	—
2	0·06	0·03	48	—	—	—
3	0·28	0·18	64	2,614	1,537	59
4	0·23	0·11	48	2,175	939	43
5	0·06	0·03	52	467	132	28
6	0·07	0·04	53	550	215	39
7	0·02	0·01	19	74	4	5
8	0·02	0·01	15	112	5	4

The study of pond communities seems never to have attracted the serious attentions of professional limnologists (other than those interested mainly in fish culture) to quite the same extent that the study of lake and running-water communities has. On the other hand, most amateur aquatic biologists, it is probably true to say, have found pond study more attractive, and certainly most books on the aquatic biota and written for more or less popular consumption emphasize the biology of ponds more so than that of lakes. One of the reasons for this general state of affairs is that the ecology

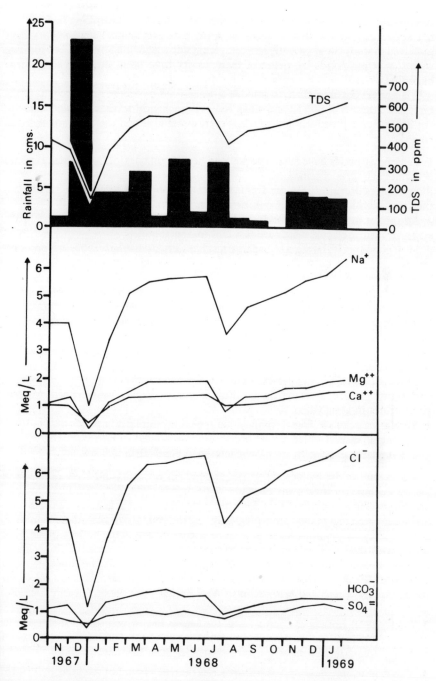

Fig. 9:3 Fluctuations in rainfall and concentration of total dissolved solids (T.D.S.) and major ions in Ellalong Lagoon, New South Wales. *From Timms (1970b)*

of ponds is complex,[3] with the occurrence of great species diversity, complicated community relationships, and highly developed matrices of controlling factors. It is not surprising, in the first instance at least, that professional limnologists have preferred to study the apparently less complex systems generated by lakes and running waters, whereas ponds by virtue of their variety have been attractive to amateur biologists.

It is not yet possible, then, to provide a generalized comprehensive account of the biological features of ponds, and many recent limnological textbooks do not attempt any at all. Some broad features are, however, discernible and these are summarized below.

1 Temporary ponds have fewer species than permanent ones, although many species occur in both sorts of ponds.
2 Branchiopod crustaceans are frequently characteristic of temporary ponds, but they may occasionally occur also in permanent ponds.
3 All species in temporary ponds, and many but not all in permanent ponds, have stages in their life-cycle which are resistant to desiccation.
4 Most pond species have good dispersal mechanisms, particularly those in temporary ponds.
5 A large part of the biota of permanent ponds occurs also in the littoral regions of lakes, especially when these regions are not subject to excessive wind or wave action.
6 Ponds with extensive amounts of macrophytes may have no truly limnetic plankton; where this is not the case a limnetic plankton develops, composed of species which also occur in lakes and some which are generally restricted to the limnetic region of ponds. Pond plankton is referred to as *heleoplankton* (tychoplankton is an incorrect term used in this context).
7 Much of the fauna of temporary ponds displays a characteristic successional pattern, and many planktonic forms in permanent ponds display well-defined seasonal periodicities in abundance.
8 Secular variations in faunal composition appear to be greater in ponds than in lakes.
9 In general, insects which are air-breathing or associated with the surface film, and plants similarly associated, are more common in ponds than in lakes.

No account of the biological features of ponds in Australia and New Zealand can yet be written that makes any pretence of being complete. Pertinent investigations are still small in number, and they deal with ponds that occur in geographically diverse and widespread regions and are ecologically very different from each other. Moreover, many studies understandably emphasize restricted aspects of the biology of the pond under investigation. The best that can be done at present is to summarize the available data on a largely separate basis—and express the wish that the situation were otherwise! *We draw attention to the fact that farm dams in particular are ubiquitous throughout Australia*[4] *and many parts of New Zealand, are generally easily accessible*

---

3. Macan (1963) remarks of Charles Elton, the famous English ecologist, that although he was once attracted to the study of ponds because he thought their communities to be controlled by a limited set of factors, he was repelled from such a study when he found as many distinct communities as ponds he had visited!
4. This chapter, appropriately, was written during the time a farm dam was being constructed less than 100 m from the writer's desk. Some of the literary disjointedness may be attributable to this.

for study, and would form admirable objects for research projects at a variety of intellectual levels. The taxonomic hurdles have been somewhat lowered recently by the publication of Williams's (1968a) book dealing with the Australian aquatic invertebrate fauna, and that of Marples (1962) on the New Zealand aquatic biota.

Temporary ponds in Australia, somewhat surprisingly for so arid a continent, are almost the least studied of Australian standing waters, although there is some work in which emphasis is given to particular taxonomic groups occurring in them. Generally, conchostracans, anostracans, and notostracans appear to be characteristic members of the fauna, as they are in temporary pools (see p. 177), but a variety of other animals has been recorded, many of which, unlike these branchiopods, are also frequently to be found in permanent ponds. Of some interest in this respect is an analysis of Brand's (1967) data on the limnetic plankton of numerous coastal (but athalassic) ponds in eastern Victoria. Several of these are noted by Brand as being temporary under normal conditions (i.e. in non-drought years). All such temporary ponds contained many forms which also occurred in similar but permanent ponds nearby. Thus, for example, he recorded *Boeckella triarticulata* from temporary as well as permanent waters. This is in line with Bayly's (1964b) statement that this species is the commonest and most widely distributed of the Australasian species of this genus, and also Timms's (1970a) evidence that it had the best powers of dispersal of the several calanoids that he studied. In addition, Brand found that *Hemiboeckella searlei* occurred mainly in temporary ponds and that several species occurred in the permanent ponds but not at all in the temporary ones, for example *Calamoecia tasmanica* and *Boeckella symmetrica*.

With respect to the littoral insect fauna of Australian temporary ponds, special mention may be accorded the Odonata (Watson 1968). Many species of this order have shortened their nymphal lives to a period that synchronizes with the duration of the pond water. Species from several families have become adapted in this way. Some, for example *Hemianax papuensis*, can even reach adulthood in 2–3 months given suitable temperatures. Many species of frog, likewise, have adapted their life-cycle so that use can be made of the littoral region of temporary ponds for breeding (Martin 1967b). Perhaps the most interesting example is provided by Fletcher's frog, *Lechriodus fletcheri*, an eastern species. Not only have the tadpoles of this frog an extremely foreshortened life (31 days), they are also facultative carnivores; in conditions of abundant plant food the tadpoles are vegetarians, but when plant food is scarce they become carnivorous and even cannibalistic. Some frogs, we note, are known to utilize both temporary *and* permanent ponds as breeding loci (e.g. *Limnodynastes tasmaniensis*).

New Zealand temporary ponds as whole entities have been studied by Stout (1964, 1969b) and Barclay (1966). The accounts of both authors are particularly interesting, because Barclay made a comparison with nearby permanent ponds and Stout (1964) studied four ponds each with a slightly different time during which water was present. The faunal results of this latter study are summarized in Table 9:2. It can be seen that a few species were found only in the very temporary pond (the chydorid cladoceran, and '*Cyclops*'), some were found in all ponds (*Cypris, Lepidurus, Prionocypris,* and *Daphnia*), most only in the three least temporary ponds, and several were restricted to the most permanent pond. Only crustaceans occurred in the very temporary pond, and an established flora (*Callitriche stagnalis, Lemna minor, Azolla rubra, Spirogyra,* and *Oedogonium*) developed best in the most permanent pond.

TABLE 9:2
DISTRIBUTION OF PRINCIPAL SPECIES OF MACROSCOPIC FAUNA IN
FOUR TEMPORARY PONDS IN CANTERBURY
Relative abundance of each species is indicated by the number of plus signs. Pond 1 was
the most temporary pond; ponds 2, 3. and 4 were in sequence less temporary.
Rearranged from Stout (1964)

Taxon[a]	Pond 1	Pond 2	Pond 3	Pond 4
chydorid	++++	+		
'Cyclops'	+++	+		
Cypris	++++	++++	++++	++++
Lepidurus	+++	++++	++	+
Prionocypris	++	++	+	+
Daphnia	+	+	++++	++++
Culex		+	+	++
Rhantus		+	+	++
Sigara		+	++	+++
Anisops		+	++	+++
Microvelia		+	++	+++
Chironomus		+	++	+++
Simocephalus		+	+	+++
tubificid		++	+++	++++
Boeckella		++	++++	++++
Candonocypris			++	++
Limnoxenus				+
Hydrachna				+
Planorbis				+
Austrolestes				++
Physastra				+++
Hyla				+++

[a] With rare exceptions, only one species per genus.

Stout (1969b) later provided more specific details concerning the most permanent pond,[5] but it is unnecessary to reiterate these in full here. We merely note that the zooplankton was very dense for most of the year, consisted mainly of *Daphnia carinata*, and also present were *Boeckella triarticulata*, *Asplanchna*, and *Cypris kaiapoiensis*. This composition contrasts with that of the zooplankton of a rather more permanent nearby pond (dry about once in 10–20 years) reported on by Stout (1969b). In this latter pond *B. triarticulata* was the dominant zooplanktonic species and rotifers were also important, including the genus *Keratella*. No truly planktonic cladocerans occurred. Many other details of the biota of this pond are given by Stout from which it is clear that the diversity of species present is even greater than that of the most permanent of the ponds reported on in 1964.

That even greater variation in zooplankton composition between temporary ponds may occur is shown by a comparison of the zooplankton in the ponds studied by Stout near Christchurch and the temporary pond near Auckland studied in detail by Barclay (1966). In Barclay's pond no calanoid copepod occurred, nor apparently any

---

5. She does not actually note that the pond described in 1969 is the pond referred to in 1964 as Bottom Pond, the most permanent one then reported on, but it is clear from her account that this is so.

rotifers; the dominant forms were ostracods (*Cypris kaiapoiensis, Cypricercus sanguineus,* and *Cypretta* sp.), and two quite different genera of swimming Cladocera occurred (*Simocephalus, Alona*). A cyclopoid (*Acanthocyclops vernalis*) was also present. Some of these differences were no doubt due to the different degrees of macrophyte development and collecting methods involved, but they do suggest that considerable faunal variation may occur between one temporary pond and another in different geographical areas.

Less variation between temporary ponds within the same geographical area is indicated by Barclay's study of three localities near Auckland. Further, with regard to her comparison of the faunas of temporary and permanent ponds within a restricted area, she noted that in general there was relatively little difference between permanent and temporary ponds of similar size. Nevertheless three faunal groups could be distinguished: a poorly represented one with a marked preference for temporary ponds (*Cypris, Cypricerus,* and *Eucypris*), a large one which occurred in both permanent and temporary ponds (Cladocera, cyclopoid copepods, oligochaetes, *Physa, Hyla*), and a small one with a preference for permanent ponds (most gastropods, *Chlorohydra*).

A comparison of the faunas of New Zealand and Australian temporary ponds either with each other or with the faunas of temporary ponds elsewhere is not justified at the present time. One difference, however, is quite clear, namely the paucity of branchiopods in New Zealand temporary ponds *vis-à-vis* those in Australia. *Lepidurus* is the one genus that occurs in these localities in both countries. The explanation most certainly involves the general rarity of conchostracans in New Zealand and the absence there of indigenous anostracans, phenomena no doubt reflecting the very much less arid climate and long-standing isolation.

No comprehensive studies of the biota of Australian permanent ponds have yet been made; the only major faunal assemblage given any attention is the plankton (Brand 1967, Geddes 1968, Timms 1967, 1970a, 1970c), and even this attention, it should be stressed, is restricted mainly to southeastern Australia. As with temporary ponds, some taxonomic groups have been the subject of particular emphasis, but such studies we can consider only incidentally apart from Weatherley's (1958a) interesting study of Tasmanian farm dams. In this, although emphasis is given to the culture of fish, brief accounts are also given of the phytoplankton, zooplankton, and bottom fauna.

Zooplankton studies indicate, not surprisingly, the absence of uniformity in species composition within the range of localities here referred to as permanent ponds. Some of the variation would undoubtedly be less, were discussion to be restricted to localities of a more specific type (e.g. farm dams of a given size), but even then some variation could be expected (cf. Fig. 9:4). Summarizing the available data, the dominant limnetic species are *Boeckella minuta, B. triarticulata, B. fluvialis, Calamoecia lucasi, Mesocyclops leuckarti, Daphnia lumholtzi, D. carinata, Ceriodaphnia cornuta, Eucyclops* sp., *Alona rectangula, Chydorus* sp., and *Asplanchna*, and, in coastal ponds only, *Calamoecia tasmanica*. Not all of these of course occur contemporaneously, and congeneric occurrences in particular are not very common. The momentary species composition (all entomostracan species considered) in seventy-two of the smaller ($<250 \times 10^3$ m^3 capacity) localities in northeastern New South Wales investigated by Timms (1970a) was on average two copepods and one cladoceran. Typically, as noted by Timms, the limnetic zooplankton of all of the localities

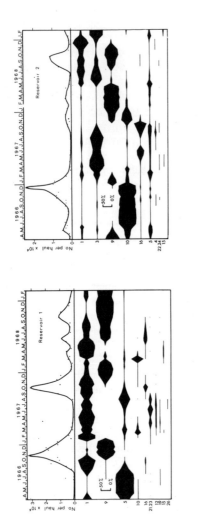
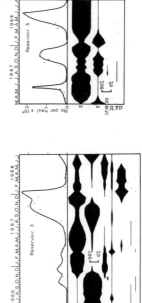

Fig. 9:4 Seasonal abundance and species periodicity of Entomostraca in four dams (reservoirs) in New South Wales. Reservoir 1, Richmond Main Colliery dam, capacity 910 x 10³m³; reservoir 2, Maitland Power Station dam, capacity 890 x 10³m³; reservoir 3, Abernethy Colliery dam, capacity 185 x 10³m³; reservoir 5, Morisset dam, capacity 125 x 10³m³. Index to species code: 1: *Boeckella fluvialis*; 2: *B. minuta*; 3: *Calamoecia lucasi*; 4: *Gladioferens spinosus*; 5: *Mesocyclops leuckarti*; 6: *Eucyclops serrulatus*; 7: *Macrocyclops albidus*; 8: *Daphnia carinata*; 9: *D. lumholtzi*; 10: *Ceriodaphnia cornuta*; 11: *Simocephalus acutirostratus*; 12: *S. elizabethae*; 13: *Moina micrura*; 14: *Ilyocryptus spinifer*; 15: *Diaphanosoma excisum*; 16: *Bosmina meridionalis*; 17: *Alona kendallensis*; 19: *Alona* sp.; 20: *Camptocercus similis*; 21: *Chydorus eurynotus*; 22: *Chydorus* sp.; 23: *Dunhevedia crassa*; 24: *Leydigia acanthocercoides*; 25: *Neoynhamia fenestrata*; 26: *Cypridopsis australis*. From Timms (1970c) in Proceedings of the Linnean Society of New South Wales and reprinted with permission

investigated by him, nearly all of which have capacities less than 1×10^6 m^3, consisted of a species of *Boeckella*, perhaps a species of *Calamoecia*, *Mesocyclops leuckarti*, and one or two species representing the genera *Daphnia*, *Ceriodaphnia*, *Bosmina*, or *Moina*. Re-analysis of his data[6] on species occurrences in localities with capacities greater or smaller than 1×10^6 m^3 indicates that there were *no* limnetic zooplankton species restricted to the larger localities, most of the calanoid copepods could exist in both sorts of locality (*Boeckella major* and *B. montana* are the only ones to occur in the smaller), and most of the cyclopoid copepods and cladocerans and all of the ostracods occurred only in the smaller localities.

In any comparison of the zooplankton composition between ponds, seasonal differences within any one pond must also be taken into consideration, for these are frequently pronounced, as indicated by Fig. 9:4. Moreover, within a particular pond, pronounced differences occur between the composition of the limnetic and littoral zooplankton. In the Brisbane pond studied by Timms (1967), for example, whereas the limnetic zooplankton was dominated by *Boeckella minuta*, in the littoral region the dominant zooplanktonic forms were, depending upon season, *Eucyclops* sp., *Macrothrix spinosa*, and *Chydorus* spp. Finally, we note, differences in the limnetic zooplankton at a given stratum occur according to the time of day for, despite the shallowness of some of the ponds investigated, there is good evidence that vertical migration of at least some of the plankton components takes place.

Little of significance can be written at present about other elements of the fauna of Australian permanent ponds. We may note, however, that a large number of algal species have been recorded in the phytoplankton, many of which belong neither to the Volvocales nor Chlorococcales, the two characteristic algal orders found in ponds. Nevertheless, cosmopolitan genera in these two orders and typical of ponds worldwide also occur frequently (e.g. *Chlamydomonas*, *Volvox*, *Pediastrum*, *Scenesdesmus*). Despite the high turbidity of most Australian farm dams, as previously noted, algal blooms are still recorded in these localities. Fish are often introduced but this is usually on a rather *ad hoc* basis. In southeastern Australia the most popular species for introduction appears to be the non-indigenous rainbow trout. In southwestern Western Australia, the marron (*Cherax tenuimanus*), an edible freshwater crayfish, has been introduced to many dams. A somewhat far-removed relative of this species, a freshwater crab belonging to the Potamonidae, *Paratelphusa* spp. (mostly *P. transversa*), may occur in farm dams throughout most of northern Australia and inland New South Wales. Its occurrence is important for it burrows into dam walls and may penetrate up to 1 m, causing significant damage.

Permanent ponds in New Zealand have been comprehensively studied only by Byars (1960), who investigated a pond near Dunedin. She has provided a long list of animal and plant species that occurred in this pond, but for present purposes it is sufficient to note only that in the zooplankton four species dominated: *Boeckella triarticulata*, *Chydorus sphaericus*, *Synchaeta pectinata* (Rotifera), and *Conochiloides coenobasis* (Rotifera). All displayed large seasonal variations in abundance and the rotifers were completely absent during the colder months. Ostracods, in contrast to the situation in the ponds studied by Barclay (1966) on the North Island (see above), were always unimportant.

---

6. Given as unpublished appendices to his paper and available on application to the Editor-in-Chief, Editorial and Publications Section, C.S.I.R.O., 372 Albert Street, East Melbourne, Victoria 3002, Australia.

SOME GENERAL COMMENTS

Two related topics not properly discussed so far in this chapter, but of obvious importance in any consideration of the biota of isolated standing bodies of water, concern the problems of dispersal and occurrence of unfavourable dry periods. Apart from some work by Timms (1970a) on the efficiency of dispersal by limnetic entomostracans in New South Wales, little can be written on these topics based on Australian or New Zealand work, but they are so important that it is appropriate to summarize pertinent information derived from studies elsewhere. The reviews of Gislén (1948), Talling (1951), and Macan (1963) are particularly relevant in this connection.

Problems of dispersal, that is difficulties relating to transport between water bodies, are least amongst microorganisms. Most of these can encyst or are otherwise able to withstand at least short periods free of water. They are therefore relatively easily dispersed passively by the wind, birds, insects, and some mammals. Maguire (1963) has reviewed the dispersal of such forms, and her own work provided a long list of microorganisms that had been passively transported to her experimental loci; prominent on this list were Protozoa, algae (including blue-green algae), and Rotifera.

Problems of dispersal are also not great in many aquatic insect groups, for the *adults* of all groups either live a completely terrestrial life or are able to fly or move away from the water when occasion demands. Many groups with terrestrial adults are easily dispersed, for example the Odonata and many families of the Diptera, and it comes as no surprise to find these common inhabitants of small, discrete, water bodies. In general in such groups one can predict a correlation between the degree of permanence of the water in the larval habitat and the extent of adult dispersal; that is, the more temporary the water in a larval habitat, the greater the extent of adult dispersal, and the converse. This is clearly indicated by Watson's (1969) work on dragonflies in the northwest of Western Australia. All 9 species with temporary larval habitats had wide adult dispersal, whereas of the 21 species with larvae in permanent habitats only 5 had wide adult dispersal and most (13) had limited adult dispersal. Other insect groups, for example the Plecoptera, have terrestrial adults which are not so easily dispersed and these groups are not commonly found in ponds, rain-pools, and similar localities. A somewhat substantial literature has arisen which records the occurrence away from water of adult insects which are typically aquatic (cf. Fernando 1958, 1959). Such peregrinations may be significant too for some small macroscopic animals that are not insects: hydracarines and small molluscs, for example, have frequently been recorded attached to the bodies of adult corixids and aquatic beetles.

Other macroscopic elements of the fauna, like microorganisms, are passively dispersed by various animals or by the wind. For those with a stage in their life-cycle that is resistant to desiccation—especially when this is small, as are most eggs—few difficulties can be envisaged in such transport. However, it must be admitted that some animals that do not have such stages, for example fish, amphipods, isopods, and atyid prawns, occur in isolated bodies of water (especially permanent ponds). These animals are probably transported as adults, but our knowledge of the actual transport mechanisms is fragmentary for the most part.

Particularly characteristic amongst these groups which have a resistant stage are the branchiopod crustaceans as well as many copepods and cladocerans. The resistant stage of these is usually the egg, but it may be a juvenile or other stage in certain copepod species. Frequently, two kinds of eggs are produced: a resting, resistant one, and one that has neither of these features. Some resting eggs are known to remain viable under conditions of desiccation for several years, and it is of some interest to note that the foundations of entomostracan taxonomy in Australia and New Zealand were provided by eggs sent in dried mud to Europe at a time well before the advent of air-travel. In Europe the eggs hatched and the adults were subsequently described (e.g. Sars 1889).

It is not easy to measure the efficiency of dispersal mechanisms, for so many factors difficult of assessment arise to complicate the issue. One such factor is the element of chance, and as a corollary to this the phenomenon of competitive exclusion may be involved too. Thus, if by chance one species reaches a locality before another with a similar ecological niche, prior occupancy may determine which species survives. On the other hand, of course, the first species to arrive may actually be superior in dispersal ability but perhaps inferior in competitive ability and will then be replaced when the second species eventually arrives; it is prevented from becoming extinct only by its superior dispersal abilities (such species are termed fugitive species).

In spite of the difficulties, under certain restricted conditions, a few authors have attempted to assess dispersal efficiency. One attempt of special interest in the present context is that by Timms (1970a). This author, on the basis of three lines of evidence, ranked fifteen species of Entomostraca occurring in two areas in New South Wales according to their ability to disperse. Not unexpectedly the two species most highly ranked were the cosmopolitan cyclopoid *Mesocyclops leuckarti* and the widespread cladoceran *Daphnia carinata*. The calanoid copepods, however, were ranked—unexpectedly—higher than the remaining cyclopoid and cladoceran species, and this, Timms suggested, may be partly the result of better adaptation by calanoids in Australia to astatic conditions.

Methods of overcoming periods of unfavourable desiccation are mostly related to the means whereby species can also be dispersed. That is, the stage dispersed is the stage that exists during the dry period. Nevertheless this is not always the case. The resistant larval stage of certain chironomids mentioned earlier in our discussion of the fauna of puddles and rock-pools, for example, is not likely to be the stage that is most responsible for dispersing the species. Likewise, adult freshwater mussels which bury themselves with tightened shells in mud to overcome a dry period are probably not the usual individuals to bring about distribution of the species to other localities.

part five
NON-MARINE OR ATHALASSIC SALINE WATERS

Salt lakes, South Australia. *Photograph by D. Darian Smith*

chapter ten
NON-MARINE OR ATHALASSIC SALINE WATERS

TYPES OF DRAINAGE SYSTEMS OR HYDROLOGICAL REGIONS
In regions where much of the rainfall finds its way back to the sea through river systems (*exorheic* regions), atmospherically supplied ions are truly cyclic in nature, and there is no marked tendency for the production of high salinity waters.[1] There are, however, two situations which interrupt the hydrological cycle and result in the accumulation of atmospherically supplied ions, and hence the production of highly saline waters. First there are *arheic* regions, in which little or no precipitation occurs or in which none is carried off superficially. In other words, no rivers or other bodies of running water arise in such regions, which are mostly absolute desert. The Gibson Desert of Western Australia, and Salinaland surrounding Kalgoorlie in the same State, might be quoted as examples. In the latter region ephemeral bodies of saline water occasionally exist. Second there are *endorheic* regions, in which any precipitation that occurs never reaches the sea. Rivers may arise in endorheic regions, but all the water is lost by evaporation before it can reach a river mouth. The rivers become lost in dry water courses or enter closed, terminal lakes. Examples of this type of system are discussed below. All three main types of drainage system are found in Australia (Fig. 10:1). Details of their distribution in Victoria are shown in Fig. 10:2. A fourth type, also found in Victoria, consisting of a system of water channels that are hidden underground, usually in limestone country, is sometimes recognized. This is the *cryptorheic* system.

Not all salt lakes are dependent on the atmosphere and ultimately the sea for a source of ions; the salts of some are mainly derived from surrounding rocks and soils by weathering processes. This usually seems not to be true of the Australian scene, but this is not the opinion of all authors—Wofner and Twidale (1967), for example, think that most of the salts in Lake Eyre have been derived in this way. These salt lakes, like those discussed above, are found predominantly in endorheic and arheic regions.

---

1. Most Australian authors describe atmospherically supplied salts as 'cyclic' under all conditions, yet, throughout the greater portion of the country, this is precisely what they are not, unless one takes a very long-term view of things.

194   INLAND WATERS AND THEIR ECOLOGY

Australia is the driest continent in the world except for Antarctica. Some indication of the extent of its aridity is given by the following simple tabulation:

	Entire Australian Continent	All Land Areas of the World
Average annual rainfall (cm)	41·9	66·0
Average annual run-off (cm)	3·3	24·9

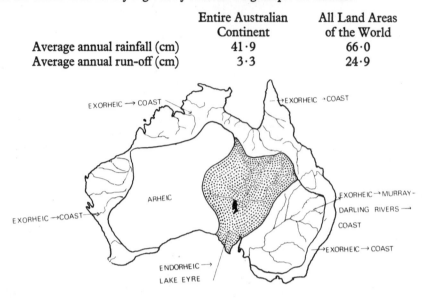

Fig. 10:1 Map showing the main hydrological regions of Australia. For explanation of terms, see text

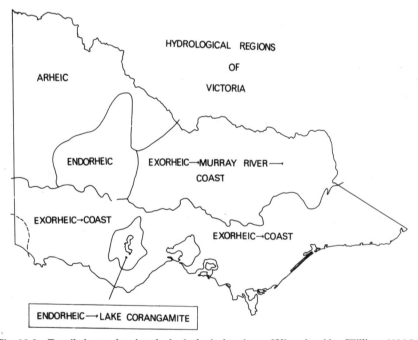

Fig. 10:2 Detailed map showing the hydrological regions of Victoria. *After Williams (1964)*

NON-MARINE OR ATHALASSIC SALINE WATERS

Eighty-seven per cent of the total area of Australia receives less than 76 cm of rainfall per annum on average and 68 per cent receives less than 50 cm. Not only is rainfall low, but evaporation is high, as suggested by the fact that the ratio of run-off to rainfall, which may be derived from the above tabulation, is much lower for Australia (0·08) than that for the world (0·38). In fact annual average loss of water by evaporation and transpiration exceeds the corresponding rainfall over almost 95 per cent of the total area of the Australian mainland.

To add to this picture there are extensive areas of arheic and endorheic drainage. Thus the largest lake in Australia, Lake Eyre (Plates 10:1 and 10:2), when it exists, is a saline lake and represents the terminal lake of a huge (*ca.* $0·5 \times 10^6$ sq ml or

Plate 10:1   The northern shore of Lake Eyre, South Australia, looking eastward from 27° 57′ S, 137° 13′ E at a time when the lake was full. Temporary pools to the north of the lake are also visible. Photograph taken 11 May 1950 from a height of 4,600 m. *By courtesy of the R.A.A.F.*

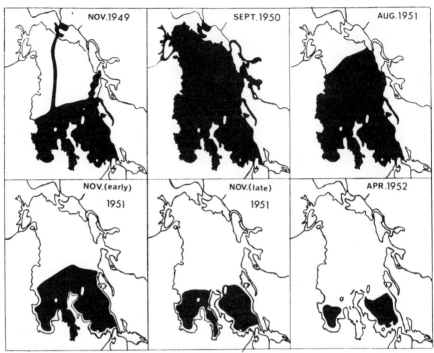

Fig. 10:3 The filling (1949–50) and drying (1951–52) of Lake Eyre, South Australia. *After Bonython and Mason (1953)*

Plate 10:2 The coastline of portion of Lake Eyre (Level Post Bay, southeast of Madigan Bay, looking southwest) when full, October 1950. On the left is Goyder Channel; in the centre is Scalloped Bay, on the right extremity of which is Prescott Point. *Photograph by C. Warren Bonython*

$1 \cdot 3 \times 10^6$ sq km) endorheic system (Fig. 10:1).[2] Lake Eyre is of course usually a dry salt-pan, but it filled in 1890–91, and again in 1949–50, and has contained some water on a few other occasions this century. The evaporation rate at Lake Eyre is about 230 cm per annum. As shown in Fig. 10:3, the lake was full in September 1950 but was almost empty again early in 1952. In October 1950 the salinity was about $39\%_{00}$, but by December 1951 it had risen to about $210\%_{00}$. The largest lake in Victoria, Lake Corangamite (Plate 10:3), is also a salt lake situated in the centre of an endorheic region (Fig. 10:2). Unlike Lake Eyre, however, it is permanent, although the salinity has fluctuated a good deal this century (see below). On average the annual amount of water finding its way into Lake Corangamite from rainfall both directly on the surface and within its catchment area is almost exactly balanced by the annual loss due to evaporation. It is this approximate equality that permits permanent existence.

SOME MORPHOLOGICAL FEATURES OF AUSTRALIAN SALT LAKES

All of the larger salt lakes are shallow. In 1950, when Lake Eyre was full, a maximum depth of about 4 m was recorded in the southeastern portion (Bonython and Mason 1953). The mean depth at this time was only about 2 m. Likewise the mean depth of Lake Corangamite even at its exceptionally high level in 1953 was only $3 \cdot 4$ m. The nearby Lake Beeac is considerably less than 1 m in depth when full. Lakes Eliza and St Clair in South Australia rarely have more than about 20 cm of water in them. In addition, there are an enormous number of very shallow salinas in Australia that never contain more than a few cm of water. However, a perhaps unique Australian feature is the existence of a few deep salt lakes which are quite permanent and do not display anything like the large degree of salinity variation which characterizes most salt lakes. These are exemplified by a few maar lakes in the western districts of Victoria. Thus Lake Bullenmerri, which currently has a salinity of about $8-9\%_{00}$, has a maximum depth of 64 m and Lake Gnotuk ($50-60\%_{00}$) has a maximum depth of about 19 m. Lake Keilambete ($55-60\%_{00}$, $z_m = 11 \cdot 5$ m), also in Victoria, is another example. Despite the fact that these depths considerably exceed those of several meromictic lakes in other parts of the world, none of these three Australian salt lakes is meromictic. This seems to be a consequence of the fact that the prevailing westerly winds are usually of considerable strength. Recently, however, a saline meromictic lake has been found in Victoria (B. V. Timms, unpublished). This is a well-sheltered maar known as West Basin Lake, in which the salinity of the mixolimnion is about $90\%_{00}$ and the maximum depth is 14 m.

PHYSICAL AND CHEMICAL FEATURES

The temperature range in a shallow salina only a few cm deep is extreme, often being in excess of 30°C seasonally, and even in the course of a day may be almost 20°C. On a hot summer afternoon temperatures of up to 40°C may be recorded, but these may fall to almost 20°C and sometimes slightly less by the following dawn after strong cool winds during the night. The minimum winter temperature in such localities in Victoria may be considerably less than 10°C but almost invariably above 0°C. A small salt lake near Douglas, Victoria, had a temperature of 34°C at midday in mid-January

---

2. Hutchinson (1957) calls it an arheic system, but since rivers that arise within it do sometimes flow (Plates 7:1–7:4), it seems more appropriate to consider it endorheic.

1962, but at 11 a.m. on 20 May 1962 it was only 9°C. Large bodies of saline water of course have smaller variations than this. Nevertheless eurythermy must be regarded as a characteristic and essential feature of all successful salt-lake organisms.

Because of their scarcity little has been published on vertical changes in the physico-chemical conditions of highly saline lakes which are deep enough to stratify in summer and yet are not meromictic. Figure 10:4 shows temperature-depth and oxygen-depth curves for Lake Gnotuk in mid-summer and mid-winter. It can be seen that the lake stratifies in summer in the same way as a freshwater lake, having a thermocline at a depth of about 13-14 m. The hypolimnion is relatively small in volume but is completely lacking in oxygen. In winter there is complete circulation at about 10°C and water in the immediate vicinity of the bottom is still 75 per cent saturated with oxygen.

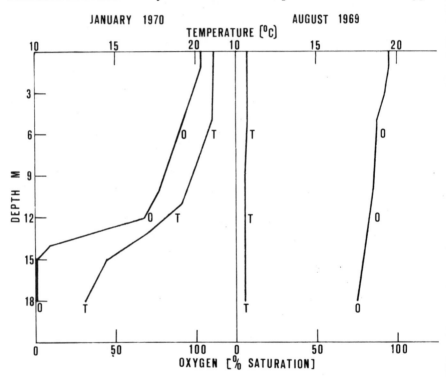

Fig. 10:4 Temperature-depth and oxygen-depth graphs for Lake Gnotuk, Victoria, for August 1969 and January 1970. *From unpublished data of B. V. Timms*

The salinity of inland saline waters is very variable and much more so than that of fresh waters. The extent of this variation in six localities in the Lake Corangamite area of Victoria from March 1964–July 1965 is shown in Fig. 10:5. Lakes Corangamite and Gnarpurt are relatively constant because of their large areas (23,400 ha and 2,230 ha respectively) and Lake Gnotuk is fairly stable because of its considerable depth (20 m). The remaining three lakes are either small or very shallow and show great fluctuations. If for each lake the maximum concentration is expressed as a percentage increase from the minimum concentration, then the mean percentage change for these six lakes during the period of study was 330 (Bayly and Williams 1966a). T.D.S. concentrations for several lakes in the Beachport Robe area, South Australia, from March 1964–March 1965 and in September 1966 and December 1969 are shown in Fig. 10:6. The mean percentage increase from minimum to maximum concentration for seven of these lakes from March 1964–March 1965 was 495. The most violently fluctuating of seventeen localities studied by Bayly and Williams (1966a) was a small lake near Douglas, Victoria. In March 1964 this contained a saturated brine with a T.D.S. concentration of 339‰. Six months later, however, the concentration was only 17·2‰, the increase from minimum to maximum being 1,970 per cent. Four freshwater lakes studied contemporaneously with the seventeen salt lakes from March 1964–March 1965 had a mean percentage increase from minimum to maximum concentration of only 53.

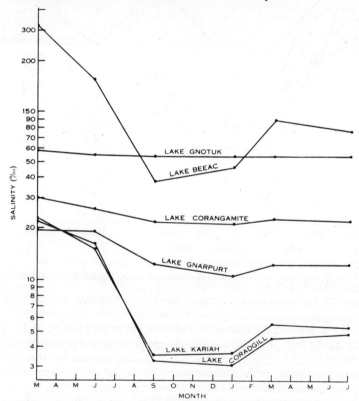

Fig. 10:5   Seasonal variation in the concentration of total dissolved solids in six lakes in the Lake Corangamite region of Victoria during the period March 1964–July 1965. *After Bayly and Williams (1966)*

In addition to seasonal variation, secular (long-term) changes occur in the salinity of salt lakes. These correlate with and are undoubtedly caused by secular fluctuations in climate. Figure 10:7 shows secular as well as seasonal variation in T.D.S. concentration for Lake Corangamite from 1958–65. In 1937 the salinity of this lake was 115‰, but in 1953 and 1960 values of only 12·5‰ and 6·5‰ respectively were recorded.

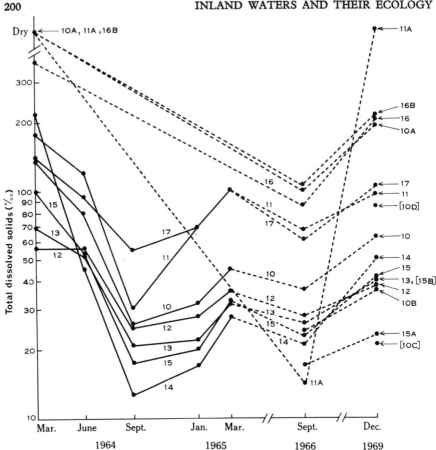

Fig. 10:6 Seasonal and secular variation in the concentration of total dissolved solids in sixteen lakes in the Beachport-Robe area of South Australia. The lakes include Lake Robe (10), Lake Eliza (10A), Lake Fellmongery (10C), Lake Amy (10D), Lake St Clair (16B), Salt Lake, Beachport (17), and ten other small unnamed lakes in the same region. *From Bayly (1970a)*

Dahl (1956) stressed the ecological importance of salinity change as distinct from absolute salinity value, but restricted his discussion to diluted or concentrated bodies of sea water found in coastal regions. Bayly (1967b) pointed out that this was an unnecessary restriction of the concept since most inland saline waters undergo great salinity fluctuations which must be of no less significance ecologically than those occurring in the waters discussed by Dahl. The term poikilosaline (Bayly 1967b) may thus be used to refer to all waters of variable salinity irrespective of mean absolute values. Such waters are subject to substantial and often rapid changes in salinity and this places a premium on osmoregulation or the tolerance of large changes by tissues. Sometimes strong regulation of the concentration of body fluids is involved. This is so with the atherinid fish, *Taeniomembras microstomus* (Lee 1969), the isopod *Haloniscus searlei* (Bayly and Ellis 1969) (see Fig. 1:5B), and also the anostracan *Parartemia zietziana*. Sometimes, however, there is almost complete conformity between body fluids and external medium over a wide salinity range and

NON-MARINE OR ATHALASSIC SALINE WATERS 201

it must be assumed that either cellular osmoregulation or cellular tolerance is well developed. This is so with the calanoid copepod *Calamoecia salina* (Bayly 1969a, G. W. Brand, unpublished data). The known salinity tolerance of this species in the field is 28–131‰. In the laboratory this species has been kept in media with a freezing-point depression range of 0·8–9·1°C. Over this range the greatest differential found between Δ_e and Δ_i was 0·26°C. It may be assumed that the ability to osmoregulate at either the organ or cellular level or the possession of exceptional cellular tolerance is a physiological necessity for success in athalassic saline waters.

Fig. 10:7 Seasonal and secular variation in the concentration of total dissolved solids in Lake Corangamite over a six-year period. *After Bayly and Williams (1966)*

As pointed out in Chapter 1, the ionic composition of Australian athalassic waters, especially those that are saline, strongly resembles that of sea water. Table 10:1 shows in quantitative terms how closely the relative proportion of major ions in Victorian and South Australian salt lakes agrees with that of sea water. The cationic order of dominance is usually Na > Mg > K > Ca, whilst that for anions is Cl > SO_4 > HCO_3. Sometimes the positions for K and Ca are transposed as are those for SO_4 and HCO_3 but the positions for Na, Mg, and Cl hardly ever change. The only

Australian saline waters with orders of dominance substantially different from these are found in a series of about six lakes near Red Rock, Victoria. In these the orders are Na > K > Mg > Ca and Cl > (HCO$_3$ + CO$_3$) > SO$_4$ (Bayly 1969c). Elsewhere in the world, for example in the Rift Valley of Africa, the Lahontan Basin of Nevada, and the lower Grand Coulee of Washington State, carbonate-dominated lakes occur. Sulphate-dominated lakes also occur, Little Manitou Lake in Saskatchewan providing an example. Salt lakes which, like those in Australia, are dominated by chloride include the Great Salt Lake, Utah, and the Dead Sea in the Middle East. Sodium is also the dominant cation in the former, but magnesium dominates the cations in the latter.

TABLE 10:1
MEAN PROPORTIONS OF MAJOR IONS IN SEA WATER AND IN ATHALASSIC SALINE WATERS OF VICTORIA, SOUTH AUSTRALIA, AND NEW ZEALAND
All values are equivalent percentage of total cations or anions except Na$^+$/Cl$^-$ ratio

Water	Na$^+$	K$^+$	Mg^{2+}	Ca^{2+}	Cl$^-$	SO$_4^{2-}$	HCO$_3^-$	Na$^+$/Cl$^-$
Sea water	77·1	1·6	17·7	3·5	90·3	9·3	0·4	0·85
South Australia								
Beachport–Robe series[a]	76·5	1·9	18·9	2·7	87·0	12·5	0·6	0·88
Beachport–Robe lakes[b]	78·8	1·5	18·0	1·5	91·8	8·0	0·5	0·86
Victoria								
Centre Lake series[a]	84·1	0·5	14·8	0·5	94·9	4·9	0·2	0·89
Corangamite series[a]	85·7	0·9	12·7	0·7	95·6	—	4·9[d]	0·90
Lake Corangamite[c]	83·3	3·5	12·9	0·2	96·1	2·8	1·4[d]	0·87
Salt Lakes of Linga[e]	96·7	<0·1	6·7	0·7	95·1	4·9	—	1·02
New Zealand								
Salt Lake, Sutton[f] (Otago)	91	2	5	1	ca. 100	—	3	0·91

[a] From Bayly and Williams (1966a). The Beachport–Robe series included eight lakes, the Centre Lake series three, and the Corangamite series six.
[b] Calculated from Jack (1921: 74). Mean data for four lakes; samples taken 27 May 1920.
[c] Mean data calculated from Anderson (1945, Table 13) for analyses of samples taken in 1938 and 1939.
[d] HCO$_3^-$+CO$_3^{2-}$.
[e] Calculated from Cane (1962: 87). There is uncertainty as to whether the data apply to liquid brine or to a sample of moist crystalline salt.
[f] From Bayly (1967a).

THE ZOOPLANKTON AND BENTHIC MICROFAUNA

In this and the following three sections comments are largely restricted to athalassic saline waters in the southeast and southwest of Australia, and are mostly based on the work of the present authors and associated students. Although the biota of salt lakes in general displays considerable cosmopolitanism (Macan 1963), our restricted treatment may be justified on the ground that the fauna of Australian salt lakes is largely endemic. Cole (1968) has recently provided a comprehensive review of the biota of salt lakes on a world basis, and reference may also be made to Hedgpeth's (1959) wide treatment of this subject.

In lakes with a salinity greater than about 200‰ the only animal to occur in any numbers in the plankton is the anostracan *Parartemia zietziana*. This may still be present in concentrations approaching saturation (> 300‰).

Within the salinity range 50–200‰ the zooplankton usually consists of one or other —exceptionally both—of the calanoid copepods *Calamoecia clitellata* and *C. salina*, together with a small ostracod of the genus *Diacypris* and a large ostracod belonging to a new genus as yet undescribed. The truly planktonic nature of these ostracods is stressed by Bayly (1970a). Sometimes the ostracod *Platycypris* may also occur in this salinity range. In addition, the cyclopoid copepod *Halicyclops ambiguus* is frequently present, but occasionally another cyclopoid, *Microcyclops arnaudi*, occurs instead. Another somewhat surprising component of the zooplankton of highly saline lakes is insect larvae belonging to the family Ceratopogonidae.

In the salinity range 10–50‰ most of the above genera or species are still represented, although the lower limit of some is probably closer to 20‰ than 10‰. The large new ostracod referred to above seems to be an obligate high-salinity species, having not yet been collected from a salinity of less than 38‰. This is probably also the case with some species of *Diacypris*, although the genus certainly tolerates salinities of less than 50‰. Nevertheless there is a distinct rise in the number of species. Thus the cladoceran *Daphniopsis pusilla* may be seasonally abundant in salinities of up to about 50–60‰. One or other and sometimes both of the rotifers *Brachionus plicatilis* and *Hexarthra fennica* may now be present. The former was almost continuously present throughout a two-year (1968–70) study of Lake Werowrap, Victoria (Plate 10:3), and during this time the range of salinity variation was 25·2–57·5‰ (K. F. Walker, pers. comm.). When a considerable amount of (bi)carbonate is present and the salinity is not too high, *Hexarthra jenkinae* may occur instead of *H. fennica*.

Plate 10:3 A view from the top of Red Rock, Victoria, looking to the southwest. In the left foreground is Red Rock 'Tarn' behind which is Lake Werowrap. Behind this again, to the right, are Eastern and Western Twin Lakes. Lake Corangamite, the largest lake in Victoria, is located in the background of the picture. *Photograph by I. A. E. Bayly*

In salinities below about 30‰ *Mytilocypris* may occur, apparently in place of the large undescribed ostracod characteristic of more highly saline waters. As already mentioned in Chapter 1, there is evidence that the calanoid *Boeckella triarticulata* occurs in mildly saline waters (up to about 20‰) if there is an adequate amount of carbonate or carbonate plus bicarbonate. There is now evidence (G. W. Brand, unpublished data) that the amount of (bi)carbonate has an influence more marked on egg development than on the adult animal. The medusae of the hydrozoan *Australomedusa baylii* may be found in the plankton of coastal athalassic saline waters. This species has also recently been found in the brackish water of an open estuarine system in Victoria (I. Neale, unpublished).

Halobiont species of *Calamoecia* are very important throughout the salinity range 10–140‰. However, although the salinity ranges of *C. clitellata* and *C. salina* overlap to a very large extent, there is evidently some degree of differentiation, *C. clitellata* being able to tolerate lower salinities than *C. salina* and vice versa. The known ranges on the basis of field data are 8–113‰ and 28–131‰ respectively. The factors that permit the coexistence of these two similarly sized calanoids have still to be elucidated. It may be of some significance that the only two known localities where they regularly coexist (Lakes Beeac and Cundare, Victoria) are both temporary and show great oscillations in salinity.

A salinity of 3‰ may somewhat arbitrarily be regarded as constituting the lower limit of saline waters, so that the range 3–10‰ forms the last of the rather arbitrary subdivisions to be considered. In this range there is a very great increase in the number of species that may be present. Most, however, are essentially freshwater species whose upper limit of tolerance is perhaps only slightly higher than usual. It is not appropriate to discuss them in detail in this chapter.

With respect to the benthic microfauna, the harpacticoid copepod *Mesochra baylyi* is of very common occurrence throughout a wide range (at least 21–94‰) of salinities. In athalassic saline waters that are quite close to the sea the harpacticoids *Onychocamptus bengalensis*, *Robertsonia propinqua*, and *Heterolaophonte* sp. nov. may also occur.

The known salinity ranges for all of the above-mentioned species are given in Table 10:2.

Other elements of the benthic microfauna, for example the Protozoa, remain largely uninvestigated. In view of the high degree of cosmopolitanism displayed by the free-living Protozoa, it seems unlikely on *a priori* grounds that significant differences will occur between the protozoan faunas of Australian salt lakes and those of salt lakes in other continents. The Great Salt Lake, Utah, has been most investigated in this respect, and reference may be made to Flowers and Evans's (1966) summary of results since this was published after Coles's (1968) review had been completed. They noted the presence of at least seven species of ciliate, two species of amoebae, and seven flagellate species. Eleven species of bacteria were also noted.

THE NEKTON AND BENTHIC MACROFAUNA

Fish are absent from all the highly saline lakes. The highest salinity at which a species has been recorded in Australia is 70‰; this record relates to a species of *Taeniomembras*,[3] an atherinid occurring in salt lakes of southeast South Australia (Lee 1969).

---

3. Previously referred to (Bayly and Williams 1966a) as *Craterocephalus* sp.

NON-MARINE OR ATHALASSIC SALINE WATERS

TABLE 10:2
SALINITY TOLERANCES OF ZOOPLANKTON AND MICROBENTHIC
FAUNA OCCURRING IN NON-MARINE SALINE WATERS IN SOUTHEAST
AUSTRALIA
Based on field records

Species	Salinity Range (‰)
COPEPODA	
Boeckella triarticulata	fresh–22
Calamoecia salina	28–131
C. clitellata	8–113
Mesochra baylyi	21–94
Onychocamptus bengalensis	19–93
Robertsonia propinqua	23–62
Heterolaophonte sp. nov.	57–94
Halicyclops ambiguus	8–62
Microcyclops arnaudi	6–93
OSTRACODA	
Diacypris (small)	21–131
Cyprinae gen. nov. (large)	38–113
Mytilocypris	fresh–36
Platycypris	20–180[a]
ANOSTRACA	
Parartemia zietziana	40–300[a]
CLADOCERA	
Daphniopsis pusilla	6–59
INSECTA	
Ceratopogonidae	57–95
ROTIFERA	
Brachionus plicatilis	15–58
Hexarthra fennica	9–58
H. jenkinae	25–41
HYDROZOA	
Australomedusa baylii	22–34

[a] M. Geddes (unpublished data).

The range of salinity within which the species has been collected in the field is 21–70‰ (Table 10:3), but laboratory investigation of its salinity tolerance gave a range of 3·3–108‰ for salinities required to kill 50 per cent of populations. Slight meristic differences occur between isolated populations of this species (Lee and Williams 1970) and these, it has been suggested, are induced by salinity.

In lakes of rather low salinity ($<ca.$ 10‰), a number of more typically freshwater fish may be found (omitted from Table 10:3). In Victoria, for example, the quinnat salmon, *Oncorhynchus tschawytscha*, lives (but does not breed) in Lake Bullenmerri, which has a salinity of 7–8‰ and, in Lake Modewarre, which has a salinity of *ca.* 3·7‰ (Lim and Williams 1971), the following species occur (Lim 1964): *Salmo gairdneri*, *S. trutta*, *Perca fluviatilis*, *Galaxias attenuatus*, *Retropinna victoriae*, *Philypnodon grandiceps*, *Anguilla* sp. The life-history of *Galaxias attenuatus* in this closed lake is noteworthy, for adults spawn in spring after migrating up an intermittently inflowing creek. The normal life-history involves an adult migration downstream to spawn in estuaries (Pollard 1966).

Other elements of the fauna that may be considered as nektonic include two species of the amphipod genus *Austrochiltonia* and several species of Hemiptera and Diptera (Culicidae). Further details are given in Table 10:3, which indicates the salinity or range of salinities at or within which each species has been recorded. For the most part the details are based on the data of Bayly and Williams (1966a) for Victorian and South Australian localities, but the data on Hemiptera are those of Ettershank, Fuller, and Brough (1966), who found three species in two localities in western New South Wales.

Table 10:3 also lists southeastern species which are macrobenthic in habitat,

TABLE 10:3
SALINITY RECORDS FOR NEKTONIC AND MACROBENTHIC FAUNA
IN NON-MARINE SALINE WATERS IN SOUTHEAST AUSTRALIA
Based on field records

Species	Salinity (‰)
PISCES	
Taeniomembras sp.	21–70
DECAPODA	
Halicarcinus lacustris (Hymenosomatidae)	0·1–10
AMPHIPODA	
Austrochiltonia subtenuis	fresh–29
A. australis	fresh–25
ISOPODA	
Haloniscus searlei ·	30–159
INSECTA	
Anisops gratus (Hemiptera)	53
A. thienemanni (Hemiptera)	7
Procorixa sp. (cf. *eurynome*) (Hemiptera)	53
Micronecta gracilis (Hemiptera)	7
Ischnura aurora (Odonata)	21
Enochrus (Methydrus) andersoni (Coleoptera)	21
Bagous sp. (Coleoptera)	21
Necterosoma sp. (Coleoptera)	50–93
Leptoceridae, species a (Trichoptera)	21
Leptoceridae, species b (Trichoptera)	50–62
Pyralidae (Lepidoptera)	21
Ephydrella (Ephydridae:Diptera)	53
Tipulidae (Diptera)	53
Tanytarsus (Tanytarsus) barbitarsus (Chironomidae:Diptera)	53–93
Procladius sp. (Chironomidae:Diptera)	22
Aedes (Ochlerotatus) vigilax (Culicidae:Diptera)	22
A. (Halaedes) australis (Culicidae:Diptera)	125
MOLLUSCA	
Coxiella striata (Gastropoda)	25–112
ANNELIDA	
Ceratonereis erythraeensis (Polychaeta)	23–62
Capitella capitata (Polychaeta)	36
NEMATODA	
Nematoda	fresh–62
TURBELLARIA	
Mesostoma sp.	13
COELENTERATA	
Cordylophora sp.	29

NON-MARINE OR ATHALASSIC SALINE WATERS

together with their known salinity ranges. Some of the information given is worth further comment. Undoubtedly the most interesting species listed is *Haloniscus searlei* for, apart from a doubtful Russian record, completely aquatic oniscoid isopods are otherwise unknown from salt lakes. Ellis and Williams (1970) have discussed its biology in some detail. It is, as indicated in the table, extremely euryhaline. The crab *Halicarcinus lacustris* is of considerable interest too in that most members of the large family to which it belongs, the Hymenosomatidae, are marine or estuarine (Holthuis 1968); only this species is recorded from inland saline waters where one of the adaptations it has developed is the complete suppression of a free larval stage in its life-history (Walker 1969). Several of the salinities listed in Table 10:3 represent the highest values at which the groups they refer to are known to occur. This applies to the leptocerid species (Trichoptera), the tipulid, the lepidopteran, and the gastropod, *Coxiella striata*. Bayly and Williams's (1966a) record of *Batillariella estuarina* (Gastropoda) from a South Australian salt lake is omitted from the table since the validity of this record now seems doubtful (Bayly 1970a). Also omitted is the large assemblage of insects known from localities with salinities less than *ca.* 5‰.

Nektonic and macrobenthic animals known to occur in southwest Western Australian salt lakes include *Austrochiltonia* sp., *Haloniscus searlei*, *Coxiella striatula*, and other species in this genus (but not *striata*), several genera and species of Odonata (in mildly saline waters), various dytiscid and hydrophilid beetles, Ephydridae, *Tanytarsus (Tanytarsus) barbitarsus*, and *Symphitoneuria wheeleri* (Trichoptera). Generally, precise salinity records are not available for these. The palaemonid shrimp, *Palaemonetes australis*, may occur in the nekton of coastal saline lakes (Serventy 1938), but it is not certain that these are truly athalassic.

Apart from the presence of unusual or atypical species, the composition of the nektonic and macrobenthic fauna of Australian salt lakes is of interest on other counts. Thus some groups, in particular the ephydrids and corixids, seem to be far less common than they are in similar localities outside Australia. Some genera, for example *Coxiella*, are apparently monospecific in eastern Australia yet multispecific in Western Australia. Certain species are common to both sides of the continent, for example *Haloniscus searlei*; yet others are restricted to eastern or western sides. The occurrence of congeneric groups of species is itself a phenomenon of interest, for the ephemeral nature of salt lakes would seem likely to be a feature operating *against* such speciation. Williams (1970) has suggested that rapidity of evolution, a consequence of the shallow survivorship curve of the fauna of salt lakes, may be an important factor in this connection favouring speciation.

THE FLORA

Highly saline lakes have no submerged macrophyte flora and, as noted by Sculthorpe (1967), most freshwater macrophytes are in general very susceptible to increasing salinity. A few freshwater genera are tolerant, for example *Myriophyllum* and *Potamogeton*, and these may extend into slightly saline waters. In addition, there are a few species which are characteristic of moderately saline lakes. In southeast Australia these are the angiosperms *Ruppia maritima*, *Lepilaena preissi*, and *Potamogeton pectinatus*, and the green alga *Enteromorpha*. Salinity data are almost entirely lacking, but *Ruppia maritima* appears to be the most tolerant, having been recorded from Lake Gnotuk at a salinity of *ca.* 56‰. *Enteromorpha* can also tolerate considerable salinities and has been recorded from Lake Corangamite at a salinity of *ca.* 27‰

(Hussainy 1969). Both taxa, as well as *P. pectinatus*, have been recorded too from coastal brackish lakes, but other angiosperm macrophytes characteristic of such situations in Australia, for example *Zostera* and *Halophila* (Higginson 1966), do not seem to occur in inland waters.

A wider variety of macrophytes is associated with the terrestrial margins of Australian salt lakes over the whole range of salinity of these. Many of them also grow in coastal situations; Chapman (1960) has recently discussed this sort of flora in great detail. Willis (1964) recorded sixteen such halophyte species in inland situations in western Victoria, including *Salicornia australasica*, *Suaeda australis*, *Pratia platycalyx*, *Wilsonia rotundifolia*, and *Selliera radicans*. On drying saline mud, *Angianthus preissianus* and *Cotulus vulgaris* occur ephemerally. *Suaeda australis*, the austral seablight, is perhaps the commonest and most characteristic halophyte growing marginally at southeast Australian salt lakes; it was formerly regarded as conspecific with *S. maritima* of almost world-wide distribution, but is now recognized as a separate and endemic species.

A reasonably large number of phytoplanktonic algae have been recorded from salt lakes, although in general the diversity is clearly and inversely correlated with salinity. The most characteristic form in highly saline lakes world-wide is undoubtedly *Dunaliella*, a chlorophycean, and species of this may occur at salinities approaching saturation. It was the only alga recorded by Hussainy (1969) in three lakes in Victoria whose salinity was mostly in excess of $100‰$. At lower salinities the algae most frequently recorded in southeast Australia include the diatom *Chaetoceros*, and the blue-greens *Nodularia spumigena*, *Anabaena circinalis*, *A. spiroides*, *Spirulina subsalsa*, and *Microcoleus lyngbyaceus*. Undoubtedly more occur, especially those $<60\mu$ in size, that is smaller than those normally caught by conventional phytoplankton nets. The substantial contribution to total production in Lake Werowrap, Victoria, by *Gymnodinium aeruginosum*, which is considerably smaller than 60μ, has already been mentioned (Chapter 2).

Nodularia spumigena appears to be one of the more important species in terms of algal biomass in most moderately saline Victorian lakes (Hussainy 1969); it has been recorded at salinities between 6 and $60‰$ in these lakes. Despite decreased algal diversity, it may be noted, chlorophyll *a* values—which serve as *estimates* of phytoplankton standing crops (see Chapter 2)—in Victorian saline lakes are as high or higher than are those found in many freshwater eutrophic lakes. The range of values recorded by Hussainy in seven lakes with salinities $> 10‰$ was $1-970 mg/m^3$.

As has been clearly shown by Wetzel's (1964) work, the importance of periphytic algae in shallow salt lakes must not be underestimated. A wide variety of such algae is known (cf. Castenholz 1960), but we have little specific knowledge concerning those which occur in Australian salt lakes. We note, however, that in a short (five-week) experiment conducted by one of our students, Miss C. Kleemeier, in 1968 and involving the suspension of glass microscope slides at various depths in Lake Gnotuk, Victoria (salinity $59‰$), the following diatom genera occurred: *Navicula*, *Amphora*, *Cocconeis*, *Amphipleura*, and *Cyclotella*. The curve relating vertical distribution and total periphyton biomass predictably showed the same general shape as that in Fig. 2:2. The actual results are reproduced in Table 10:4. In the slightly less saline but more alkaline Lake Werowrap Dr K. F. Walker (pers. comm.) has recorded as periphytes the diatoms *Navicula* and *Nitzschia* and (rarely) the chlorophycean *Cladophora*, and as benthic algae the blue-greens *Oscillatoria* and *Schizothrix briessii*.

TABLE 10:4
VERTICAL DISTRIBUTION OF PERIPHYTON BIOMASS ON GLASS SLIDES IN LAKE GNOTUK, VICTORIA, WINTER 1968

Depth (m)	1	2	3	4	5	6	7	8	9	10	11	12	13	17	18
Periphyton dry wt (mg cm^{-2})	0·6	1·1	1·5	1·2	0·4	0·1	0·1	<0·1	<0·1	0·1	0·2	<0·1	0·1	<0·1	0·1

DERIVATION AND AFFINITIES OF THE FAUNA

To many it may come as a matter of surprise to learn that most animals in athalassic saline waters are more closely related to freshwater forms than to marine or marine-brackish ones, yet this is undoubtedly the case, as was pointed out by Beadle (1943) many years ago. This is closely and quantitatively documented with respect to the calanoid copepods of non-marine saline waters in southeast and southwest Australia (Bayly 1967b, table 3:2; Bayly and Arnott 1969, table 1). *Calamoecia clitellata* and *C. salina* must be interpreted as the end products of a morphological reduction series that has accompanied physiological evolution, enabling the successive occupancy of marine through brackish and fresh to athalassic saline environments. In order to read the series in the opposite direction it would be necessary to postulate that Australian athalassic saline waters constitute the original home of the calanoid family Centropagidae—a most improbable if not absurd proposal in view of the specialized and physiologically difficult nature of such environments. But irrespective of the direction of evolution it is clear that the closest relatives of the two salt-lake species of *Calamoecia* are the more numerous freshwater species in the same genus. Similarly, all the ostracods, which are so prominent and frequently dominant in highly saline lakes, belong to the predominantly freshwater subfamily Cyprinae. Likewise, most if not all of the insects, the amphipods, *Microcyclops arnaudi*, and the rotifers of inland saline waters are of freshwater descent.

Next there is a smaller group of salt-lake forms that have close affinities with marine or marine-brackish water species. In Australia all of the harpacticoid copepods and the cyclopoid *Halicyclops ambiguus* belong to this category. This group is of greatest importance in non-marine waters that are quite close to the sea (Bayly 1970a). Several of these essentially marine species, including the polychaete *Ceratonereis erythraeensis* and the calanoid copepod *Acartia clausii*, are not capable of establishing self-perpetuating populations in athalassic saline waters, being susceptible to extinction when extreme cond˙ions occur. They cannot be re-established without re-introduction occurring from a marine source. Sea-birds probably play an important part in this process.

Finally, there is a very small group of only one or two species of terrestrial or semi-terrestrial descent which have secondarily become adapted to athalassic saline waters. The isopod *Haloniscus searlei*, which is unique to Australia, is almost certainly of this type. *H. searlei* belongs to an almost exclusively terrestrial taxon, the Oniscoidea, but is truly aquatic in athalassic saline waters. There is both morphological (Chilton 1920, Ellis and Williams 1970) and physiological (Bayly and Ellis 1969) evidence that this species is descended from a terrestrial ancestor. The isosmotic value for the haemolymph (Fig. 1:5B) is in close agreement with that found in most terrestrial isopods. The gastropod *Coxiella striata* may be another example.

This belongs to the family Truncatellidae and is closely related to marine littoral forms such as *Acmea*. The direction of evolution within this family may have been as follows: marine littoral → marine supra-littoral → (semi-)terrestrial → athalassic saline supra-littoral → athalassic saline.

THE FAUNA OF SALT LAKES IN NORTHERN AUSTRALIA

Little investigation has been made of salt lakes in Queensland, the Northern Territory, and northern parts of Western Australia, although many such lakes do occur in these regions. However, the little information that is available suggests that their fauna is much less distinctive than that in the southeast and southwest of Australia and that its affinities are perhaps more with the northern hemisphere.

In January 1965 the present authors visited Lake Buchanan, a large salt lake in Queensland between Aramac and Charters Towers. The salinity at this time was $87 \cdot 6\%_{00}$, the cationic order of dominance was Na > Ca > Mg > K, and chloride completely dominated the anions—almost negligible amounts of HCO_3 and SO_4 were recorded. These previously unpublished chemical findings are in very close agreement with those of Dunstan (1921), except that at the time of his examination the salinity was only $29 \cdot 6\%_{00}$.

In January 1965 the fauna consisted of an enormous number of small ostracods (as yet unidentified) and in addition the copepod *Apocylops dengizicus*. *Parartemia* was also common. This record of *A. dengizicus* has been previously documented by Kiefer (1967). A large number of empty shells of the gastropod *Coxiellada gilesi* (cf. *Coxiella* in southern Australia) were collected from Lake Galilee, another large lake south of Lake Buchanan, which was dry in January 1965 but is slightly saline when it exists.

A. dengizicus is an halobiont species that is very widely distributed in the Palaearctic region. Rylov (1963) mentions its occurrence in Iraq, India, Haiti, Egypt, and U.S.S.R. He also lists California but this record refers to the occurrence of the possibly distinct species *A. dimorphus* in the Salton Sea. Rylov also states that this species is peculiar to saline waters of desert, semidesert, and, in part, steppe, and records it from water on the Muganskaya steppe with a T.D.S. concentration of $25 \cdot 9\%_{00}$. *A. dimorphus* occurs in the Salton Sea during summer and autumn months and during the years 1954–56 inclusive tolerated a salinity range of about 30–$33\%_{00}$ (Carpelan 1961a, 1961b).

NON-MARINE SALINE WATERS IN NEW ZEALAND

For the most part New Zealand is well watered and athalassic saline waters are very scarce. There seems to be only one natural non-marine salt lake worthy of the name. This is a small (A = 3·7 ha), shallow, temporary lake (Salt Lake) located in the double rain-shadow of the Southern Alps and the Rock-and-Pillar Range near the township of Sutton in Otago (Bayly 1967a) (Plate 10:4). There are also a few temporary saline ponds and pools to the north of this, near Patearoa. In addition, some of the pools and springs associated with the geothermal activity in the Rotorua district appear to be slightly saline.

On 25 January 1966 Salt Lake had a salinity of $15 \cdot 2\%_{00}$. The ionic composition at that time is shown in Table 10:1. As is usually the case in Australia, the cationic order of dominance was Na > Mg > K > Ca and the anions were strongly dominated by chloride. The fauna in January 1966 consisted of the rotifer *Brachionus plicatilis*,

NON-MARINE OR ATHALASSIC SALINE WATERS

the cyclopoid copepod *Microcyclops monacanthus*, the ostracod *Diacypris* sp., and the dipteran *Ephydrella* sp. *Microcyclops monacanthus* is very closely related to but probably distinct from the Australian species *M. arnaudi*. Kiefer refers both species to the genus *Metacyclops* but there seems to be no good reason for separating *Metacyclops* from *Microcyclops* (Bayly 1967a).

The saline ponds and pools near Patearoa are highly alkaline and owe their salinity mainly to sodium (bi)carbonate. In January 1966 the fauna of one of these consisted of the three hemipterans, *Sigara arguta*, *Anisops wakefieldi*, and *A. assimilis*, and the dipteran *Ephydrella novaezealandiae*. Dumbleton (1969) described another species of *Ephydrella*, *E. thermarum*, from hot (40–50°C) 'mineralized' springs in central North Island. Anion data only were given for one of its habitats but on the basis of this it seems likely that it is sometimes slightly saline. Dumbleton also recorded two other ephydrids, *Neoscatella vittithorax* and *Scatella* sp., a culicid, *Culex rotoruae*, and a chironomid, *Chironomus* sp., from 'mineralized' waters of normal temperature. It is not at all clear, however, what the salinity range for any of these species is.

A large number of artificial salt lakes occur at Lake Grassmere near Cape Campbell in the South Island in connection with the commercial production of salt by solar evaporation of sea water. Apparently *Artemia salina* occurs in many of the evaporation ponds at Lake Grassmere, but this species is not a New Zealand native.

SALT LAKES AS ECOSYSTEMS

Understandably, most investigations of the biota of salt lakes have been concerned with either the physiological adaptations of the fauna or the general nature and relationships of the biota. However, it has recently been suggested (Williams, 1969c, 1970, 1972) that salt lakes and their biota are as interesting, and in certain respects more interesting, when approached as examples of total ecosystems. As such they have several significant features which make studies of them of wider importance.

Many of the foregoing sections in this chapter have—at least implicitly—indicated that as the salinity of a salt lake rises, the diversity of the species present decreases. This phenomenon, in conjunction with the decreased degree of habitat heterogeneity, the usual absence of macrophytes, the general shallowness, and, as a consequence of being the termini of endorheic drainage basins, the discrete and closed nature of the basins of salt lakes, generates ecosystems which compared to most freshwater ones are *relatively* simple to study. This has important consequences for those ecologists interested in trophic-dynamic and related aspects of aquatic ecosystems.

Additionally, because salt lakes are subject to various degrees of seasonal and longer-term fluctuations in salinity, they also represent a spectrum of ecosystems subject to a wide range of environmental perturbation. If, then, such perturbations (which clearly have profound biological repercussions) are regarded as a form of 'exploitation' (*sensu* Margalef 1968) tending to maintain the ecosystem upon which they operate in a state of ecological simplicity, a series of salt lakes spanning a range of salinity may be regarded as a series of ecosystems spanning a range of ecosystem complexity. The most saline will represent a simple ecosystem, the least a complex one; or, in the terminology of Margalef (1968), the range is from 'immature' to 'mature' ecosystems respectively.

The range of ecosystem complexity may also provide, it has been suggested, a range of suitable ecosystems from which to withdraw known components for laboratory, gnotobiotic study of micro-ecosystems. The contribution that such studies in general

Plate 10:4   A view of Salt Lake, near Sutton, South Island, New Zealand, looking westward. The Rock-and-Pillar Range is in the background. *Photograph by I. A. E. Bayly*

may make is potentially very high (Taub 1969a, 1969b), although so far they have mostly used components withdrawn from rather complex natural ecosystems, with obvious consequential difficulties of extrapolation and prediction from results. Nixon (1969), it may be noted, has already been able to generate a laboratory microecosystem using a small number of biotic units from a salt lake.

part six
MAN AND INLAND WATERS

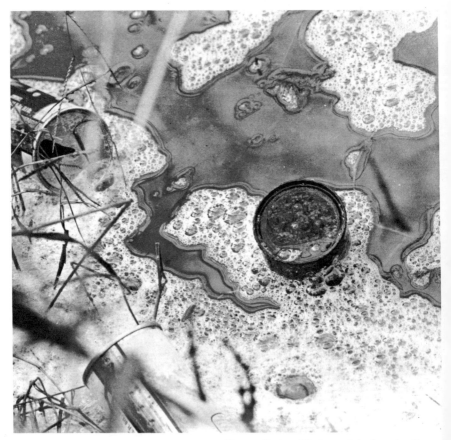

River pollution. *Photograph by Lorraine Symes*

chapter eleven
USES OF INLAND WATERS

In many books like this applied aspects of limnology are not fully or are scarcely discussed. Since, however, inland waters represent such an important basic resource, particularly in a country so arid as Australia, and since in any event the division between pure and applied ecology is an unnatural one (Williams 1969b), this chapter and the next are not considered out of place. Whether the volume of inland waters will become a limiting basic resource in Australia is uncertain, but there is no doubt about the present and continued importance of inland waters to mankind.

Inland waters are used for nine main purposes: for domestic, industrial and agricultural consumption, for generating hydro-electricity, for navigation, fishing and recreation, for aesthetic purposes, and for the harvesting of salt.

DOMESTIC SUPPLIES

Presumably for a considerable time after evolution as a distinct species, *Homo sapiens* made no provision for the storage of drinking water and like almost all other animals was directly dependent for supplies upon local pools, lakes, streams, and rivers. With increased communal living such sources frequently became insufficient or too unreliable, and at some time in prehistory water began to be stored. At first storage was confined to easily transported containers, but later was effected in much greater volumes in suitably altered natural basins or in entirely new constructions. In time, wells and aqueducts also appeared; classical Rome, for example, had some 600 km of aqueducts to provide each citizen with about 200 l of water daily (Peterson 1966). Man is now long past this early stage of domestic water supply, and the recent claim is perhaps not unjustified that 'The provision of fresh water is almost certainly the most important technical problem of our time. To provide it cheaply presents the greatest challenge to the world's scientists and engineers . . .' (Kronberger 1966).

There are six sources of domestic water: perennially flowing streams and rivers, lakes or reservoirs, artificially collected and stored rain water, ground water in river and stream substrata, subartesian and artesian bores, and by desalination of saline (usually marine) water. The extent to which these sources individually contribute, if at all, to the total supply varies a great deal according to local conditions. Thus in southeastern England a substantial contribution is derived more or less directly from the River Thames, whereas in Australia most of the big cities derive their main supply from large reservoirs. Privately stored rain water makes an insignificant contribution to domestic supplies in western Europe, but is of considerable importance in many other parts of the world and is of some significance in Australia; the domestic supply for one of us and his family was based for three and a half years entirely upon rain water collected from the roof of his house. In some parts of Western Australia large sloping areas of ground are covered with an impermeable material and run-off is used for domestic and other supplies (Fernie 1930).

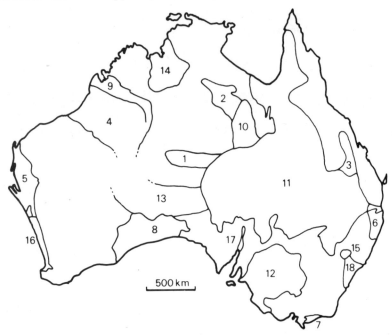

Fig. 11:1 Major subsurface water basins in Australia. 1: Amadeus Basin; 2: Barkly Basin; 3: Bowen Basin; 4: Canning Basin; 5: Carnarvon Basin; 6: Clarence Basin; 7: East Gippsland Basin; 8: Eucla Basin; 9: Fitzroy Basin; 10: Georgina Basin; 11: Great Australian Artesian Basin; 12: Murray Basin; 13: Officer Basin; 14: Ord-Victoria Region Basins; 15: Oxley Basin; 16: Perth Basin; 17: Pirie-Torrens Basin; 18: Sydney Basin. *Redrawn and modified from* Official Yearbook of the Commonwealth of Australia (*1962*)

Ground waters contained in river and stream substrata provide important sources in some of the drier areas of Australia; nineteen towns in Queensland, for example, derive their domestic supplies from such sources. Sometimes sources of this nature may be very large even in very dry areas, and the four pools of the Millstream group of springs in the arid Hamersley area of Western Australia (average annual rainfall ca. 30 cm), for example, are said to discharge 836 × 10^6 l of very fresh water daily into the sands of the Fortescue River (Simpson 1928).

Subartesian waters (with non-flowing bores) and artesian waters (with flowing bores) occur widely, and in Australia we know relatively a good deal about them (e.g. Reports of Interstate Conferences on Artesian Water 1913, 1914, 1921, 1925, 1928). The major Australian subsurface basins are plotted in Fig. 11:1. The most

USES OF INLAND WATERS

extensive is the Great Australian Artesian Basin, which is said to be the largest in the world. Many subartesian and artesian basins are not suitable for domestic purposes, or are only partly so if, for example, they yield hot but mineralized water. This appears to be the case more frequently in Australia than in countries of higher rainfall. Even within a given basin the quality of the water is often widely variable.

Desalination of salt water for domestic supplies is an economically feasible procedure in certain parts of the world, for example the Middle East, where alternative supplies are lacking. Nevertheless high production costs make it an uneconomic process where other supplies are available. The advent of cheap nuclear power would of course alter this balance.

Large storage reservoirs or impoundments[1] along the course of rivers are the most important source of domestic water supplies both in Australia and other countries. Because of their fundamental similarity to natural lakes it is worthwhile enumerating some of the main factors which are considered in their siting, construction, and maintenance; many such factors are derived from limnological principles discussed in Part II of this book.

Essentially water supply engineers are concerned to produce waters with low amounts of dissolved and particulate matter. Such waters occur in oligotrophic lakes. It follows that in the construction of reservoirs engineers should seek to reproduce as far as is feasible in any given situation conditions conducive to the development and maintenance of oligotrophy. Knowledge of the causative factors of oligotrophy are important here, and these all underpin the fact that oligotrophic lakes are those with low productivity. In the siting and construction of reservoirs, consideration should therefore be given to:

1 surface area/volume relationships—clearly the smaller this fraction, the less productive the reservoir will be in proportion to the volume of stored water;
2 siting the reservoir in non-agricultural areas which, apart from precluding the entry of excess fertilizers (and weedicides!), is also likely to mean that the catchment area itself will contribute only small amounts of nutrient material; and
3 clearing the actual basin of most vegetation before it is filled with water, as was done at Warragamba Dam, N.S.W., but not, with consequent deleterious results on the oxygen content of the outflow during the first years, at Tinaroo Falls Dam, Queensland.

In the maintenance of oligotrophy, minimizing nutrient inflow is the cornerstone of good water supply engineering. Afforestation of the catchment area is important here, for not only does this reduce erosion, that is nutrient inflow, but it also means that run-off is more even because of the greater water-absorptive powers of forest soils with consequently less likelihood of floods. Catchment grazing is an activity likely to lead to increased nutrient inflow and is therefore best minimized; grazed pasture is more subject to erosion than forest, and of course there is also the question of faecal contamination. Periodic cropping of reservoirs, especially by the removal of fish, is a practice that has a sound ecological basis and although not widely undertaken in

---

1. Many Australian government departments concerned with water storage officially refer to impoundments as lakes. Thus some five years ago the decision was made in Victoria to change the name of all impoundments previously called 'reservoirs' to 'lakes'. It seems that lakes are politically more 'glamorous' than reservoirs!

Australia is sometimes done elsewhere. Settling tanks on reservoir inflows, and the drawing-off of different strata of water according to season and the nature of chemical stratification, are two other measures of some importance in the maintenance of oligotrophy.

We have inferred imprecisely above that domestic water supplies contain low amounts of dissolved and particulate material. Table 11:1 contains some quantitative information on this matter with respect to dissolved materials. It refers to the standards used by the New South Wales Department of Health and derived from World Health Organization figures. With reference to this table, however, it should be noted that, apart from certain poisonous solutes (Pb, Se, As, Cr, cyanides), the quality of drinking water depends not so much upon the values of single solutes but rather upon the values of all in combination, so that some solutes in acceptable domestic water frequently exceed recommended values. The limits for total dissolved solids are likewise rather arbitrary. Many country towns in New South Wales, for example, have domestic water supplies containing far above 500 mg/l T.D.S., and about 2 per cent of Queensland's population receives water having more than 1,000 mg/l T.D.S. The highest T.D.S. concentration in water which can be consumed by humans (with salts in the same proportion as in sea water) appears usually to be about 6,000 mg/l, but may be a little higher in special circumstances. The human taste threshold between fresh and salty (sodium chloride) water is about 1,000 mg/l, but can be as low as 200 or as high as 3,400 (Richter and MacLean 1939, Bayly 1967a).

TABLE 11:1
DRINKING WATER STANDARDS OF THE NEW SOUTH WALES DEPARTMENT OF HEALTH[a]
All data except those of pH expressed as mg/l

Solute	Permissible	Excessive
Total dissolved solids	500	1,500
Calcium	75	200
Magnesium	50*	150
Sulphate	200*	400
Chloride	200	600
Iron	0·3	1·0
Manganese	0·1	0·5
Copper	1·0	1·5
Zinc	5·0	15
Phenolic substances	0·001	0·002
Lead	—	0·05
Selenium	—	0·01
Arsenic	—	0·05
Chromium	—	0·05
Cyanide	—	0·2
pH	7·0–8·5	<6·5 or >9·2

[a] According to Griffin (1964). However, some values given by Griffin (1964, Table 2) are clearly erroneous and have been replaced directly by those published by the World Health Organization (1963) (replacement values indicated by *). Additionally, there are differences between the values given by Griffin for lead, selenium, arsenic, and cyanide (none for chromium), and those given by the World Health Organization (1963) for these solutes. Only the latter are tabulated and refer to maximum allowable concentrations.

USES OF INLAND WATERS

Two major sorts of particulate matter in domestic waters are of especial concern to water supply authorities: certain sorts of bacteria and freshwater algae. All fresh waters contain free-living bacteria, but sometimes there also occur bacteria which are normally to be found in the intestine of man and other warm-blooded vertebrates. These can survive free in water for significant periods, and their presence in water has therefore become a standard procedure by which the degree of faecal contamination is judged. The coliform bacterium *Escherischia (Bacterium) coli* I is particularly important in this respect; Chapter 12 discusses more fully its use as an indicator of pollution.

Freshwater algae in domestic supplies give rise to concern mainly when they impart unpleasant odours or tastes or when they clog purifying filters. Occasionally, also, algal toxicity may cause concern. Some common algae implicated in the production of unpleasant odours and tastes are *Synura, Dinobryon, Asterionella*, and many blue-greens, especially *Oscillatoria, Anabaena, Anacystis*, and *Aphanizomenon. Anabaena* and *Anacystis* have also been reported as toxic to man (Palmer 1962). By and large few green algae are implicated. The most serious filter-clogging algae are diatoms and especially species of *Asterionella, Fragilaria, Tabellaria*, and *Synedra*.

The type of treatment that raw water undergoes to yield supplies for domestic use depends largely of course upon the characteristics of the original source. Here only passing mention need be made of the various types. The most important are those which result in the reduction or elimination of harmful and pathogenic bacteria and other organisms, and in this connection storage, filtration, and chlorination are the most important. Other treatments frequently involved are chemical coagulation, flocculation, sedimentation, and chemical correction. In chemical coagulation the normal coagulant used is aluminium sulphate, but sodium sulphate, ferrous sulphate, hydrated lime, and other coagulants are occasionally employed. Chemical correction includes such processes as water-softening, chemical treatment to reduce corrosive activity, and addition of activated carbon, potassium permanganate, or other substances to eliminate objectionable tastes and odours. Filtration is commonly of two sorts: rapid sand filtration (in combination with chemical coagulation) and slow sand filtration in which the film of bacteria and algae on the surface of the sand grains plays an especially important role. One of the problems associated with the management of water-treatment filters is that of clogging by algae. Sometimes algal clogging can be controlled by the use of algicides such as copper sulphate in the original source, but in some cases it is more practicable to use, temporarily at least, less contaminated water. One of the important tasks of biologists employed by water supply authorities is the monitoring, prediction, and sometimes control of phytoplankton densities in waters stored in reservoirs.

INDUSTRIAL SUPPLIES

Much of what has been written above is applicable here too, for usually no distinction is drawn between domestic and industrial supplies, the same water frequently going to both sorts of customers. Two points, however, may be accorded special mention. The first is that industry is very thirsty, thirstier in specific cases than the human population it supports. It has been estimated, for example, that about 300 m^3 of water are needed for the production of one metric ton of paper. And the second is that industrial waters have different quality criteria according to the industry involved. Some, those for instance used in food production, must have a very high degree of purity; others, such as those used primarily for cooling, need be of only low quality.

The total amount of water used for both domestic and industrial purposes varies greatly from country to country. Generally, however, there is a trend for increased relative consumption with increasing affluence (in turn associated with industrialization). Data for Melbourne, a city of some 2 million inhabitants, will indicate the tremendous volumes involved (Victorian Yearbook 1968). In the year 1965–66 the total consumption was 300×10^6 m^3, equivalent to a daily average consumption of $0 \cdot 8 \times 10^6$ m^3 or about 370 l per day per head of population. Sydney was slightly more profligate, the daily consumption per head of population being about 410 l. In North America the average daily usage is about twice this figure. However, large increases in Australian usage of domestic and industrial supplies of water have been predicted (Callinan 1969).

AGRICULTURAL SUPPLIES

Water for agricultural purposes serves two uses: stock consumption and irrigation. It is convenient to discuss these separately.

Table 11:2 provides information on the approximate upper salinity limits for drinking water tolerated by various stock species. Clearly, sheep are the most tolerant, and all except pigs and poultry can drink water of greater salinity than can man. As a result in Australia, if not elsewhere, bore waters figure prominently as a source for stock supplies. Indeed it is only the presence of large artesian and subartesian basins containing water of relatively high salinity that has permitted the extensive development of pastoralism in the drier regions of Australia. The usual procedure in the case of subartesian bores is for the wind to provide the pumping power, and for the water to be temporarily stored in surface tanks or ponds from where it is reticulated to stock drinking troughs. Bore windmills and associated storages are characteristic features of the landscape in much of outback Australia. Artesian bores discharge directly into storage tanks or troughs. Stock water tanks often have a rich fauna, but have been little investigated by freshwater biologists.

TABLE 11:2
ARBITRARY UPPER LIMITS OF TOTAL SALTS IN WATER SUITABLE FOR DRINKING BY STOCK[a]
All data as mg/l

Stock	Salinity
Pigs	4,000
Poultry	4,000
Horses	7,000
Cattle	10,000
Sheep on dry feed	12,000
Sheep on green grass	15,000

[a] Based mainly on Griffin (1964).

Apart from total salinity the effect of individual ions must also be considered in determining the suitability of water for stock. In particular, magnesium should not exceed 500 mg/l for sheep and 300 mg/l for horses; the safe upper limit for fluoride is regarded as being about 6 mg/l although higher concentrations can be tolerated; and for nitrate a limit has been set of 30 mg/l although as much as 120 mg/l may still be satisfactory for some stock. If too much nitrate is present, then stock may succumb to methaemoglobinaemia, a blood disorder.

USES OF INLAND WATERS

Especially in the drier temperate regions of the world, including large areas in Australia (*ca*. 13,000 km²), irrigation is an important part of agricultural practice and uses significant amounts of water. It has in fact been claimed that the development of the kind of agriculture made possible by irrigation played a crucial role in the evolution of cities and the founding, therefore, of the social structure of civilized man. Large areas of the world are subject to irrigation, using either ancient methods and animal (including human) motive power in the absence of gravity, as in the so-called developing countries, or scientific methods and mechanical power, as, for example, in North America and Australia. Often irrigation is effected by direct withdrawal from rivers, but in Australia the high climatic variability, with consequent effects on river flows, has resulted in the construction of large numbers of storage reservoirs to provide more even flows (Plate 11:1).

Plate 11:1 Adaminaby Dam and Lake Eucumbene—a major feature of the Snowy Mountains Scheme in the Australian Alps. Lake Eucumbene has a capacity of $4 \cdot 7 \times 10^9$ cu m and a surface area of 145 sq km

The nature of crops grown under irrigation varies of course according to country. In Australia the principal crops grown under irrigation are (in general order of importance): pastures, cereals and related crops including rice and sugar cane, lucerne and fodder, vines and fruits, vegetables. One problem now associated with the production of these under conditions of intensive irrigation is increased soil salinity. Only passing mention of this phenomenon can be made here; briefly, it involves the gradual accumulation of salts originally present in the irrigating water, and is partly controlled by using water additional to the crop requirement and evaporation losses, this extra water being referred to as the *leaching requirement*.

HYDRO-ELECTRICITY

Moving water as a source of power has been used for many centuries, but the conversion of its kinetic energy to a readily transportable form, electricity, dates only from about the middle of the last century when the hydraulic turbine was invented. Since that time a great number of dams have been built upon rivers specifically to harness the available energy and, additionally, many turbines have been placed to use overflows from impoundments containing water stored principally for other reasons. The world potential hydro-electric capacity has been estimated at about $2,900 \times 10^6$ kW (Hubbert 1969).

Hydro-electricity is relatively much more important in New Zealand than Australia, and in absolute terms there is little difference between the two countries; about as much hydro-electricity is generated in New Zealand as in the whole of Australia. In New Zealand the total installed capacity for hydro-generators in 1967 was 2,257,000 kW. This represented 84 per cent of the total installed capacity for generators of all types. The corresponding values for Australia were 2,893,000 kW and 26 per cent. The mean loading factor in New Zealand would be higher than that in Australia, so that the actual amount of hydro-electricity generated would probably be greater. The Waikato River is almost totally harnessed for the production of hydro-electricity in the sense that almost the entire fall that occurs between the source (Lake Taupo) (Plate 11:2) and sea level is thus exploited; the eight generating stations along its length together have a static head of 292 m, whereas the altitude of Lake Taupo is 357 m. The ultimate combined capacity of these eight stations will be 1,045,000 kW. In the South Island the Waitaki River is also well exploited, where three stations (Waitaki, Benmore, and Aviemore) will eventually have a combined capacity of 865,000 kW. In addition, Roxburgh station on the Clutha River has a capacity of 320,000 kW while the first stage of the huge Lake Manapouri scheme is rated at 400,000 kW.

In Australia, Tasmania (total installed capacity of 860,000 kW in 1967) and New South Wales (1,561,000 kW) are the only two States which generate large amounts of hydro-electricity. That produced in New South Wales comes almost entirely from the Snowy Mountains Scheme (Plate 11:1), although this scheme, it should be noted, was designed to a significant extent to serve the needs of irrigation (Fig. 11:2). The full hydro-electric potential of the latter region and the northeastern highlands of Victoria has not yet been realized; plans relating to the Snowy Mountains and Kiewa schemes call for the installation of a total generating capacity of more than $4 \cdot 0 \times 10^6$ kW by 1975. After that date the relative importance of hydro-electricity in Australia will almost certainly undergo a continuous decrease. In Tasmania the major power stations are Poatina (250,000 kW), which is fed with water from the Great Lake, and

USES OF INLAND WATERS   223

Tungatinah (125,000 kW), fed by the Nive, Little, and Pine Rivers. Exploitation of the power potential of the Derwent River and its tributaries is virtually complete. The first stage of the Gordon River scheme which is scheduled for completion in 1975 will have a capacity of 240,000 kW.

Unlike water for domestic, industrial, and agricultural purposes, the quality of water driving hydro-electric turbines need not necessarily be of high quality. When, however, water is piped considerable distances, its quality may then become an important consideration. Thus one of the significant problems encountered by the Hydro Electric Commission of Tasmania results from the manganese content of some waters and the subsequent appearance of manganese deposits. These deposits may form a lining of up to about 1 cm thick which then causes significant friction in the pipes, thereby reducing the available kinetic energy.

The problem of manganese deposition is certainly not peculiar to Tasmania and appears to be world-wide. Most published reports on the problem implicate sheathed 'iron bacteria' (order Chlamydobacteriales) as the oxidizing agents. Recently, however, some careful work and exacting microscopy by Tyler and Marshall (1967a, 1967b, 1967c) indicate that stalked, budding bacteria (*Hyphomicrobium*) are of far greater significance in this role in Tasmanian pipelines, and, they suggest, this may also prove to be the case elsewhere. Their examination of slime from Kareeya pipelines in northern Queensland certainly supports this suggestion.

Plate 11:2   A view of Lake Taupo looking southwest from Taupo township. The source of the Waikato River is seen in the foreground and the snow-capped peaks in the background represent Mounts Tongariro, Ngauruhoe, and Ruapehu. *Photograph by Whites Aviation*

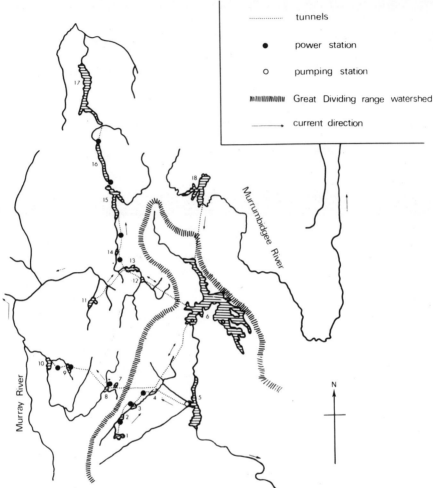

Fig. 11:2 Snowy Mountain Scheme, southeastern New South Wales. 1: Kosciusko Reservoir; 2: Guthega pondage; 3: Munyang pondage; 4: Island Bend pondage; 5: Jindabyne Reservoir; 6: Eucumbene Reservoir; 7: Windy Creek pondage; 8: Geehi Reservoir; 9: Murray Two pondage; 10: Khancoban pondage; 11: Tooma Reservoir; 12: Happy Jacks pondage; 13: Tumut pond reservoir; 14: Tumut Two pondage; 15: Tumut Three Reservoir; 16: Tumut Four Reservoir; 17: Blowering Reservoir; 18: Tantangara Reservoir

NAVIGATION

Man's use of inland waters as navigable highways ranks in antiquity next to their use for domestic purposes. This use must not be underestimated as a significant factor in man's history, for frequently it was along river valleys that human migrations occurred and in certain countries it was rivers that provided almost the sole pathways for exploration. Most of the early exploration of Canada was possible only by following rivers closely. Rivers were not so important in the exploration of Australia or New Zealand, but the Murray-Darling system formerly served as an important inland shipping route in Australia.

USES OF INLAND WATERS

Much faster land and aerial transport has now largely replaced this use of inland waters in most industrialized countries, especially so far as passenger transport is concerned, but in the less industrialized countries inland waters still play an important role as routes of communication and transport for goods and people. Nevertheless rivers and lakes in many industrialized countries still function for the transport of certain goods and to some extent as internal pathways for ships: the recently opened St Lawrence Seaway in North America is especially noteworthy. In some industrialized countries also, particularly England, the former extensive use of inland waters for navigation has left a legacy in the form of a network of linked canals—a remarkable, man-made limnological environment.

FISHING

Man fishes in inland waters for two reasons: for pleasure and to procure food. By and large these partly correlate with economic status. Fishing for pleasure is the major reason in more affluent countries; subsistence fishing is more often the reason in poorer countries. In the latter, fish may contribute significantly to available food resources, as for example in many parts of Southeast Asia, and in certain parts of Africa and South America. The greatest contributions occur associated with the tropical and semi-tropical rivers having an annual pattern of flooding. The fish production from the Grand Lac area of Indo-China has been estimated as about 150,000 metric tons *per annum* (Hickling 1961), no mean contribution to the total food resources of the region's inhabitants. Even in affluent countries the contribution of freshwater fish to human diets is not insignificant. The Canadian salmon, for example, is renowned as a table delicacy, and in Australia as well as New Zealand there are a number of freshwater fish of commercial interest. Table 11:3 lists these for Australia.

A variety of methods is used in fishing (readers are referred to Hickling [1961] for details). The simplest is found when fish become trapped in shallow lagoons which dry out; the fish are simply caught by hand. Usually, however, supplementary tools are used. For capturing single fish they include nooses, hooks, spears, and harpoons; in multiple fish capture various sorts of entangling nets, trawls, barriers, traps, and 'filtering' devices are employed. Poisons and explosives are also occasionally involved (illegally so in most countries[2]). Capture methods are to some extent determined by the physical nature of the waters containing the fish; obviously, rather different techniques are needed in running and in still waters.

A topic of interest in relation to subsistence fishing is *fish farming*. By this we mean the raising of edible fish in special (and generally artificial) impoundments in a manner analogous to the cultivation of other forms of meat. Fish farming is an extremely ancient human activity: it is probably older than 3,000 years. Apparently its origins lie in China, but it is known to have been undertaken in Europe from at least the Middle Ages. Nowadays fish farming is practised widely, particularly in Central Europe, the Middle East, Russia, China, and the Far East. In many of these regions it contributes valuable protein to human diets, but in the more affluent parts of Central Europe it continues mainly because it produces fish which have become

---

2. One of us well remembers the dismay after a long journey to some remote water-holes in the Macdonnell Ranges near Alice Springs on discovering that some local inhabitants (Caucasian) had used explosives to catch the fish in the holes.

TABLE 11.3
AUSTRALIAN NATIVE INLAND FISH OF COMMERCIAL OR RECREATIONAL IMPORTANCE[a]

Common name	Order	Family	Species	Importance	
				Commercial[b]	Recreational
Short-finned eel	Anguilliformes	Anguillidae	*Anguilla australis*	++	
Tasmanian whitebait	Clupeiformes	Aplochitonidae	*Lovettia sealii*	++	
Freshwater catfish	Cypriniformes	Plotosidae	*Tandanus tandanus*		+++++++
Macquarie perch	Perciformes	Serranidae	*Macquaria australasica*	++	
Golden perch			*Plectroplites ambiguus*	++	
Murray cod			*Maccullochella macquariensis*	++	
Spangled perch		Theraponidae	*Madigania unicolor*		
Silver perch			*Bidyanus bidyanus*	+	
River blackfish		Gadopsidae	*Gadopsis marmoratus*		
Australian bass		Macquariidae	*Percalates colonorum*		
Barramundi		Latidae	*Lates calcarifer*	+	

[a] Partly after Lake (undated).
[b] For year 1961–62 total freshwater fish production (excluding Queensland for which no data were available) was 572 metric tons.

esteemed as traditional delicacies, particularly on festive occasions. We may assure readers that the Anglo-Saxon snobbery concerning the eating of coarse fish has no reasonable basis.

In Europe and other temperate regions the principal fish cultivated is the carp, *Cyprinus carpio*, but the rainbow trout, *Salmo gairdneri*, and the tench, *Tinca tinca*, are also frequently cultivated. In the more tropical parts of Asia the principal species favoured belong to the genera *Tilapia*, *Puntius*, and *Ctenopharyngodon*. *Ctenopharygodon idella*, the grass carp, and several *Tilapia* species are important, as they are vegetarian and hence shorten the food chain, thereby increasing the total amount of energy fixed in a form consumable by man. In Israel, where fish farming has reached an extremely high degree of efficiency and organization, and in other parts of the Middle East a wide variety of fishes is farmed, including both *Cyprinus* and *Tilapia*. The Israelis, it may be mentioned, have had considerable success in farming fish in inland waters almost a third as salty as sea water.

With suitable techniques fish farming is at least as efficient—and in many cases much more efficient—a use of land as typical farming involving crops and livestock. In certain subtropical regions with poor soils the superiority of fish farming over normal agriculture may be as much as 20:1. Broadly, the productivity of fish ponds shows an inverse correlation with latitude. Some figures (from Hickling 1968, partly recalculated) to illustrate this and relating to unfertilized fish ponds are given in Table 11:4.

TABLE 11:4
PRODUCTIVITY IN UNFERTILIZED FISH PONDS AT THREE LATITUDES
From Hickling (1968), partly recalculated

Region	Annual Fish Production (kg/ha)
Northern Europe, Siberia	70–80
Southern Europe	200–250
Malacca (Malaysia)	314

Equivalent values for natural fresh waters in temperate regions may be expected to be *ca.* 20–30 kg/ha. With fertilization of fish ponds, threefold or fourfold increases in production occur, and in certain special environments higher values still have been recorded; for example, carp grown in the treated sewage effluent of Kielce (Poland) produced a crop of 1,300 kg/ha; and in Java, in raw sewage, a fish crop of 3,000–4,000 kg/ha has been recorded.

Fish farming on anything more than a limited scale does not occur in Australia or New Zealand. One of the reasons for this is undoubtedly the conservative eating habits Australians and New Zealanders have inherited from their (mainly) Anglo-Saxon forbears. The apparent absence of suitable native fish is another. In Australia *Cyprinus carpio*, although present, has been declared a noxious fish on account of its habits of deleteriously disturbing its environment, and its cultivation is now forbidden by law (see *Report of the State Development Committee on the Introduction of European Carp into Victorian Waters*). *Salmo gairdneri* has also been introduced to Australia, but despite its euryoecious nature (it can survive oxygen tensions down to 3 mg/l, and continue to feed up to 28°C) it is not farmed on a commercial basis. The tench, likewise introduced, has a rather patchy distribution in southeastern Australia, and appears to encounter certain spawning difficulties in the Australian environment (Weatherley 1962).

A discussion of fish farming leads naturally to a discussion of fish hatcheries. Formerly it was the widespread practice to raise artificially large numbers of fish of angling importance for subsequent release into natural waters. In this practice was the implicit assumption that there was a straightforward relationship between the numbers of fish available for anglers to catch and the initial number of young fish present. Such has now been shown to be *not* the case in most circumstances. Any given natural water can support only a certain weight of fish, and the addition of further amounts as hatchery-bred material is simply wasted effort; almost all of the additional fish are quickly lost to natural predators or are otherwise eliminated. One of the early Australasian investigations illuminating the futility of much hatchery practice—and one which is seldom referred to—was that of Hobbs (1948), who worked on New Zealand streams. Investigations elsewhere have shown that his conclusions are widely applicable. Nicholls (1958), referring to northwestern Tasmanian rivers, concluded that of every fifty-one trout taken by anglers, only one resulted from stocking activities: in general, stocking streams with hatchery-reared fish has an insignificant effect upon fish populations.

Although much hatchery work has now been thoroughly discredited, there are still some quite valid functions fulfilled by hatcheries. In Australia, for example, there are many inland waters which support large populations of salmonids and are consequently well-known and popular angling localities. They do not, however, possess suitable spawning regions, or for other reasons the fish population is not self-reproducing. The stocking of these localities is clearly one legitimate function of hatcheries. The hatchery operated by the Victorian Department of Fisheries and Wildlife is said to be the largest in the southern hemisphere.

How long man has engaged in fishing as a recreational activity is uncertain, although we may be sure that small boys have always! However, at least from the eleventh century onwards, angling has been recognized in Europe as an activity in which a man with some leisure and of contemplative bent may reasonably indulge. With increasing urban population densities and concomitant social pressures, and with increasing amounts of leisure time and ease of travel, the number of such men—both absolutely and relatively—has shown a marked increase in the past two decades in most countries of the western world. This has had a number of effects. One is that the provision of suitable tackle, bait, boats, and clothing now constitutes an exceedingly healthy industry in most western countries. Secondly, angling pressure has sharply increased at the more accessible waters and, whilst this has had some obvious direct effects, largely reversible, upon fish stocks, of perhaps greater consequence are the indirect effects upon the total environment: trampling of surrounding vegetation, litter, noise, disturbance of wildlife, pollution, and so on. Fortunately, overall population densities of man in Australia and New Zealand remain low in spite of the relatively high degree of urbanization, so that such indirect effects are not yet significant here.

Expectedly, the insecurity of Australia's first colonists created for them the urgent need for some feelings of connection with home, England, and it is not surprising therefore that they found in the unique Australian environment little to please them by way of animals and plants. At all events, despite the 'great Plenty of Fish' noted by

Sir Joseph Banks in 1770 (*Journals of the House of Commons* 1778–80), not many years elapsed before attempts were made to introduce fish of more familiar breeds (Weatherley and Lake 1967). The first to be introduced successfully were *Perca fluviatilis* (1862), *Salmo trutta* (1864), and *Carassius carassius* and *C. auratus* (ca. 1876). Later, *Salmo gairdneri*, *Tinca tinca*, *Rutilus rutilus*, and *Cyprinus carpio* were also successfully transplanted to the Antipodes. The most recent introductions to Australia have been of the mosquito fish, *Gambusia affinis*, now very widespread, which was introduced supposedly to control mosquitoes (1925), and the Quinnat salmon, *Onchorynchus tschawytscha*, which exists only in some closed Victorian lakes where it is continually re-stocked. Apart from these successful introductions, there have been many failures, notably the successive attempts to introduce the Atlantic salmon, *Salmo salar*. In addition, various people have tried (unsuccessfully) to introduce over fifty species of aquarium fish. Most of the successful Australian introductions have also been successfully introduced into New Zealand, which has a few species more as well (e.g. *Salmo salar*, which has had, however, only limited success) (Allen 1956).

From the recreational viewpoint, the most important Australasian transplant was *Salmo trutta*, the brown trout. Its introduction from Britain was finally achieved, after failures in 1852, 1860, and 1862, by packing eggs in moss and charcoal in a huge icebox on board the ship *Norfolk* which sailed from Falmouth on 28 January and berthed at Melbourne on 15 April 1864 (Wilson 1879). Subsequent hatching took place at Plenty, Tasmania, where the original hatchery is still functional, and from where continental Australia and New Zealand were initially stocked. Despite the number of species introduced, outside of Tasmania and the highlands of eastern Australia the most popular sport fish in Australia are native species. In general they belong to the perch-like group (Table 11:3).

Any discussion of fishing in inland waters is incomplete if it is restricted only to discussing representatives of the class Pisces: a wide variety of other animals is removed from fresh waters by man, whose activities in so doing are broadly referred to as 'fishing'. In Australia such animals include freshwater crayfish, crocodiles, and, formerly, the duck-billed platypus (see also Chapter 8). Freshwater crayfish occur widely in Australia and a few of the larger species are of some commercial importance as food. In the east the species principally involved is *Euastacus armatus*, the Murray River 'lobster', which may reach 2·7 kg in weight. *Astacopsis gouldi*, the giant Tasmanian crayfish, however, attains a slightly greater maximum weight; individuals may weigh 3·6 kg. In the southwest the principal species is the marron, *Cherax tenuimanus*. Both crayfish and palaemonid shrimps are fished for on a subsistence basis in New Guinea.

Two species of crocodiles occur in Australia, of which one, Johnstone's crocodile, *Crocodilus johnstoni*, is entirely restricted to fresh waters. It is found throughout northern Australia, where it is protected by law from fishermen in Western Australia and the Northern Territory, but not in Queensland. In this latter State its numbers are decreasing rapidly as a result. Its skin is of some value as a type of 'leather', and preserved juveniles, 'stuffers', are sold as souvenirs to tourists whose only sense of taste lies apparently in their mouth.

Following Australia's colonization the platypus was subjected to severe pressure because of the value of its pelt in rug-making. There is no doubt that it was quite common in eastern Australia at the beginning of the nineteenth century—it was certainly numerous at the time of Governor Macquarie's journey across the Blue Mountains in 1815—but by the middle of the same century its numbers had plunged, as recorded by Gould (1863). Now, although less abundant than formerly, because of complete protection its continued existence is no longer seriously threatened.

RECREATIONAL AND AESTHETIC USES

A discussion of the recreational and aesthetic uses of inland waters follows naturally from the preceding account of angling, with which indeed there is logically some overlap. Fishing apart, the main recreational uses of inland waters are for: bathing, skin-diving, water-skiing, rowing, sailing, powered boating, and duck-shooting. Little need be said of these except to emphasize their popularity, and to note their stimulation, like angling, of a large equipment supply industry. The popularity of water-based recreation shows a marked rise with increased affluence, so that in the United States over 40 per cent of the population is now said to prefer such recreation over any other (Mackenthun, Ingram, and Porges 1964). Increased public awareness of the use of inland waters for leisure finds expression in the apparent increase of government benevolence in many countries towards wildlife conservation, which frequently involves conservation of what are termed 'wetlands'. In Victoria it has justifiably been claimed that the initial impetus for the acquisition of wildlife reserves came from the urgent need to conserve natural inland waters.

Recreational and aesthetic uses are not easily discussed separately, for many people (including the authors) find an aesthetic appreciation of water bodies, whether still or flowing, sufficient 'recreation' in itself. At all events, there is no doubt that natural waters in certain situations are exceedingly appealing to humans or are otherwise emotionally evocative. Judging from the high value of real estate on lake edges in many parts of the western world, summer cottages, picnicking, and camping are all more attractive near water. And it was certainly not isolation alone which led John Ruskin, one of the greatest of English landscape connoisseurs, to spend, from 1871 onwards, almost thirty years of his life in the English Lake District. For those who must spend most of their lives in the city, ornamental ponds and lakes in municipal gardens provide welcome visual relief from pre-stressed concrete. The claim that the Royal Botanic Gardens in Melbourne is one of the finest landscaped gardens in the world is largely based upon W. R. Guilfoyle's skilful integration of lakes, trees, and lawns. There will be many of us in cities who know what W. B. Yeats meant when he wrote:

> 'I hear lake water lapping with low sounds by the shore;
> While I stand on the roadway, or on the pavements gray,
> I hear it in the deep heart's core.'

The aesthetic appeal of lakes may even be more important than we have hitherto thought. Hutchinson (1963: 688) had this to say: 'all lakes can play an increasing role in providing an aesthetically satisfactory background to human life, and it is by no means unlikely that, great as the immediate practical value of an increasing number of lakes may be, the less obvious role may prove to be the more important. This aesthetic aspect of limnology may well prove to be inversely associated with the increasingly significant problems of mental health that develop in highly industrialized and urban societies.'

USES OF INLAND WATERS

SALT SUPPLIES

Common salt, that is sodium chloride or halite, is used in large quantities for many purposes: for example, in the manufacture of other sodium salts, chlorine, hydrochloric acid, dyes, and some organic and other chemicals; in food industries such as meat-packing, canning, preserving, fish-curing; in leather, hide, and textile production; in the diet of man and livestock; and in various metallurgical processes. Commercially important sources are rock salt, lake-bed deposits in arid areas, and salt solutions (including the sea). In Australia, which has large salt resources, salt is produced from the solar evaporation of sea water (the biggest proportion) and from the harvesting of salt deposits from the beds of inland saline lakes which are dry either seasonally or for longer periods. Such deposits in Australia are almost entirely sodium chloride (Williams 1967a), although this is frequently not so in the salt lakes of other continents.

According to Barrie (in McLeod 1965), the principal Australian lakes from which salt is gathered are: *Queensland*—Lake Buchanan; *Victoria*—Spencer's Lake, Mystic Park, Lake Boga, Lake Tyrrell, Pink Lakes (Underbool); *Tasmania*—small lakes near Tunbridge and Ross; *South Australia*—Lake Fowler and other lakes on Yorke Peninsula, Lake Lochiel, Lake MacDonnell, Pernatty Lagoon, Lake Dutton outlet, Lake Yaninee; *Western Australia*—Pink Lake (Esperance Bay area), Lake Lefroy, lakes on Rottnest Island. Significant inland commercial sources of salt are not present in the Northern Territory or New South Wales although some inland deposits do occur. New Zealand likewise has no significant sources of salt from inland lakes (see Chapter 10).

Considerable expansion of salt production from Australian lake deposits is currently underway (Williams 1969b). Further information on this topic can be gained from Bain (1949), Betheras (1950), Dunstan (1921), Jack (1921), Kalix (1962), Barrie (in McLeod 1965), and Willington (1959).

chapter twelve
THE EFFECTS OF MAN ON INLAND WATERS

Man has greatly affected inland waters and the total limnological environment in a wide variety of ways. With exponentially increasing human populations the day is not far removed when there will be no inland water which will not to a greater or less extent have been altered from its natural condition by man. Concern for such a state of affairs is largely responsible for 'Project Aqua', an international effort to conserve as scientific reference localities a number of natural waters scattered throughout the world. Those localities in Australia and New Zealand provisionally proposed for such conservation on the basis of their scientific value are listed in Table 12:1. Already, however, some (e.g. Lake Pedder, Tasmania; Bayly 1965) are doomed to early death as natural environments, or have undergone modification (e.g. Red Rock lake complex).

It is convenient to discuss the effects of man under the arbitrary headings of direct effects, pollution, eutrophication, and indirect effects.

Direct Effects

The most obvious direct effect is the conversion by man of many naturally formed lakes into storage reservoirs and the creation of numerous artificial lentic water bodies loosely referred to as reservoirs, dams, weirs, ponds, tanks, or impoundments. The complete drainage of some lakes and other localities—often the most interesting or ecologically important—has tended to decrease the volume of natural waters, but in general the overall result has been to increase the *total* volume of inland waters. The additional water bodies, however, differ in many characteristics from natural ones. First, many have been created in regions where no natural lentic waters of any size exist; the Aswan Dam in Egypt springs immediately to mind and, in Australia, the numerous stock tanks and dams in central Australia which draw their water from bores. Even where natural waters do exist, these usually differ strongly in morphometry from artificial ones, since the latter are designed whenever possible to have a low area/volume ratio. In Victoria, for example, most of the larger lakes are exceedingly shallow (Lakes Hindmarsh, Albacutya, and Corangamite are all less than 7 m deep) whereas all reservoirs of comparable area are much deeper. There may also be chemical differences between natural and artificial waters, especially in drier regions, since stored water to be useful must be fresh. Certainly in Australia the total number, proportion, and degree of permanency of lentic waters containing less than 500 p.p.m. T.D.S. is much greater now than before the advent of Caucasian man, and many such waters exist in regions which naturally possess only saline lakes or ephemeral bodies of fresh water.

TABLE 12:1
LIST OF AQUATIC SITES IN AUSTRALIA AND NEW ZEALAND PROVISIONALLY PROPOSED FOR CONSERVATION ON THE BASIS OF THEIR SCIENTIFIC VALUE[a]

Position	Site
Australia	
Victoria	Lake Tarli Karng; anabranch lakes of Hattah Lakes system; Lake Gnotuk; Lake Corangamite; Red Rock lake complex
South Australia	Lake Eyre; salt lakes between Robe and Beachport
New South Wales	Little Llangothlin Lagoon; Lake Hiawatha; Lake Minnie Water; The Gap Lagoons; Kosciusko plateau lakes
Queensland	Lake Eacham; Lake Barrine; Lake Euramoo; The Crater
Tasmania	Lake Seal; Lake St Clair; Lakes Sorell and Crescent; Lagoon of Islands; Lake Pedder; Lake Fergus or Lake Ada; Lake Judd
New Zealand	
North Island	Lake Rotokawau; Lake D; Lake Rotowhero; Lake Rotomahana; Waimangu Thermal Reserve; Lake Okataina; Lake Rotopounamu
South Island	Waikoropupu Twin Springs; Lake Marion; Vagabonds Inn; Lake Christabel; Lake Mueller

[a] From compilation of H. Luther and J. Rzóska (1971): *Project Aqua*. Further details concerning the sites are given in this publication.

Secondly, the patterns of fluctuation in levels are frequently quite different in artificial and in natural standing waters. In impoundments created specifically for hydro-electric purposes there may be marked diurnal fluctuations resultant upon diurnal changes in demands for power; in those created to store irrigation water marked draw-offs may occur in the dry season (generally speaking, winter in northern Australia, summer in southern Australia), or, as occurs in many reservoirs in north-eastern Victoria forming part of a reticulated system, there may be little or no fluctuation in level since attempts are made to keep them always at 'full supply level'. Unnatural fluctuations in level may also occur in certain lakes whose levels are partly controlled to mitigate or prevent flooding, for example Lakes Hindmarsh and Corangamite in western Victoria. Hydro-electric authorities may induce similar effects upon lakes forming part of a single drainage system; a large number of lakes on the central Tasmanian plateau, including especially the Great Lake, provide a good example of such a system.

The direct effects of man upon the physical nature of many natural *lotic* waters have been equally as profound: the creation of unnatural flow regimes, straightening, dredging, and alteration of drainage patterns are some of these effects. Alteration of temperature may be another if impoundments occur on a river; the water in these takes longer to heat up in spring and to cool in autumn, with consequent effects upon the temperature of waters released downstream. Release of water from different levels within an impoundment may also result in unnatural temperatures, particularly when water is drawn off from the hypolimnion in summer. Size provides no escape and even some of the largest rivers can now hardly be considered truly natural waters. The River Murray, flowing through three Australian states, has several major dams across it and the pattern of flow has now become so altered that only the severest flood can replenish many lateral anabranch systems (e.g. the Hattah Lakes) which formerly depended for their existence upon semi-regular replacements of water during high winter flows.

In addition to alterations to natural running waters, man has created many new lotic environments mostly in the form of canals for navigation or the transport of water. They are usually quite different in character from natural watercourses and many are merely exposed earthen aqueducts. On the other hand, with time some may assume at least a quasi-natural appearance, as for instance the British system of canals which was laid down in the immediate post-Industrial Revolution era.

The biological effects stemming from these direct changes wrought by man upon the inland aquatic environment have been far-going. Unnatural fluctuations in level or flow frequently adversely affect the biota, either directly or indirectly by preventing colonization or breeding. Hynes (1961b) has shown how severely altered was the biological character of Llyn Tegid, Wales, following large-scale changes in water level. The absence of fringing emergent vegetation from most reservoirs and controlled lakes is usually due to fluctuating water levels. In Australian rivers it has been suggested that alterations of natural flow regimes have inhibited the breeding of many animals, including for example *Plectroplites ambiguus*, a fish whose reproduction is dependent upon natural floods (Lake 1967c). It has also been suggested that *Maccullochella macquariensis*, the Murray cod, has been largely replaced by introduced trout in certain waters of southeastern Australia because of lowered water temperatures resulting from upstream impoundments (Butcher in Williams 1967b).

Where canals link separate drainage systems, genetic introgression, competition, or predation may occur in populations that were formerly allopatric. Perhaps the best-known biological effect resulting from the creation of a canal is provided by the entry into the North American Great Lakes of the sea lamprey, *Petromyzon marinus*, via the series of locks which provided a route for this species around Niagara Falls (see Beeton and Chandler [1963] for a more complete account). Since its entry into the Great Lakes about 1921, populations of this parasitic fish have exploded, to the detriment of native species, particularly lake trout, *Salvelinus namaycush*. Commercial fishing for this has collapsed.

The direct effects of man on the inland aquatic environment have undoubtedly had important consequences for the distribution and abundance of much of the aquatic biota, consequences which, moreover, are probably still a long way from full attainment.

Pollution

Most readers will be intuitively aware of what is meant by the term 'water pollution'. It is, however, very difficult to give precision to such intuitive awareness and, since this book is not designed for use in courts of law, we shall not try. As a basis for discussion—no more—we provide the definition that pollution is a significant and deleterious change in the natural character of a water resulting from the addition of material or heat by man. The broad and imprecise nature of this definition will not escape the notice of critical readers. Other definitions are not so broad. Tarzwell (1965: 15) ties the definition to usage: 'water pollution is the addition of any material or any change in character or quality of a water that interferes with, lessens, or destroys a desired use.' Simmonds (1962: 76), referring to water pollution in Queensland, has an even narrower definition (despite his claim that it is a broad one): pollution is 'the presence in a water of any substance, physical, chemical or biological, which renders that water unsuitable or unfit for some domestic purpose.' Legal and official definitions of pollution, stripped of verbiage, are generally as imprecise as the one we provide.

Ever since men aggregated in settlements of any size, pollution has probably been of some consequence to natural waters. At first its extent was relatively minor but with increasing size of human aggregations, that is the advent of towns and cities, sewers were constructed and are known even from the ancient cities of the Indus valley— the centres of the Harappa civilization (2500–1500 B.C.) possessed an intricate series of drains (Allchin and Allchin 1968). The Romans, too, had an elaborate sewerage system in many of their towns. However, following the demise of their empire such systems fell into disrepair and subsequent disuse, and from then until about the middle of the nineteenth century pollution of European rivers and streams occurred with scant regard for either hygiene or aesthetics. Accounts, for example, of sanitary conditions in early London are horrendous.

In many parts of the world industrialization and consequent trade effluents greatly compounded the problem in the nineteenth century, and in Europe, England, and North America by the end of that century it was clear that authorities needed to pay attention to pollution control. In England this took the legislative form of several Parliamentary Acts beginning with the Gas Clauses Act (1847). Of especial note was the formation in 1898 of a Royal Commission on Sewage Disposal which sat for fifteen years and issued ten reports. The recommendations within these, whilst not spawning any specific Act, were generally accepted. The reports were concerned *inter alia* with methods of assessing degrees of pollution, and proposed as one such method what is now known as the five-day Biochemical Oxygen Demand test ($B.O.D._5$). This is a measure of the amount of oxidizable organic matter in the water. It has been widely adopted and is discussed in more detail later.

The precise form of control taken elsewhere depended upon the political structure of the country involved and the nature of the pollution. In Germany a number of regional authorities were formed to control both sewerage and drainage; the earliest of these was the Emschergenossenschaft (Emscher River Commission) formed and constituted under Prussian State Laws in 1904 (Mulvany 1951). Regional approaches were also a feature of early United States efforts to control pollution, although controls, for obvious reasons, appeared later; it was not until 1948, for instance, that the Ohio River Valley Water Sanitation Commission was set up to deal with the grossly polluted Ohio River. The recent (1966) Conservation Bill of the United States Government is a much more far-reaching anti-pollution effort.

History, demography, and geography have favoured Australia and New Zealand, and pollution of their inland waters until fairly recently was *relatively* unimportant. A feature of major significance was the coastal location of almost all major cities. Now the situation is rapidly deteriorating and marked environmental degradation has begun. At present in Australia pollution control is vested in various State departments, although the actual policing of legal requirements may be by local authorities. This has clearly been inefficacious and a Parliamentary Senate Select Committee has now recommended (June 1970) that a national water body concerned *inter alia* with water pollution be set up (*Report from the Senate Select Committee on Water Pollution* 1970).

A number of arbitrary classifications of pollutants have been proposed based either on source or ecological effects or both. We recognize the following: domestic refuse, organic trade residues, inert material, inorganic poisons, organic poisons, hot water, non-poisonous salts, radioactive wastes. Frequently, of course, waters contain pollutants of several sorts, so this list may not be applicable as a scheme of classification for the waters themselves. Even so, it is possible to distinguish waters affected predominantly by one or other sort of pollutant and so to recognize specific effects. We discuss first the general nature of each of the various types of water pollutants, then the major effects of pollutants on inland waters, and finally the ways in which the degree of pollution is assessed.

DOMESTIC REFUSE The two principal pollutants in this category are sewage and detergents. The former consists mainly of faecal matter and urine and is often subjected to varying degrees of purification before discharge as effluent. Complete treatment can yield water of drinking quality, but treatment is usually not carried so far. Commonly, three levels of sewage treatment are recognized: primary, secondary, and tertiary. Primary treatment is merely mechanical removal of solid material. Secondary treatment is mostly biological and may result in the removal of up to 90 per cent of degradable organic wastes. Tertiary treatment involves removal of plant nutrients and other contaminants. Untreated sewage is termed 'raw' sewage. Since it consists mainly of dissolved or suspended organic matter it provides a rich source of energy for bacteria and heterotrophs and this is responsible for its major effect upon receiving waters, that is deoxygenation, discussed in more detail later. Oxidation and mineralization are the two fundamental features of sewage treatment, so that treated sewage effluent contains less potential energy.

Detergents, too, are organic compounds but many, unlike sewage, are not easily broken down. This persistence has unfortunate effects and so new detergents with straight carbon-chain molecules as opposed to branched ones have been synthesized; these are as efficient but are far less stable. Many countries now insist upon the use of these less stable, 'soft', detergents.

Apart from large volumes of organic material, domestic refuse may contain significant numbers of pathogenic bacteria, helminth eggs, and other disease-producing organisms.

ORGANIC TRADE RESIDUES Many industries give rise to waste effluents containing organic material, particularly the 'food' industries such as dairies, abattoirs, breweries, and sugar-beet, canning, and fish-meal factories. Laundries, paper mills, tanneries, and textile and petrochemical factories are also important. The chemical stability of their effluents varies considerably; those from the food industries are relatively quickly open to bacterial attack, as is sewage, whereas others, notably from paper mills, may be much more stable. Carbohydrates, proteins, fats, amino-acids, cellulose, and wood sugars are the principal constituents but, according to the industry involved, admixtures of other substances, both inorganic and organic, may be present. Sulphites, for example, are often present in paper-mill wastes, and phenol in effluents from chemical factories. Oil, motor-exhaust wastes, and organic chemicals used to decrease evaporation from reservoirs are further organic pollutants that may be considered here. Oil is chiefly derived from road-washings, garages, and various sorts of factory machinery. Motor-exhaust wastes are derived from powered boats and may be important when they give rise to tainting of fish flesh. Chemicals to inhibit evaporation are used chiefly in Australia and other arid areas; they are fatty alcohols (cetyl and stearyl) and are spread upon the water surface in extremely thin films. It is claimed that they have an insignificant biological effect, but it is known that they can hinder insect emergence (United States Department of the Interior 1957).

INERT MATERIAL More or less chemically inert waste that is insoluble and either finally divided and particulate or in liquid form is found temporarily suspended in effluents from many sources. Coal mines may contribute coal dust in their washings; paper mills may give rise to wood pulp and fibrous material; petrochemical industries may give rise to heavy oil and tar fractions. In some countries china-clay is a well-known pollutant of this sort.

INORGANIC POISONS Above certain low concentrations, acids, alkalis, sulphides, cyanides, and salts of chromium, arsenic, lead, copper, zinc, and many other elements are poisonous to aquatic life. This notwithstanding, they are added to inland waters in many effluents. Gas-works, petrochemical factories, certain engineering industries, and mining are sources especially implicated. A noteworthy Australian example of pollution from mining wastes is provided by the now-closed zinc mine at Captain's Flat, about 48 km south of Canberra. Tailings, spoil, and washings from this mine have had a profound effect upon the adjacent river, whose pollution has received considerable publicity since it flows into 'Lake' Burley Griffin in Canberra. Pollution from the mine has been described by Weatherley, Beevers, and Lake (1967) under four separate headings: *i* acidity; *ii* base metals in solution [chiefly zinc]; *iii* thiocyanates, sulphides, cyanides, cresols, phenols, etc.; and *iv* finely divided tailings material. Pollution of type *ii* is the most important.

ORGANIC POISONS Of increasing significance as poisons are those organic compounds referred to as weedicides, herbicides, pesticides, insecticides, and so on. They are either refined from natural sources (e.g. pyrethrum—now also manufactured—rotenone, nicotine alkaloids) or manufactured specifically as poisons. Although most are used primarily in terrestrial situations, spray contamination, run-off, biological transference mechanisms, atmospheric transfer, and so forth soon extend their effect to the aquatic environment. This has been clearly illustrated by the careful investigation under the auspices of the Fisheries Research Board of Canada of fish and invertebrate populations in an area of New Brunswick which was sprayed with DDT[1] to control spruce budworm. The results have been summarized recently by Kerswill (1967). Briefly, there was extensive loss of wild fish within days of DDT spraying, exposed caged fish also died, aquatic insects generally ceased emerging, fish diets changed, there was delayed fish mortality, and adult stocks of anadromous fish declined in quantity. Even in Victoria, with low human population densities and large tracts of unsprayed land, it has proved impossible to find freshwater fish *without* pesticide residues in their bodies (Butcher 1965). A few samples of fish from supposedly DDT-free waters in New Zealand have shown tissue levels of DDT from nil to 0·7 p.p.m. (Hopkins, Solly, and Ritchie 1969).

Organic poisons are of two main structural types: chlorinated hydrocarbons and alkyl or organic phosphates. Examples of the former are DDT, chlordane, heptachlor,

---

1. DDT = dichloro-diphenyl-trichloro-ethane.

and the chlorinated naphthalenes, dieldrin, aldrin, and endrin; examples of the latter are parathion, malathion, and most systemic insecticides. Two other important sorts of organic poisons are the 'dinitro' compounds (e.g. DNOC[2]) and the carbamates (e.g. sevin). 2,4-D,[3] dalapon, trichloracetic acid, CMU,[4] to name just a few, are further compounds which must be mentioned since they have been used directly to control aquatic weeds. The State Rivers and Water Supply Commission of Victoria, for example, uses or has used all four as weedicides (Dunk 1957).

Methyl mercury, which results from the action of microorganisms on inorganic mercury discharged into natural waters, is a further organic poison that of late has caused concern about its effect in both the inland aquatic and marine environments.

The toxicity of organic poisons to aquatic life is awe-inspiring, and particularly vulnerable are many insect larvae and nymphs and small crustaceans. Table 12:2 illustrates the potency of various insecticides to the cladoceran, *Daphnia magna*, and a small North American fish, the fathead minnow (*Pimephales promelas*). The concentration of organic poisons actually within animal tissues may become many times greater than their initial concentrations in water. A classic example of such concentration or 'biological magnification' within the ecosystem is supplied by Clear Lake, California (a summarized account of published and unpublished data is given by Rudd 1964). DDD[5] (a compound related to DDT) was added to the lake in an effort

TABLE 12:2
TOXICITY OF VARIOUS INSECTICIDES TO *DAPHNIA MAGNA* (CLADOCERA) AND *PIMEPHALES PROMELAS* (PISCES)
Based on Anderson (1960) and Henderson, Pickering, and Tarzwell (1960)

Pesticide	D. magna[a]	P. promelas[b]
Aldrin	0·0292	0·033
DDT	0·0014	0·032
Dieldrin	0·33	0·016
EPN[c]	0·0001	0·2
Heptachlor	0·0577	0·094
Malathion	0·0009	12·5
Methoxychlor	0·0036	0·064
Parathion	0·0008	1·4–2·7

[a] Estimated concentration (p.p.m.) required to immobilize in 50 hours.
[b] Median tolerance limit (p.p.m.) after 96 hours' exposure.
[c] EPN = ethyl p-nitrophenyl benzenephosphonothionate.

to control a midge nuisance, *Chaoborus*; in 1948 it was applied at a concentration of *ca.* 0·014 p.p.m. and in 1949 and 1954 at *ca.* 0·02 p.p.m. After some years extensive mortality occurred amongst grebes on the lake and an examination of these and various other lake inhabitants revealed DDD residues in extremely high concentrations. The range in fish was generally 40–2,500 p.p.m., whilst grebes had about 1,600 p.p.m.! The explanation suggested is that concentration is brought about via the food chain, although with regard to fish it has recently also been suggested

2. DNOC = dinitro-ortho-cresol.
3. 2,4-D = 2,4-dichlorophenoxyacetic acid.
4. CMU = 3-(p-chlorophenyl)-1, 1 dionethylurea.
5. DDD = p,p¹-dichlorodiphenyldichloroethane.

(Mellanby 1967) that concentration occurred because of the need to 'breathe' large volumes of water.

HOT WATER Thermal pollution, that is the artificial elevation of water temperatures, is chiefly brought about by the use of river and stream waters for cooling purposes in electric power stations. The pickling of steel, washing of wool, cooling of gas, and other industrial activities may also produce heated effluents.

The extent to which water temperatures are raised by electric power stations depends principally upon the size and efficiency of the generating plant and the initial quantities of water available. According to Jones (1964), referring to British power stations, older inefficient plants may produce twice as much waste heat energy as more modern plants. When large quantities of water are available, the system might be designed for a rise in temperature of the coolant of approximately 6°C; when less water is available the system might be designed for a rise of the order of 8°C. A modern electric power station in Britain of 100,000 kW capacity whose effluent is raised 5°–6°C would use about 19×10^6 l of water per hour. Estimates of future requirements of cooling water by industrialized nations predict major problems in waste heat management (Christianson and Tichenor 1968).

Although an appreciable proportion (> 10 per cent) of the power generated in Australia is derived from hydro-electric sources (particularly in Tasmania) needing no cooling of the sort required by thermal power stations, most is derived from thermal sources and as a result thermal pollution of river waters does occur. Thus in Victoria the present generating plant of the State Electricity Commission located at Yallourn uses both cooling towers and water direct from the Latrobe River. There is a statutory temperature limit of 35°C at a checkpoint 4 km downstream and the cooling towers are used when the Latrobe River flow alone is insufficient for operation below this value (State Electricity Commission of Victoria, pers. comm. 25 February 1969). The smaller power station at Richmond, Melbourne, uses water from the River Yarra for cooling. Two major power stations on the Brisbane River in Queensland, Bulimba and Tennyson, with a combined capacity of 440,000 kW, also rely solely on river water for cooling. Nevertheless it seems that, generally speaking, thermal pollution in Australia is comparatively unimportant.

NON-POISONOUS SALTS Certain industrial activities and mining, notably for oil, may result in the production of brine wastes. These, whilst not directly poisonous at low concentrations, may so elevate salinities in inland waters that natural communities are destroyed or, according to species tolerance limits, significantly altered in composition.

RADIOACTIVE WASTES In certain parts of the world radioactive waste material from research laboratories, nuclear power stations, and other sources is of increasing significance as a pollutant. High- and medium-energy wastes are generally disposed of either at sea or by being buried, but low-energy wastes are more usually removed in aqueous effluents. Strontium-90 and cesium-137 are among the more common of such wastes, but an extremely large number of radionuclides (of varying length of life) may be disposed of in this way. Over sixty radionuclides have been identified in the effluents of the Hanford reactors to the Columbia River (Washington, U.S.A.), and of these twenty have been measured in river organisms collected within 1·6 km downstream (Davis, Perkins, Palmer, Hanson, and Cline 1958).

In general we may regard two sorts of radionuclide as being of biological importance: those which are isotopes of essential elements, for example cobalt-60, iodine-131, sodium-22, phosphorus-32; and those which have chemical behaviours analogous to essential elements and are therefore subject to accumulation by aquatic biota, for example strontium-90 and cesium-137. The half-lives[6] of these are shown in Table 12:3.

TABLE 12:3
HALF-LIFE OF SOME RADIONUCLIDES OF BIOLOGICAL IMPORTANCE
After various sources

Radionuclide	Half-life
Iodine-131	8·04 days
Phosphorus-32	14·3 days
Sodium-22	2·6 years
Cobalt-60	5·27 years
Strontium-90	28 years
Cesium-137	30 ± 3 years

In Australia, where thus far nuclear power stations have not been built, such radioactive wastes that do occur come mainly from hospitals, research laboratories, mining, and uranium production. Their disposal is governed by the individual State radioactive substances Acts. The major governmental institution concerned with radioactive material is the Australian Atomic Energy Commission whose research establishment is located at Lucas Heights about 40 km southwest of Sydney. Low-activity liquid wastes from this establishment are discharged into the Woronora River not far from its estuary. These consist almost entirely of mixed fission products, cobalt-60, tritium, and traces of activation products (e.g. zinc-65) (Australian Atomic Energy Commission, private communications). The amounts discharged are strictly controlled according to the formula approved by the Maritime Services Board and the Radiological Advisory Council of New South Wales. Regular environmental and faunal sampling ensures that no health hazards occur.

Two characteristics of atomic waste disposal are noteworthy in the present context. Radioactivity may be concentrated by the aquatic biota, sometimes by very high factors. For example, larvae of *Hydropsyche cockerelli* (Trichoptera) collected within 4·8 km downstream of the effluents from the Hanford reactors commonly have a gross beta radioactivity 1,400 times greater than that of the water (Davis 1965). The extent of such concentration is dependent upon many factors including trophic level within the ecosystem. Second, there is a differential accumulation of radionuclides (cf. Fig. 12:1), the accumulation factors for a given radionuclide being widely variable (Table 12:4).

MAJOR EFFECTS OF POLLUTANTS ON INLAND WATERS Having considered the broad nature of water pollutants and their sources, we now need to consider in more depth the important topic of effects. The effects of *organic material* have been found to be more or less similar irrespective of source and nature—domestic or industrial—of the pollutant; obviously, many differences occur between

---

6. The period of time taken for the activity of a radioactive substance to decay to half its initial value.

THE EFFECTS OF MAN ON INLAND WATERS

localities, but by and large it is possible to compose a *generalized* picture of effects that is widely applicable. The major difference in expression is determined by whether the receiving water is still or flowing. The following discussion is confined to the effects on rivers and streams; the effects on lakes, loosely referred to as cultural eutrophication, are more conveniently discussed separately (see below). It should be noted also that the following discussion of necessity leans heavily on accounts of organic pollution of northern hemisphere running waters (e.g. Hynes 1960, Ingram, Mackenthun, and Bartsch 1966).

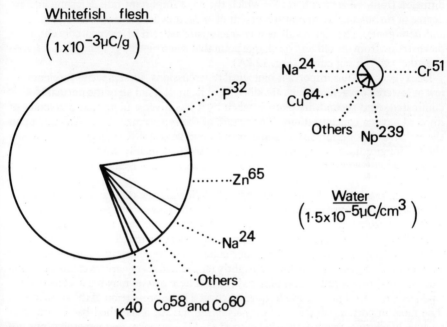

Fig. 12:1 Radionuclide composition of flesh of whitefish (*Prosopium williamsoni*) and Columbia River Water, Washington, U.S.A. Both sets of data were obtained in autumn 1961 and from the same location. The areas of the circles are proportional to total radioactivity. *After Foster and McConnon (1965)*

TABLE 12:4
RADIONUCLIDE ACCUMULATION FACTORS FOR *HYDROPSYCHE COCKERELLI* LARVAE (TRICHOPTERA), COLUMBIA RIVER, WASHINGTON
From Davis (1965)

Radionuclide	Accumulation factor (radioactivity/g larvae, live weight) (radioactivity/ml water)
Phosphorus-32	100,000
Zinc-65	20,000
Chromium-51	3,000
Copper-64	400
Sodium-24	80
Neptunium-239	30

Following the discharge of organic material (and no poisons) the most significant event that occurs is bacterial attack on the material. Its swiftness, rate, and effectiveness depend upon many factors including temperature, prior presence of a suitable bacterial flora, and the nature of the organic material. The result is always the same: bacteria commence the breakdown of the complex organic molecules into simpler ones and inorganic salts, and in the process remove oxygen from the water. If large amounts of organic material are present all oxygen may be removed; conditions become anaerobic. Reoxygenation occurs downstream by photosynthesis and by physical diffusion from the atmosphere, for which the most important rate determinants are degree of turbulence, temperature, extent of oxygen deficit, and presence of oil films and detergents. The net result is a characteristic pattern of oxygen concentration downstream from the effluent discharge point that when plotted graphically is referred to as the 'oxygen sag curve' (Fig. 12:2A).

Deoxygenation has important biological repercussions. First, species intolerant to low or lowered oxygen tensions are killed; that is, most if not all of the normal biota is eliminated. And second, its place is taken by a less diverse biota that is tolerant of lowered oxygen concentrations. The nature of the replacement biota depends upon the severity of pollution and to some extent upon geography. Its major components show a longitudinal downstream sequence correlated mainly with the form of the oxygen sag curve and by and large exhibit the sort of arrangements summarized in Fig. 12:2B, C.

Bacteria reach high densities immediately following the discharge of the effluent and gradually decrease in density downstream as the organic material is broken down and they become subject to predation. Although numerous, they themselves are generally not obvious. A noticeable component, especially in strongly polluted rivers, is 'sewage fungus', a term used to refer collectively to a variety of organisms that characteristically associate to form unsightly ragged masses of various colours ranging from white through yellow and pink to brown. These masses may reach considerable sizes, may break off pieces, and frequently cover a large proportion of the substratum. The most important constituent is *Sphaerotilus natans*, a sheathed bacterium with various life-forms, many formerly referred to different taxa, and generally, but not always, filamentous. Also involved is *Beggiatoa alba*, a sulphur bacterium, several true fungi which may include *Apodya lactea*, *Fusarium (Nectria) aqueductus*, *Mucor*, and *Geotridium*, and various genera of sessile ciliate protozoans, especially *Carchesium*, *Epistylis*, and *Vorticella*. 'Sewage fungus' reaches its peak of abundance in the most severely deoxygenated parts of the river or stream, short of those that are completely anaerobic. A similar pattern of distribution is displayed by those Protozoa not intimately part of 'sewage fungus', for example *Bodo*, *Colpidium*, *Claucoma*, *Coleps*, although flagellates like *Bodo* occur in more heavily polluted regions than do the ciliates.

With reoxygenation downstream, certain species of algae replace 'sewage fungus' as the major attached element (Fig. 12:2B). These exhibit a more or less characteristic succession, the species composition at any one point being dependent upon local conditions and often different from that only a little distance upstream or downstream. Large, dense growths of *Cladophora*, a green alga, are frequently characteristic, but many algal genera have been recorded, including particularly the cyanophyceans *Oscillatoria* and *Phormidium*, the chlorophyceans *Ulothrix*, *Euglena*, *Stigeoclonium*, and *Spirogyra*, and the bacillariophyceans *Gomphonema*, *Nitzchia*, *Navicula*, and *Surirella*.

THE EFFECTS OF MAN ON INLAND WATERS

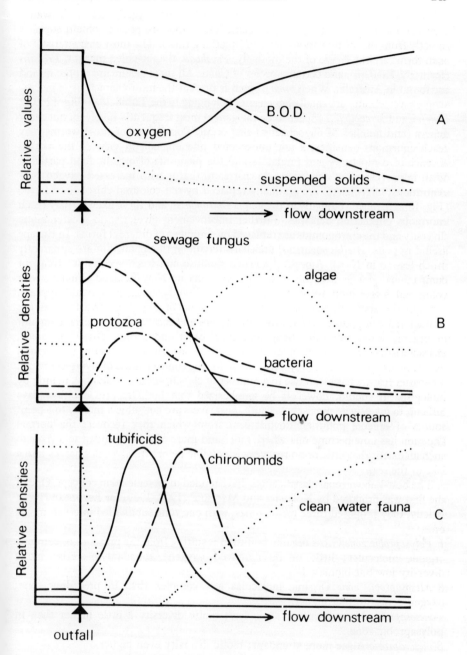

Fig. 12:2 General form of longitudinal downstream changes in some physico-chemical and biological features following discharge (outfall) of organic material. A: physico-chemical parameters; B: floral and microbial changes; C: changes in macroinvertebrates. *Redrawn after Hynes (1959)*

When pollution is so severe that extremely little or no oxygen is left in the water, very few macroscopic invertebrates occur. Those that are present obtain supplies directly from the air by various sorts of breathing tubes. The most characteristic of such forms are the larvae of the moth-fly, *Psychoda*, the rat-tailed maggot, *Eristalis* (formerly *Tubifera*), and certain species of *Culex*. All three genera are dipterans and are found in Australia. When *some* oxygen is present the most characteristic macroscopic invertebrates are small oligochaetes belonging to the Tubificidae (Fig. 12:2C); *Tubifex* and *Limnodrilus* appear to be the genera most frequently involved, but other genera (and families of oligochaetes) also occur. Populations of these worms may reach enormous densities, a not unexpected phenomenon in view of the virtual absence of competitors and predators and the plenitude of organic food particles. With improvement of conditions downstream, that is with increased oxygen concentrations, oligochaetes are gradually replaced by red-coloured chironomid larvae (Fig. 12:2C), most frequently of the genus *Chironomus*, and these likewise may reach enormous population densities. Further improvement gives rise to greater biotic diversity and the chironomids are replaced by a number of different forms. In Europe asellid isopods (*Asellus aquaticus*) are characteristic at this point, but are apparently much less so in North America.[7] Certain mollusc and leech species are frequent companions. *Asellus* does not occur in Australia or New Zealand. Finally, with continued oxygenation, the biota typical of unpolluted conditions gradually appears.

From the above discussion and a consideration of Fig. 12:2, two features are characteristic of polluted rivers: they can accept and recover from a certain amount of organic waste run into them, and their biota and certain physico-chemical characteristics display a more or less well-defined longitudinal succession. The former, naturally, has provided a ready part-excuse for the continued use of rivers as sewers. The latter feature has given rise to a number of classifications (zonations), mainly by biologists. It should, however, be emphasized that both features are nowadays difficult to investigate precisely because most rivers are not subject merely to a point source of organic pollution downstream from which they recover; the normal situation has now become one where more and more effluent is added to a river to such an extent that little or no recovery is possible before the river discharges into a lake or the sea.

The best-known zonal classifications have tended to be regional in concept. One of the first was proposed by Kolkwitz and Marsson[8] (1908, 1909) for European rivers. Briefly, they proposed three major zones, with one zone subdivided into two minor ones:

1 Polysaprobic zone Gross organic pollution resulting from the presence of complex organic molecules; little or no dissolved oxygen; abundant bacteria; biotic diversity low but biomass large.

2 Mesosaprobic zone Organic molecules less complex than in preceding zone; oxygen concentration higher.

α-*mesosaprobic* Many bacteria; algae few; biotic diversity a little higher than in polysaprobic zone.

β-*mesosaprobic* Algae more abundant; biotic diversity even higher.

---

7. Somewhat surprisingly, little precise ecological information is available on *Asellus* in North America.
8. Translations appear in Kemp, Ingram, and Mackenthun (1967: 47–52, 85–95).

3 *Oligosaprobic zone* Mineralization complete or almost; oxygen concentration near normal; biota similar to unpolluted conditions.

At the basis of this classification, known as the *Saprobiensystem*, is the assumption that certain animals or plants characterize or act as indicator species for a given degree of pollution or recovery. This assumption is not wholly justified, as emphasized by a number of critics (e.g. Hynes 1960). The Saprobiensystem nevertheless has some merits and has been altered, modified, and added to by several European workers, especially Fjerdingstad (1965), Liebmann (1951, 1962), and Zelinka and Marvan (1961). Sládeček (1965) has recently summarized the arguments for and against the system. The multiplicity of new terms introduced to expand the original concepts need not concern us, and in view of the distinctiveness of the Australasian aquatic biota it is inappropriate to list the biological indicators for the system.

By and large the Saprobiensystem was not widely accepted outside continental Europe. In Britain, Butcher (1946), on the basis of his study of the River Trent, proposed a system of four zones characterized as follows:

1 Zone of foul pollution Sewage fungus abundant; biotic diversity low.

2 Zone of pollution Oligochaetes and chironomids numerous but few other macroscopic invertebrates; sewage fungus still common, but algae more abundant.

3 Zone of mild pollution Many species of algae present; more macroscopic invertebrates present; sewage fungus rare.

4 Repurified zone Biota similar to that of unpolluted rivers.

Several American classifications have been put forward. One of the most widely accepted recognizes the following four zones: degradation, active decomposition, recovery, cleaner water.

Hynes (1960) has cogently argued against attempts to evolve formal classificatory zonations and his arguments have added force for Australasian biologists dealing with a distinctive biota and environment. At all events, no schemes of zonation have yet been applied in Australia, although some of the *general* features of zonation exhibited by organically polluted rivers in the northern hemisphere seem to be shown also by Australian rivers. This is indicated by the recent study by Jolly and Chapman (1966) of the effects of (mainly) organic pollution on two confluent rivers in New South Wales. They investigated in detail samples from three stations on Farmer's Creek below the outfall of the sewage treatment plant from the town of Lithgow, and in less detail from one station upstream of this outfall, and from five stations on Cox's River (two above the confluence with Farmer's Creek and three below). The most pertinent results are shown in Tables 12:5 and 12:6 and Figs 12:3 and 12:4. Their data have been partly simplified in that only those relating to a linear sequence are included (i.e. data for the two stations on Cox's River above the confluence with Farmer's Creek are omitted). For clarity, their stations have also been relettered so that station A is about 3 km above the sewage-plant outfall, and stations B–G are below the outfall, B about 0·5 km, C about 2 km, D about 12 km (and just above confluence of the two rivers), E about 12·2 km (and just below confluence), F about 28 km, and G about 44 km.

The tables and figures will illustrate the authors' summary (p. 186):

It was found that the association of oligochaetes, chironomids, snails, sewage fungus and small growths of *Stigeoclonium* and diatoms immediately below the outfall was rapidly succeeded by extensive growths of *Stigeoclonium*, high counts of Protozoa, and an increase in the number of snails. Recovery from the effects of pollution in Farmer's Creek before the confluence of the two streams was marked by the appearance of stoneflies, dragonflies, mayflies, caddis, and riffle beetles, although snails, limpets and chironomids were the dominant organisms. In the lower reaches of Cox's River most of the wide range of species found in the clean water above the confluence was recorded and variations between the fauna at the two stations were regarded as habitat differences. It was found however that the nutrients produced by mineralization of the sewage had induced a more extensive growth of algae in Cox's River. No 'indicator species' of pollution was found.

Hirsch's (1958) study of the macroscopic invertebrates of several organically polluted streams in New Zealand is also of interest at this point. He found that in grossly polluted areas the macroscopic invertebrates largely comprised tubificid worms but sometimes included naid worms and certain chironomid larvae. When pollution was less, a molluscan fauna commonly occurred as well. With further decrease in organic pollution, caddis larvae and various other organisms survived until finally ephemeropteran nymphs and other sensitive species survived too.

TABLE 12:5
ALGAE FROM FARMER'S CREEK AND COX'S RIVER
Based on monthly collections, 1961–62; data (rearranged) from Jolly and Chapman (1966, Table 5)[a]

Genus[c]	Station[b] A ↓	B	C	D	E	G
	outfall		— downstream →			
Batrachospermum	+					
Nitella	+				+	+
Closterium	+		+	+	+	+
Oscillatoria	+	+	+	+	+	+
Dinobryon		+	+			
Scenedesmus		+	+			+
Stigeoclonium		+	+	+	+	+
Cosmarium		+			+	
Staurastrum			+		+	
Spirogyra			+	+	+	+
Mougeotia					+	
Melosira					+	
Coleochaete					+	+
Gomphonema						+
Merismopedia						+
Cladophora						+
Ankistrodesmus						+
Tribonema						+
Rivularia						+
Draparnaldia						+

[a] In addition to the genera recorded in this table, the authors also recorded (1966, Table 6), using glass slides suspended in the river, the following genera: *Amphora, Acnanthes, Cocconeis, Cymbella, Diploneis, Epithemia, Frustularia, Navicula, Neidium, Nitzschia, Rhizosolenia, Rhoicosphenia, Rhopolodia, Synedra, Tabellaria,* and *Pleurotaenium.*
[b] See text for explanation of station position.
[c] *Surirella* was apparently also recorded but data were omitted from the original table.

TABLE 12:6
MACRO-INVERTEBRATES FROM FARMER'S CREEK AND COX'S RIVER
Rearranged from Jolly and Chapman (1966, Table 9)

Taxon	Station[b]	outfall		— downstream →				
		A[a] ↓	B	C	D	E	F[a]	G
Gastropoda: *Lymnaea tomentosa*					+			
Physastra gibbosa			+	+	+	+		+
Pettancylus sp.				+	+	+		+
Diptera: Tanypodinae				+				
Chironominae					+			3
Orthocladiinae		3	3	3	3			3
Simulium ornatipes						+		+
Antocha sp.					+	+		+
Ceratopogonidae					+			
Tabanidae			+					
Lepidoptera: Nymphulinae					+		+	+
Trichoptera: Calamoceratidae					2			
Polycentropidae					+			+
Leptoceridae					+	3		+
Hydropsychidae					+	3		3
Odontoceridae						2		+
Sericostomatidae						2		
Hydroptilidae					+	2		2
Rhyacophilidae						+		+
Megaloptera: *Archichauliodes guttiferus*								+
Coleoptera: Psephenidae					+	+		+
Hemiptera: Veliidae								+
Plecoptera: *Dinotoperla* sp.			+		+	+	+	+
Trinotoperla sp.		+			+	+	+	+
Odonata: Anisoptera		+			+			+
Zygoptera					+			+
Ephemeroptera: Baetidae					+	+		+
Caenidae								+
Leptophlebiidae		+		+	+	3	+	2
'Potamanthidae'[c]		+						
Malacostraca: *Paratya australiensis*						+		+
Ostracoda: *Candonocypris candonoides*								+
Cypretta sp.								+
Ilyodromus sp.								+
Copepoda: *Boeckella* sp.		+	+	+				
Ectocyclops phaleratus						+		
Eucyclops agilis		+						+
Macrocyclops albidus					+			+
Mesocyclops leuckarti					+			
Paracyclops fimbriatus		+	+	+				+
Cladocera: *Alona abbreviata*						+		+
A. affinis					+	+		+
Ceriodaphnia quadrangula				+	+			
Daphnia carinata		+	+	+				
Ilyocryptus sordidus		+						
Macrothrix spinosa		+	+	+	+		+	
Simocephalus vetulus		+	+					+
Hirudinea: Glossiphoniidae		+			+	2		+

248　　　　　　　　　　　　INLAND WATERS AND THEIR ECOLOGY

TABLE 12:6—continued

	Station[b]	outfall		— downstream →			
Taxon	A[a] ↓	B	C	D	E	F[a]	G
Oligochaeta: Lumbriculus sp.		+	+	+	+		+
Tubifex sp.		+	+	+	+		
Naididae		+	+	+	+		+
Polyzoa: Plumatella repens				+	+		+
Nematoda		+	+	+	+		+
Turbellaria: Rhabdocoela				+	+		+
Coelenterata: Hydra sp.							+
Porifera: Spongilla sp.							+

+ Present; numbers refer to number of species present.
[a] These stations were studied in less detail.
[b] See text for explanation of station position.
[c] The Potamanthidae is not otherwise recorded for Australia.

The effects of organic wastes upon rivers are not confined to those resulting from lowered oxygen concentrations. Almost invariably associated are effects resulting from the addition of fine suspended solids and, frequently if not invariably, from the presence of detergents in domestic sewage over and above the simple contribution of these to the total organic loading. With detergents the chief noxious effect is foaming, which may appear when even mild river turbulence occurs. The foam is often persistent and foam aggregations have been aptly termed 'detergent swans'. At a recent international golf championship in Melbourne a further course hazard was added when participants had to contend with numbers of such 'swans' blown in from an adjacent creek. The use of soft detergents would largely have prevented this problem.

The effect of *inert suspended materials* discharged into rivers depends principally upon the size of the material particles and the speed of the river. If the particles are small and river speed moderate, then they may remain in suspension for a considerable time, causing more or less permanent high turbidity. This deleteriously affects the aquatic vegetation since it cuts down the penetration of light. In highly turbid situations plant life may be almost eliminated with obvious consequences for dependent invertebrates and, one step removed, fish. The latter may also be adversely affected by a decreased ability to hunt for food in the lowered light intensities and by some clogging of their gills. However, the evidence that gills become clogged and thus cause the death of fish is somewhat inconclusive and it seems that fish mortality in highly turbid rivers is more the result of a lack of food or increased attack from microorganisms.

When inert particles are larger and/or river speed slower they may soon settle on to the bottom. If present in sufficient quantity they will then completely blanket the flora and bottom fauna and cause the death of these. The invertebrates most characteristic of areas with large inert sediment deposits are chironomids and tubificids. If the particles are large and the current swift, a certain degree of abrasiveness may also adversely affect river biotas.

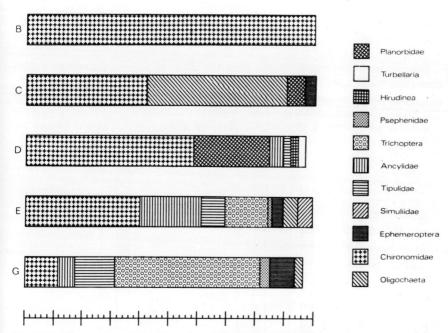

Fig. 12:3 Percentage composition of macroinvertebrates on six stones at five successive stations (B–G, see text for details) downstream of sewage effluent outfall into Farmer's Creek, New South Wales. Scale at bottom of figure indicates percentage values. *Redrawn and modified from Jolly and Chapman (1966)*

The effects of *poisons* depend upon many factors. The form of any direct effect, for example, depends largely upon the chemical nature of the poison involved. Cyanides act mainly as respiratory depressants in both invertebrates and fish (Krogh 1939, Jones 1964). Sulphides act similarly, at least in fish. The salts of the heavy metals (lead, zinc, copper, cadmium, mercury, etc.) cause fish deaths mainly as a result of asphyxiation brought about either by a reaction between the salts and gill mucus (coagulation film anoxia) or by disintegration of gill epithelia, or by both (Jones 1964). The salts of the alkali and alkaline earth metals (sodium, potassium, calcium, magnesium, barium, etc.), on the other hand, are much less toxic and lethal quantities do not affect fish gills. Some at least are toxic because of effects upon muscular and nervous activity. Others affect different body systems; potassium chromate, for instance, may cause pathological intestinal changes in fish (Fromm and Schiffman 1958).

Of the organic poisons the organic phosphates generally affect nervous systems for they inhibit certain neural enzymes. The action of other organic poisons is less clear but, again, action upon the nervous system is frequently implicated (e.g. Holan 1969). In general, we note, rather little is known about the direct pharmacological action of poisons, both inorganic and organic, on aquatic animals, and particularly on aquatic invertebrates.

Direct effects, it must be emphasized, are often delayed in expression; that is, poisons (particularly those that are accumulated rather than progressively eliminated) may not be immediately lethal to an individual, but cause sublethal effects such as reduced viability of adults or juveniles, decreased growth rates, adverse behavioural changes, and decreased reproduction. A number of organic poisons have been shown to exert this sort of effect.

Other factors involved in determining the effect of poisons include their concentration and dosage, the exposure time, the presence of other poisons and solutes, pH, temperature, salinity and oxygen tension in the receiving water, and the species and age of the animal or plant involved. At all events, for any given aquatic species it is impossible to compile a simple table showing absolute lethal concentrations for different poisons; there are too many complicating variables.

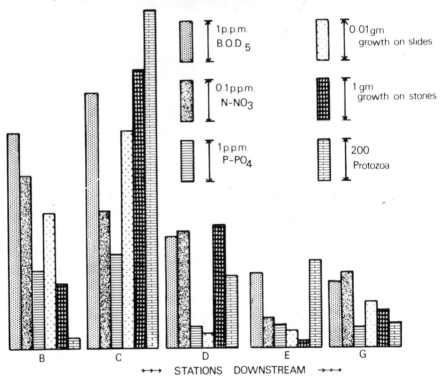

Fig. 12:4 Some chemical and biological changes at five successive stations (B–G, see text for details) downstream of sewage effluent outfall into Farmer's Creek, New South Wales. *Redrawn and modified from Jolly and Chapman (1966)*

With respect to poison concentration and dosage, Hynes (1960), Jones (1964), and others have stressed the need to distinguish clearly between the threshold dose of a poison, that is the maximum quantity that can be absorbed without resulting in death, and the concentration of a poison that can be tolerated for a specified time of continuous exposure. A table which summarizes data on lethal concentrations for various fish of numerous metallic salts at different exposure times is given by Jones (1964, table 5). The exposure time can be an important variable since many species may be affected if exposed for long periods to poison concentrations which have little if any effect during a short exposure time.

THE EFFECTS OF MAN ON INLAND WATERS

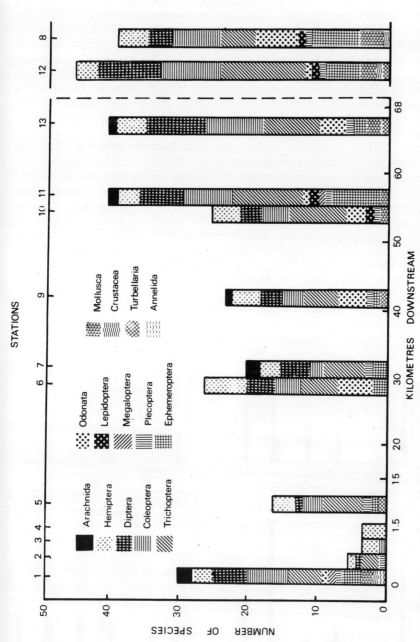

Fig. 12.5 The faunal composition (1963) of the Molonglo River and two 'control' tributaries near Canberra. *Redrawn and slightly modified from Weatherley, Beevers, and Lake (1967)*

The presence of other poisons or solutes may sometimes ameliorate the toxicity of a given poison or cause it to increase. Calcium can decrease the toxicity of heavy metals in certain situations; this is an example of *antagonism*. On the other hand, the inorganic poisons zinc sulphate and copper sulphate, when mixed, can have a combined effect which is greater than the sum of their separate effects; this is an example of *synergism*. Similar variation in toxicity may be caused by alterations in the pH, oxygen tension, and temperature of receiving waters. Generally, however, an increase in temperature or a decrease in oxygen tension increases the toxicity of inorganic poisons. Conversely, for some organic poisons, for example DDT, increase in temperature may decrease toxicity. The effect of pH is dependent upon the chemical nature of the poison.

Finally, as noted, the effect of a poison varies from species to species and according to the age of the individual organism. Certain invertebrates are much more sensitive to inorganic poisons than fish, yet others, for example chironomids, may be amongst the most tolerant of all animals to such poisons. The extreme toxicity of many organic poisons has already been referred to. The end result of the selective toxicity of poisons on animal and plant communities is that both diversity *and* biotic abundance are depressed downstream of the inflow of a poison (cf. effect of non-poisonous organic effluents). This is admirably illustrated by the work of Weatherley, Beevers, and Lake (1967) on the macroinvertebrates of the Molonglo River near Canberra. As noted previously, the major pollutant in this river is zinc. The authors investigated the fauna at thirteen stations, of which eleven (Stations 1–7, 9–11, 13) were on the Molonglo itself, and two (Stations 8, 12) were on tributaries and thus served as control situations. In part their results are summarized in Figs 12:5, 12:6. Station 1, it should be noted, is essentially upstream of the pollution source.

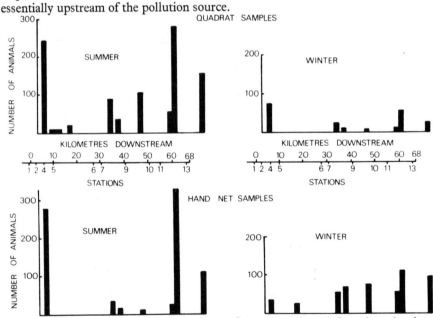

Fig. 12:6. Numbers of animals caught in $0 \cdot 1$ m² frame sampler and five minute hand net collections in Molonglo River near Canberra in 1963. *Redrawn and slightly modified from Weatherley, Beevers, and Lake (1967)*

The figures illustrate the depressing effect on both number and variety of macroinvertebrates of the zinc pollution; indeed, it was not for some 50 km downstream that the variety of macroinvertebrates in the Molonglo reached that in the unpolluted tributaries or upstream of the zinc pollution. In brief, Crustacea, Mollusca, Annelida, Ephemeroptera, Megaloptera, and Turbellaria were largely intolerant to the zinc pollution; Hemiptera, certain Plecoptera (*Dinotoperla*), and eruciform Trichoptera (all inhabitants of the slower stretches) were amongst the more tolerant. The authors observed that few if any fish had permanently inhabited this section of the Molonglo, an observation not surprising when the high zinc concentrations in the Molonglo are considered in relation to Affleck's (1952) findings that zinc concentrations of the order of 0·01 p.p.m. were lethal to ova and juveniles of brown and rainbow trout in Snob's Creek trout hatchery, Victoria. There, Affleck noted, it was sufficient for supply water merely to flow through galvanized pipes, containers, and gauze to pick up enough zinc to cause heavy mortality.

The major determinant of the effect of *thermal pollution* in rivers is of course the extent to which water temperatures are raised. Small elevations will have little effect; larger ones will have a greater effect. The effects include direct lethal consequences to the biota and alterations to breeding patterns, a reduction in the amount of oxygen dissolved, and increased chemical and biochemical reaction rates. Further, there are two associated features of thermal pollution that are important when, as frequently occurs, the water for cooling is itself polluted either by organic material or inorganic poisons. In the former case, bacteria and other plants as well as animals may be killed by heat or chlorination procedures, and this in effect increases the organic loading in the final effluent; in the latter, temperature increase often results in increased toxicity.

So little is known of the *actual* effects on river biotas of *radioactive waste*, as distinct from knowledge concerning radionuclide concentration and differential accumulation (p. 240), and apart from the predictable effects as poisons, that little can be said on this matter. Radionuclides, like all toxic substances that are mainly mutagenic in action, pose greater hazards to man than to the aquatic biota (Woodwell 1970). The disposal of such wastes, both in Australia and New Zealand and elsewhere, is subject to far more rigorous control than other pollutants, so that this lack of knowledge is of less consequence than it might otherwise be.

Lastly, in any consideration of the effects of pollutants, mention should be made of the broad overall effect of pollution on inland aquatic ecosystems, as distinct from the specific effects of pollutants. On this point, important as it may well be, only a limited amount of precise information is available. That which is available indicates that major general trends include changes from complex to simpler trophic arrangements, from high to lower species diversity, from stability to instability of biotic interrelationships, from narrow-niched to broad-niched species, from tight, closed nutrient cycles to loose and more or less open ones, and from efficient to less efficient ecosystem use of available solar energy (Woodwell 1970).

ASSESSMENT OF POLLUTION While strictly any assessment of water pollution should directly relate to the nature of the pollution and the quality of the water that is required for a given purpose, in practice a generalized approach with more or less fixed quality criteria is frequently adopted and often necessary. The following factors represent those most usually estimated in pollution assessments: temperature, pH, concentration of ammonia and dissolved oxygen, amount of suspended solids, density of coliform bacteria, extent of biochemical oxygen demand, and the nature of the fauna and flora. For many practical details of methods involved, readers are referred to standard texts such as that published by the American Public Health Association (1965). The first five factors require no further comment in a book of this sort, but some further explanation and comment is appropriate with regard to the others. Although inappropriate for further discussion, passing mention may be made of the 'saprobity indices' of several European workers; these indices are indirect measures of pollution based upon a combination of chemical, bacteriological, and biological features (e.g. Pantle and Buck 1955, Sládeček 1969).

There are a number of non-pathogenic coliform bacteria inhabiting the human intestine, and many of them survive for considerable periods freely in water. Coliform densities therefore provide a measure of the degree of human faecal contamination. Since water can carry a number of disease-causing organisms including those responsible for typhoid fever, cholera, dysentery, and over seventy enteric virus diseases[9], all of which organisms are difficult to assess directly in water, determination of coliform densities has become a standard test by which to judge the sanitary condition of a water. The higher the coliform count, the greater the health hazard presented by the water. In particular, one intestinal bacterium, *Escherichia coli* I, is widely used as a measure of faecal contamination, although it should be emphasized that this organism is not confined to humans (see Bardsley 1948), and the interpretation of observed densities requires careful consideration of several factors and much experience.

Standards for *E. coli* I densities vary from country to country and are related to water usage. In many countries drinking-water supplies with counts of <2 *E. coli* I/100 ml are regarded as satisfactory, and with counts of >100 as unsatisfactory. In waters to be used only for bathing, the American Public Health Association (in Fair and Geyer 1954) suggested the classification for total coliform bacteria/100 ml: <50, good; 50–1,000, doubtful or poor; $>1,000$, very poor. However, wide variation in coliform standards for recreational waters exists in the U.S.A. (Anon. 1963). A limit of 1,000 coliform organisms/100 ml was proposed by the Commonwealth Department of Works (1961) for Australian streams subject to occasional bathing by a relatively small population.

Little has been published on bacterial densities in Australian rivers, but available data do indicate that at least at times in certain rivers such densities may be quite high, as is shown by Table 12:7 which summarizes some unpublished data relating to three rivers in the Melbourne area (other data are added for comparative purposes). In a much more intensive investigation of bacterial pollution in ten rivers distributed throughout New South Wales, Brown (1963), though generally recording lower figures than those of Table 12:7, still concluded (p. 188) that 'undesirable levels of faecal contamination are apt to occur at any time' and that 'none of the streams can therefore be regarded as safe, from the bacteriological viewpoint, for human consumption'.

Coliform densities may be loosely correlated with organic waste levels in rivers when the wastes are primarily from humans and animals. A much better estimate,

---

9. Of especial interest here is the demonstration by Reid (1966) of a relationship between human gastro-intestinal troubles and waste-water pollution of the surface waters of Lake Maraetai, New Zealand.

THE EFFECTS OF MAN ON INLAND WATERS

irrespective of source, is given by the test known as the Biochemical Oxygen Demand test (B.O.D.). It was originally proposed by the British Royal Commission on Sewage Disposal (see p. 235), and is an entirely arbitrary measure of the amount of oxygen taken up by a sample of water incubated at 20°C for a given period. A five-day incubation period is usually adopted, the test then being referred to briefly as $B.O.D._5$. The amount of oxygen absorbed is of course dependent mainly upon the amount of organic material present—the greater the amount, the larger the demand—but it should be stressed that numerous other factors are of significance. The basic mechanism of the test is similar to that causing deoxygenation in streams polluted by organic wastes: bacteria break down the complex organic materials into simpler substances and in the process remove oxygen from the water.

On the basis of $B.O.D._5$ values, the British Royal Commission proposed the following widely accepted classification for river quality:

QUALITY	$B.O.D._5$ (mg/l)
Very clean	1
Clean	2
Fairly clean	3
Doubtful	4
Bad	10

Reflecting the results of bacteriological examination, samples from rivers in the Melbourne area frequently exceed values given in this classification. Thus $B.O.D._5$ values for six samples from the River Yarra (19 March 1970) ranged from 4–11 mg/l. Raw sewage has values from about 300–600 mg/l.

TABLE 12:7
ESCHERICHIA COLI I DENSITIES IN THREE MELBOURNE RIVERS AND IN OTHER SOURCES[a]

Sample source	Date	Number of sampling points	E. coli I/100 ml
River Yarra	12 March 1970	3	800–5,600
River Yarra	19 March 1970	6	190–>50,000
River Yarra	23 April 1970	10	1,900–150,000
Maribyrnong Creek	12 March 1970	5	500–25,000
Dandenong Creek	1 September 1969	4	60–34,100
Dandenong Creek	1, 2 October 1969	10	10–14,600
Farm Pond	1 October 1969	1	275,000
Domestic stormwater on unsewered estate	1 October 1969	2	60, 90
Melbourne tap water	1 October 1969	1	0

[a] Standard bacteriological methods were followed, entailing dilution with Ringer's solution, filtration on $0·45\mu$ membrane filter paper, incubation for two hours at 36°C on a resuscitation medium, and culturing for 16 hours at 44°C on MacConkey's medium.

Until fairly recently in many countries it was common practice to assess pollution by physical, chemical, and indirect bacteriological parameters only. This is still largely the case in Australia. Such practice neglects of course the salient point about pollution: it is essentially a biological phenomenon since its primary expression is upon the aquatic biota. Rather more attention is now directed to the use of the fauna and flora in pollutional assessments, and especially since biological assessment has many advantages over non-biological (Hynes 1964). Thus the biota in essence represents an integrated total of the effects of pollution; that is, a more accurate picture of *prevailing* conditions is likely to be obtained from a single sample of the static *in situ* biota than from a large number of samples studied by physical, chemical, and bacteriological methods alone. Intermittent pollution is much more likely to be detected by the biologist than by non-biologists. Moreover, biological assessment may be a great deal more sensitive than non-biological assessment in that pollution may be indicated when standard physical, chemical, and bacteriological methods can barely detect it.

Early attempts at biological assessments of water pollution centred upon the use of individual components of the biota as 'indicator species' (e.g. Kolkwitz and Marsson 1908, 1909). This approach is now much less emphasized, emphasis being given more to total community analyses and to diversity, absolute numbers, and species proportions within selected groups (e.g. tubificids, algae) than to single species. A number of biotic indices of pollution have been proposed, based on this approach. There is a tendency perhaps for fish still to be used as indicator species in a relatively simple manner, since a great deal more is known of fish tolerances to poisons and other factors inimical to fish life than of similar invertebrate and floral tolerances. However, fish, unlike other river biota, are capable of moving away from pollution, and there remains the difficult problem of extrapolation from laboratory results to field situations.

Eutrophication

The classical concept that lakes originate as oligotrophic water bodies, gradually become more productive and eutrophic, and then fill in is challenged by recent research. This suggests that any given lake is as productive as local conditions allow, and that with increase in age and loss of nutrients from the watershed, the trend is from eutrophic to oligotrophic conditions. This sequence, however, would only apply to lakes in natural environs. Many lakes nowadays are in areas more or less altered by man's activities so that they receive unnaturally large amounts of artificial nutrient material which cause a marked increase in lake productivity. Nutrient materials principally involved include sewage effluents, agricultural fertilizers, and detergents. The result is a relatively rapid change in the nature of the lake from oligotrophic to eutrophic conditions, and this unnatural acceleration is usually referred to as eutrophication, or more specifically (but imprecisely!) as cultural eutrophication. Basic accounts of this phenomenon are given by Hasler (1947), Sawyer (1966), Williams (1969a), and many others.

Especially implicated as specific nutrients are nitrates and phosphates, anions which in nature are frequently present in limiting amounts. Others which also probably play a part include potassium, magnesium, sulphates, trace elements (cobalt, molybdenum, copper, zinc, boron, iron, manganese, etc.), and organic growth factors (Vollenweider 1968).

What are the major effects of eutrophication which impinge upon the use of inland waters by man? First, and usually most obviously, there is a tendency for algal overproduction, a feature having a number of unfavourable aspects. In waters used for domestic and industrial supplies excessive algal growth may result in severe filter-clogging. Algal blooms frequently result in aesthetically displeasing, shiny floating masses of objectionable odour. Many algae, as indicated previously, impart unpleasant tastes and odours to water, and some even release toxic substances, as recorded apparently for the first time by Francis (1878), who noted the presence of *Nodularia spumigena*, a blue-green alga toxic to stock, in Lake Alexandrina near Adelaide. Flint (1966) has recently summarized New Zealand records of toxic algae. Large standing crops of phytoplankton may also result in murky, turbid waters.

Second, there is the tendency in eutrophic waters for extensive growth of macrophytes, and this becomes a nuisance when it interferes with boating, swimming, fishing, water skiing, and other recreational aquatic activities. Macrophytes also of course play an important role in the extinction of lake basins. Moreover, their presence may favour the greater production of such nuisance invertebrates as mosquitoes.

And third, associated with the above changes in floral composition there is a change in the nature of the fauna. So far as man is concerned the most significant change is usually taken to be a change in the composition of the fish fauna, with salmonids being replaced by coarse fish, an alteration generally displeasing to anglers. In extreme cases of eutrophication, where primary production is so high that subsequent decomposition of organic material completely deoxygenates the surrounding water, fish may be eliminated altogether.

With some exceptions the best-studied examples of lakes undergoing eutrophication are in North America. Noteworthy are the series of four lakes at Madison in Wisconsin, Lake Washington at Seattle, and (even) some of the Great Lakes—Lakes Erie and Ontario in particular (Beeton 1969, Edmondson 1969b). In Europe, Zürichsee in Switzerland and Windermere in England may be mentioned. Because most of the population live near the coast and because there is a paucity of large freshwater lakes, eutrophication is not generally of such concern in Australia as elsewhere. It is nevertheless a phenomenon that in certain regions may be expected to become increasingly important (cf. Williams 1969a). The situation is rather different in New Zealand where in a number of lakes dense macrophytic growth is of considerable nuisance value (see Chapman and Bell 1967, Fish 1963a, 1963b, Reid 1966).

It is perhaps instructive to consider one example in some detail. Eutrophication of Zürichsee first became evident about 1896 and was characterized by a bloom of *Tabellaria fenestrata* (a diatom), hypolimnetic oxygen depletion, and the disappearance of the profundal fauna (Hasler 1947). *Oscillatoria rubescens*, a blue-green alga often typifying incipient eutrophication, also appeared. The list of nuisance algae is now rather long: *Oscillatoria limosa, Spirogyra, Ulothrix, Cladophora, Zygnema,* and others. One of the most recent additions to this list is *Hydrodictyon reticulatum* which forms unsightly strand lines when thrown on to the beach (Thomas 1965). Coregonid fish no longer occur, their place having been taken by coarse fishes. And finally, swimming in the lake is now unpleasant, and its use as a source of domestic water supplies is inhibited.

Remedial measures for eutrophication centre upon the principles of limiting nutrient inflow (particularly of nitrates and phosphates), removing nutrients already present in the lake, and mechanically or chemically ridding the lake of nuisance plants by cutting or poisoning (National Academy of Sciences 1969). The removal or diversion of nutrients from lake inflows is now widely practised in the United States and elsewhere; removal procedures usually involve chemical coagulation of one sort or another. Phosphates can now be fairly easily removed, but nitrates prove more difficult to deal with. Occasionally, pre-blooming or complete distillation is resorted to in order to remove nutrients. Available evidence indicates that a lake is in danger with respect to its trophic level 'when its springtime concentration of assimilable phosphorus compounds and inorganic nitrogen compounds exceeds 10 mg P/m^3 and 200–300 mg N/m^3 respectively, and/or when the specific supply loading per unit area of lake reaches $0 \cdot 2$–$0 \cdot 5$ g P/m^2 per year and 5–10 g N/m^2 per year' (Vollenweider 1968). It should be stressed, however, that such values cannot be applied universally since lake productivity is the result of many interrelated factors, as Vollenweider points out. Much pertinent information concerning phosphates and nitrates in inland waters has been abstracted by Mackenthun (1965).

Internal remedial measures include the intensive cropping of fish, algae, or macrophytes, and the removal of hypolimnetic water rich in nutrients during the summer stagnation period. Mechanical removal of nuisance plants largely applies to the cutting of rooted macrophytes; chemical measures apply to the poisoning of macrophytes and phytoplankton as well as animals. In former times copper sulphate was chiefly used against phytoplankton and sodium arsenite against macrophytes, but nowadays organic weedicides are more frequently employed in both cases (e.g. 'aqualin', 'diquat', 'paraquat', '2,4-D'). Biological methods of macrophyte control that have been considered include the use of various aquatic herbivores such as the manatee (*Trichechus*), a mammal, and certain fish (species of *Tilapia* and *Ctenopharyngodon*), snails, insects, and pathogens (bacteria, fungi, viruses).

Indirect Effects

There are several human activities that may have considerable impact upon the inland aquatic environment but which do not involve direct physical alteration to inland waters and which have not been dealt with in our discussions of pollution and eutrophication.

The principal effect of land-clearing (deforestation), over-grazing by stock, certain sorts of agricultural malpractices, burning-off, and uncontrolled fires is to change the pattern of run-off on drainage basins and hence alter the nature of fluctuations in lake levels and river and stream flows. Invariably the trend is for such fluctuations to become more extreme. Thus formerly perennial streams may become ephemeral or approach this condition, and be more prone to damaging spates (see Chapter 8). Additionally, accelerated removal of silt and nutrient substances from drainage basins and their accession to lakes and rivers is an important associated feature, the biological repercussions of which have already been discussed. In the twelve months after the 1939 catastrophic forest fires in Victoria, the siltation rate in 'Lake' Eildon (an impoundment) exceeded that of the succeeding sixteen years (River Murray Commission 1957).

Afforestation also may have a significant effect, depending upon the previous nature of the drainage basin and the species of tree involved. Spruce, for example, has been shown to inhibit the growth of macroinvertebrates and fish in streams flowing through spruce plantations (Huet 1951). We may note here, too, our circumstantial evidence (Bayly and Williams 1966) that differences in the chemical composition of Lakes Leake and Edward in South Australia are largely the result of differences in the proportions of the drainage basin of each covered by plantations of pine, *Pinus radiata*. The replacement of the natural vegetation by cultivated crops may also cause a chemical change in run-off as indicated by Wood's (1924) work in Western Australia, where he found an increase in the salinity of run-off following replacement of the natural vegetation.

The transference by man, accidental or otherwise, of many species of aquatic biota from one water body to another is also a human activity that has affected the nature of inland waters and their biota. Such transferences have occurred on a global scale. Frequently they have had beneficial results for man, as many anglers, for example, would say with regard to the widespread introduction of the trout, *Salmo trutta*, to regions throughout the world where it is not indigenous. The increase in fish biomass caught for subsistence in many tropical countries and resulting from fish introductions is clearly another beneficial result for man (see Hickling 1961). On the other hand, many transferences have had deleterious results for mankind. Irrespective of the advantages or disadvantages to man their total impact upon inland waters and biota ranges from an apparently insignificant one to one that is highly significant.

Amongst plants that have had important and generally deleterious effects upon inland waters into which they have spread or been introduced, special mention must be made of the water hyacinth, *Eichhornia crassipes*, and the water fern, *Salvinia auriculata*. Both are natives of South America where they are not nuisance plants; both have now spread to many tropical countries throughout the world where they are frequently pestiferous, since they form huge floating masses that hinder fishing, impede navigation, impair recreation, and significantly alter water quality (Little 1966). They are now present in Australia and *E. crassipes*, a declared noxious weed, infests many waters of northern New South Wales and southern Queensland, and has even been seen in northern Victoria and South Australia (Parsons 1963). At present *S. auriculata* is less pestiferous. Other well-known examples of macrophytes that on transference have become nuisances (at least initially) include *Elodea canadensis* in temperate and warm regions and *Pistia stratiotes* and *Papyrus* in warm regions (but see Table 6:6).

Mention has already been made of the profound consequences of the accidental entry into the North American Great Lakes of the sea lamprey. Another example of this sort, of some commercial but greater biological significance, involves the transference of the fish, *Lates niloticus*, from lakes in Africa where it occurs naturally (Lakes Rudolf and Albert) to several others. Some of these have many endemic fish species—Lake Victoria, for example, has 117 endemic species of *Haplochromis*. *Lates niloticus*, a highly predatory and euryoecious fish of catholic taste in prey (Gee 1969), is now said to be exerting considerable predation pressure on many of these biologically very interesting species.

With reference to the many introduced fish species in Australia (Chapter 11), Weatherley and Lake (1967) contend that the influence of these upon native fish is much less than has been suggested. However, little precise information is available on this matter and some of their arguments are based on premises with which a great many ecologists would disagree. The importance of the introduced carp, *Cyprinus carpio*, upon the actual physical environment is also uncertain. Butcher (1962) maintains that this species causes great physical disturbance to a habitat resulting from its 'roiling' of bottom deposits and the subsequent increased turbidity and uprooted aquatic plants. Weatherley and Lake, on the other hand, doubt that such behaviour is important in Australian rivers which are naturally muddy. At all events, carp are now the subject of a vigorous extermination policy in Victoria (see also Chapter 11).

Little is known of the effects of fish introductions upon aquatic invertebrates in Australia. It has, however, been suggested that the present almost complete restriction of the Tasmanian syncarid shrimp, *Anaspides tasmaniae*, to small streams and moorland pools is a result of predation upon it by the introduced trout, *Salmo trutta*; the shrimp is absent from lakes and streams where the trout now occurs abundantly (Williams 1965a) but is still found in a few remote lakes to which trout have not been introduced, and it has been recorded in the past from lakes where it apparently does not now occur although the trout does.

With regard to the effect of introduced salmonid fish into New Zealand, Allen (1961) contended that there has probably not been any great change in the freshwater fauna, particularly so far as the benthic fauna is concerned. The almost complete disappearance of the grayling, *Prototroctes oxyrhynchus*, between 1870 and 1927 does not seem to have been caused by the introduction of salmonids, according to Allen.

Finally in this section, mention should be made of the cane toad, *Bufo marinus*, introduced in Queensland. There, at certain seasons, it becomes extremely numerous, and when individuals subsequently die in water-holes during the dry season their corpses give rise to extremely foetid conditions.

Bibliography

Abell D. L. (1961) The role of drainage analysis in biological work on streams. *Verh. int. Verein. theor. angew. Limnol.* **14**, 533–7.
Adamstone F. B. (1924) The distribution and economic importance of the bottom fauna of Lake Nipigon with an appendix on the bottom fauna of Lake Ontario. *Publ. Ontario Fisheries Res. Lab.* (Toronto Univ.) No. 24, 35–100.
Affleck R. J. (1952) Zinc poisoning in a trout hatchery. *Aust. J. mar. Freshwat. Res.* **3**, 142–69.
Allchin B. and **Allchin R.** (1968) *The Birth of Indian Civilization*. Penguin Books, Harmondsworth.
Allen K. R. (1949) Lakes. *N.Z. Sci. Rev.* **7**, 112–19.
Allen K. R. (1950) The computation of production in fish populations. *N.Z. Sci. Rev.* **8**, 89.
Allen K. R. (1951) The Horokiwi Stream: a study of a trout population. *Fish. Bull. N.Z.* No. 10, 231 pp.
Allen K. R. (1956) The geography of New Zealand's freshwater fish. *N.Z. Jl Sci. Technol.* **14** (3), 3–9.
Allen K. R. (1961) Relations between Salmonidae and the native freshwater fauna in New Zealand. *Proc. N.Z. ecol. Soc.* **8**, 66–70.
American Public Health Association (1965) *Standard Methods for the Examination of Water and Wastewater* (12th edn.). American Public Health Association, Inc., New York.
Anderson B. G. (1960) The toxicity of organic insecticides to *Daphnia*. In: *Biological Problems in Water Pollution, Trans. 1959 Seminar*. Robert A. Taft Sanit. Eng. Cent., Cincinnati.
Anderson G. C. (1958a) Seasonal characteristics of two saline lakes in Washington. *Limnol. Oceanogr.* **3**, 51–68.
Anderson G. C. (1958b) Some limnological features in a shallow saline meromictic lake. *Limnol. Oceanogr.* **3**, 259–70.
Anderson R. O. and **Hooper F. F.** (1956) Seasonal abundance and production of littoral fauna in a southern Michigan lake. *Trans. Am. microsc. Soc.* **80**, 266–307.
Anderson V. G. (1940) Old and new systems for reporting the inorganic constituents in natural waters. *J. Proc. Aust. chem. Inst.* **7**, 187–212.
Anderson V. G. (1945) Some effects of atmospheric evaporation and transpiration on the composition of natural waters in Australia. *J. Proc. Aust. chem. Inst.* **12**, 40–68, 83–98.
Andersson E. (1969) Life-cycle and growth of *Asellus aquaticus* (L.) with special reference to the effects of temperature. *Rep. Inst. Freshwat. Res. Drottningholm* **49**, 5–26.
Andronikova I. N. (1965) [The levels of productivity of zooplankton of strongly humified basins of the Karelian isthmus] (in Russian). *Gidrobiol. Zh.* **1** (4), 34–8.
Angino E. E., Armitage K. B., and **Tash J. C.** (1965) A chemical and limnological study of Lake Vanda, Victoria Land, Antarctica. *Kans. Univ. Sci. Bull.* **45**, 1097–118.
Anon. (1963) Coliform standards for recreational waters. Progress report of the Public Health

Activities Committee of the Sanitary Engineering Division. *J. Sanit. Engng Div. Am. Soc. civ. Engrs* **89**, *Proc. Pap.* 3617, 57–94.

Armitage K. B. and **House H. B.** (1962) A limnological reconnaissance in the area of McMurdo Sound, Antarctica. *Limnol. Oceanogr.* **7**, 36–41.

Bailey-Watts A. E., Bindloss M. E., and **Belcher J. H.** (1968) Freshwater primary production by a blue-green alga of bacterial size. *Nature, Lond.* **220**, 1344–5.

Bain A. D. N. (1949) Salt production in Victoria. *Min. geol. J.* **3** (6), 4–7.

Baker A. N. (1967) Algae from Lake Miers, a solar-heated Antarctic lake. *N.Z. Jl Bot.* **5**, 453–68.

Barclay M. H. (1966) An ecological study of a temporary pond near Auckland, New Zealand. *Aust. J. mar. Freshwat. Res.* **17**, 239–58.

Bardsley D. A. (1948) A study of coliform organisms in the Melbourne water supply and in animal faeces, with observations on their longevity in faeces and in soil. *J. Hyg., Camb.* **46**, 269–79.

Baylor E. R. and **Smith F. E.** (1957) In: *Recent Advances in Invertebrate Physiology* (ed. B. T. Scheer). Univ. Oregon Press, Eugene.

Baylor E. R. and **Sutcliffe W. H.** (1963) Dissolved organic matter in seawater as a source of particulate food. *Limnol. Oceanogr.* **8**, 369–71.

Bayly I. A. E. (1962) Ecological studies on New Zealand lacustrine zooplankton with special reference to *Boeckella propinqua* Sars (Copepoda: Calanoida). *Aust. J. mar. Freshwat. Res.* **13**, 143–97.

Bayly I. A. E. (1963) Reversed diurnal vertical migration of planktonic Crustacea in inland waters of low hydrogen ion concentration. *Nature, Lond.* **200**, 704–5.

Bayly I. A. E. (1964a) Chemical and biological studies on some acidic lakes of east Australian sandy coastal lowlands. *Aust. J. mar. Freshwat. Res.* **15**, 56–72.

Bayly I. A. E. (1964b) A revision of the Australasian species of the freshwater genera *Boeckella* and *Hemiboeckella* (Copepoda: Calanoida). *Aust. J. mar. Freshwat. Res.* **15**, 180–238.

Bayly I. A. E. (1965) The fate of Lake Pedder and its biota. *Aust. Soc. Limnol. Newsl.* **4** (2), 26–30.

Bayly I. A. E. (1966) Queensland's coastal dune lakes and their life. *Wildlife in Australia* **3**, 154–7.

Bayly I. A. E. (1967a) The fauna and chemical composition of some athalassic saline waters in New Zealand. *N.Z. Jl mar. Freshwat. Res.* **1**, 105–17.

Bayly I. A. E. (1967b) The general biological classification of aquatic environments with special reference to those in Australia. In: *Australian Inland Waters and their Fauna: Eleven Studies* (ed. A. H. Weatherley). A.N.U. Press, Canberra.

Bayly I. A. E. (1969a) The body fluids of some centropagid copepods: total concentration and amounts of sodium and magnesium. *Comp. Biochem. Physiol.* **28**, 1403–9.

Bayly I. A. E. (1969b) Factors influencing productivity of lakes. *Proc. R. Soc. Vict.* **83**, 11–16.

Bayly I. A. E. (1969c) The occurrence of calanoid copepods in athalassic saline waters in relation to salinity and ionic proportions. *Verh. int. Verein. theor. angew. Limnol.* **17**, 449–55.

Bayly I. A. E. (1970a) Further studies on some saline lakes of south-east Australia. *Aust. J. mar. Freshwat. Res.* **21**, 117–29.

Bayly I. A. E. (1970b) A note on the zooplankton of the Mount Kosciusko region. *Aust. Soc. Limnol. Bull.* **3**, 25–8.

Bayly I. A. E. and **Arnott G. H.** (1969) A new centropagid genus (Copepoda: Calanoida) from Australian estuarine waters. *Aust. J. mar. Freshwat. Res.* **20**, 189–98.

Bayly I. A. E. and **Ellis P.** (1969) *Haloniscus searlei* Chilton: an aquatic 'terrestrial' isopod with remarkable powers of osmotic regulation. *Comp. Biochem. Physiol.* **31**, 523–8.

Bayly I. A. E., Peterson J. A., Tyler P. A., and **Williams W. D.** (1966) Preliminary limnological survey of Lake Pedder, Tasmania, March 1–4, 1966. *Aust. Soc. Limnol. Newsl.* **5** (2), 30–41.

Bayly I. A. E., Peterson J. A., and **St John P.** (1970) Notes on Lake Kutubu, southern highlands of the Territory of Papua and New Guinea. *Aust. Soc. Limnol. Bull.* **3**, 40–7.

Bayly I. A. E. and **Williams W. D.** (1966a) Chemical and biological studies on some saline lakes of south-east Australia. *Aust. J. mar. Freshwat. Res.* **17**, 177–228.

BIBLIOGRAPHY

Bayly I. A. E. and **Williams W. D.** (1966b) Further chemical observations on some volcanic lakes in the south-east of South Australia. *Aust. J. mar. Freshwat. Res.* **17**, 229–37.

Beadle L. C. (1932) Scientific results of the Cambridge expedition to the East African lakes, 1931-1. No. 4. The waters of some East African lakes in relation to their fauna and flora. *J. Linn. Soc. (Zool.)* **38**, 157–211.

Beadle L. C. (1943) Osmotic regulation and the faunas of inland waters. *Biol. Rev.* **18**, 172–83.

Beadle L. C. (1959) Osmotic and ionic regulation in relation to the classification of brackish and inland saline waters. *Archo Oceanogr. Limnol. Roma* **11** (Suppl.), 143–51.

Beadle L. C. (1966) Prolonged stratification and deoxygenation in tropical lakes. I. Crater Lake Nkugute, Uganda, compared with Lakes Bunyoni and Edward. *Limnol. Oceanogr.* **11**, 152–63.

Beadle L. C. (1969) Osmotic regulation and the adaptation of freshwater animals to inland saline waters. *Verh. int. Verein. theor. angew. Limnol.* **17**, 421–9.

Beeton A. M. (1969) Changes in the environment and biota of the Great Lakes. In: *Eutrophication: Causes, Consequences, Correctives*. Nat. Acad. Sci., Washington, D. C.

Beeton A. M. and **Chandler D. C.** (1963) The St Lawrence Great Lakes. In: *Limnology in North America* (ed. D. G. Frey). Univ. Wisconsin Press, Madison and Milwaukee.

Beklemischew V. N. and **Baskina-Zakolodkina V. P.** (1933) [Experimental premises relating to the ecological geography of inland waters. Part II] (in Russian). *Izv. biol. nauchno-issled. Inst. biol. Sta. perm. gosud. Univ.* 8 (9–10), 361–74.

Bell R. A. I. (1967) Lake Miers, South Victoria Land, Antarctica. *N.Z. Jl Geol. Geophys.* **10**, 540–56.

Berg K. (1938) Studies on the bottom animals of Esrom Lake. *K. danske Vidensk. Selsk. Skr.* **9** (8), 1–255.

Berman T. and **Rodhe W.** (1971) Distribution and migration of *Peridinium* in Lake Kinneret. *Mitt. int. Verein. theor. angew. Limnol.* **19**, 266–76.

Berner L. M. (1951) Limnology of the lower Missouri River. *Ecology* **32**, 1–12.

Bērziņš B. (1958) Ein planktologisches Querprofil. *Rep. Inst. Freshwat. Res. Drottningholm* **39**, 5–22.

Betjeras F. N. (1950) Salt industry in South Australia. *Min. Rev. Adelaide* **91**, 187–99.

Bibra E. E. and **Mason R. G.** (1967) *Victorian River Gaugings to 1965*. State Rivers and Water Supply Commission, Melbourne.

Bicknell A. K. and **Bunt F.** (1952) The vertical distribution of certain Entomostraca in Sodon Lake. *Lloydia* **15**, 56–64.

Bird E. C. F. (1964a) The formation of coastal dunes in the humid tropics: some evidence from north Queensland. *Aust. J. Sci.* **27**, 258–9.

Bird E. C. F. (1964b) *Coastal Land Forms*. A.N.U. Press, Canberra.

Bishop J. A. (1967a) Seasonal occurrence of a branchiopod crustacean, *Limnadia stanleyana* King (Conchostraca) in eastern Australia. *J. Anim. Ecol.* **36**, 77–95.

Bishop J. A. (1967b) Some adaptations of *Limnadia stanleyana* King (Crustacea: Branchiopoda: Conchostraca) to a temporary freshwater environment. *J. Anim. Ecol.* **36**, 599–609.

Bishop J. A. (1968) Resistance of *Limnadia stanleyana* King (Branchiopoda, Conchostraca) to desiccation. *Crustaceana* **14**, 35–8.

Bishop J. A. and **Dyce A. L.** (1968) Survival of animals inhabiting temporary rainfilled rockpools near Sydney. *Aust. Soc. Limnol. Newsl.* **6** (1), 5–6.

Bishop J. E. and **Hynes H. B. N.** (1969) Upstream movements of the benthic invertebrates in the Speed River, Ontario. *J. Fish. Res. Bd Can.* **26**, 279–98.

Blum J. L. (1956) The ecology of river algae. *Bot. Rev.* **22**, 291–341.

Bonython C. W. and **Mason B.** (1953) The filling and drying of Lake Eyre. *Geogrl J.* **119**, 321–33.

Boone E. and **Baas Becking L. G. M.** (1931) Salt effects on eggs and nauplii of *Artemia salina* L. *J. gen. Physiol* **14**, 753–63.

Borutzky E. V. (1939a) [Dynamics of the biomass of *Chironomus plumosus* in the profundal of Lake Beloie] (in Russian). *Trudȳ limnol. Sta. Kosine* **22**, 156–95.

Borutzky E. V. (1939b) [Dynamics of the total benthic biomass in the profundal of Lake Beloie] (in Russian). *Trudȳ limnol. Sta. Kosine* **22**, 196–218.

Bottomley G. A. (1956a) Free oscillations of Lake Wakatipu, New Zealand. *Trans. Am. Geophys. Un.* **37**, 51–5.

Bottomley G. A. (1956b) Seiches on Lake Wakatipu, New Zealand. *Trans. R. Soc. N.Z.* **83**, 579–87.
Bowen S. T. (1964) The genetics of *Artemia salina*. IV. Hybridization of wild populations with mutant stocks. *Biol. Bull. mar. biol. Lab.*, *Woods Hole*, **126**, 333–44.
Bozniak E. G., Schonen N. S., Parker B. C., and **Keenan C. M.** (1969) Limnological features of a tropical meromictic lake. *Hydrobiologia* **34**, 524–32.
Brand G. W. (1967) Studies on some south-east Australian dune lakes, with special reference to the distribution of *Calamoecia tasmanica* Smith (Copepoda : Calanoida). B.Sc. Honours Thesis, Monash University.
Brass L. J. (1953) Results of the Archbold Expeditions. No. 68. Summary of the 1948 Cape York (Australia) Expedition. *Bull. Am. Mus. nat. Hist.* **102**, 141–205.
Brinkhurst R. O., Chua K. E., and **Batoosingh E.** (1969) Modifications in sampling procedures as applied to studies on the bacteria and tubificid oligochaetes inhabiting aquatic sediments. *J. Fish. Res. Bd Can.* **26**, 2581–93.
Broch E. S. and **Yake W.** (1969) A modification of Maucha's ionic diagram to include ionic concentrations. *Limnol. Oceanogr.* **14**, 933–5.
Brook A. J. (1965) Planktonic algae as indicators of lake types with special reference to the Desmidiaceae. *Limnol. Oceanogr.* **10**, 403–11.
Brook A. J. and **Woodward W. B.** (1956) Some observations on the effects of water inflow and outflow on the plankton of small lakes. *J. Anim. Ecol.* **25**, 22–35.
Brooks J. L. (1969) Eutrophication and changes in the composition of the zooplankton. In: *Eutrophication: Causes, Consequences, Correctives*. Nat. Acad. Sci., Washington, D.C.
Brooks J. L. and **Dodson S. I.** (1965) Predation, body size, and composition of plankton. *Science, N.Y.* **150**, 28–35.
Brothers R. N. (1957) The volcanic domes at Mayor Island, New Zealand. *Trans. R. Soc. N.Z.* **84**, 549–60.
Brown J. K. (1963) The extent of bacterial pollution in stream water in New South Wales, Australia. M.Sc. Thesis, University of New South Wales.
Brundin L. (1949) Chironomiden und andere Bodentiere der südschwedischen Urgebirgsseen. *Rep. Inst. Freshwat. Res. Drottningholm* **30**, 1–914.
Brundin L. (1956) Die bodenfaunistischen Seetypen und ihre Anwendbarkeit auf die Südhalbkugel. Zugleich eine Theorie der produktionsbiologischen Bedeutung der glazialen Erosion. *Rep. Inst. Freshwat. Res. Drottningholm* **37**, 186–235.
Brundin L. (1958) The bottom faunistical lake type system and its application to the southern hemisphere. Moreover a theory of glacial erosion as a factor of productivity in lakes and oceans. *Verh. int. Verein. theor. angew. Limnol.* **13**, 288–97.
Brundin L. (1966) Transantarctic relationships and their significance as evidenced by chironomid midges, with a monograph of the subfamilies Podonominae and Aphroteniiae and the Austral Heptagyiae. *K. svenska Vetensk. Akad. Handl.* (Fjärde Serien) **11** (1), 1–472.
Burnet A. M. R. (1969) A study of the inter-relation between eels and trout, the invertebrate fauna and the feeding habits of the fish. *Fish. tech. Rep. N.Z. mar. Dep.* **36**, 1–23.
Burns C. W. and **Rigler F. H.** (1967) Comparison of filtering rates of *Daphnia rosea* in lake water and in suspensions of yeast. *Limnol. Oceanogr.* **12**, 492–502.
Burrows G. B. (1968) A comparative study of the major zooplankton species in Lake Sorrell and Lake Crescent, Tasmania. B.Sc. Honours Thesis, University of Tasmania.
Butcher A. D. (1962) Why destroy the European carp? *Fisheries Circular No. 6*. Fisheries and Wildlife Department of Victoria.
Butcher A. D. (1965) Wildlife hazards from the use of pesticides. *Australas. J. Pharm.* **1965**, S105–9.
Butcher R. W. (1946) The biological detection of pollution. *J. Proc. Inst. Sew. Purif.* **2**, 92–7.
Byars J. A. (1960) A freshwater pond in New Zealand. *Aust. J. mar. Freshwat. Res.* **11**, 222–40.
Callinan B. J. (1969) Domestic and industrial uses of water. *Proc. R. Soc. Vict.* **83**, 37–41.
Cane R. F. (1962) The salt lakes of Linga, Victoria. *Proc. R. Soc. Vict.* **75**, 75–88.
Carpelan L. H. (1961a) Physical and chemical characteristics. In: *The Ecology of the Salton Sea, California, in Relation to the Sportfishery* (ed. B. W. Walker). *Fish. Bull. Calif.* No. 113.
Carpelan L. H. (1961b) Zooplankton. In: *The Ecology of the Salton Sea, California, in Relation to the Sportfishery* (ed. B. W. Walker). *Fish. Bull. Calif.* No. 113.

Carr J. L. (1969) The primary productivity and physiology of *Ceratophyllum demersum*. I. Gross macro primary productivity. *Aust. J. mar. Freshwat. Res.* **20**, 115–26.
Cassie V. (1969) Seasonal variation in phytoplankton from Lake Rotorua and other inland waters, New Zealand 1966–67. *N.Z. Jl mar. Freshwat. Res.* **3**, 98–123.
Cassidy N. G. (1949) Note on the composition of M.I.A. irrigation water. *C.S.I.R.O. Aust. Irrig. Research Stn, Griffith, Int. Rep.* No. 11, 4 pp.
Castenholz R. W. (1960) Seasonal changes in the attached algae of freshwater and saline lakes in the Lower Grand Coulee, Washington. *Limnol. Oceanogr.* **5**, 1–28.
Chapman D. W. (1967) Production in fish populations. In: *The Biological Basis of Freshwater Fish Production* (ed. S. D. Gerking). Blackwells, Oxford.
Chapman D. W. (1968) Production. In: *Methods for Assessment of Fish Production in Fresh Waters* (ed. W. E. Ricker). I.B.P. Handbook No. 3, Blackwell Scientific Publications, Oxford and Edinburgh.
Chapman V. J. (1960) *Salt Marshes and Salt Deserts of the World*. Interscience Publishers, New York.
Chapman V. J. and **Bell C. A.** (1967) *Rotorua and Waikato Water Weeds*. Dept of University Extension, University of Auckland, Auckland.
Chappuis P. A. (1927) Die Tierwelt der unterirdischen Gewässer. *Binnengewässer* **3**, Stuttgart.
Cheeseman T. F. (1907) Notice of the occurrence of *Hydatella*, a genus new to the New Zealand flora. *Trans. N.Z. Inst.* **34**, 433–4.
Cheeseman T. F. (1925) *Manual of the New Zealand Flora* (2nd edn). N.Z. Govt Printer, Wellington.
Cheng D. M. H. (1968) Comparative limnological studies on Lake Sorell and Lake Crescent, Tasmania. B.Sc. Honours Thesis, University of Tasmania.
Chilton C. (1894) The subterranean Crustacea of New Zealand, with remarks on the fauna of caves and wells. *Trans. Linn. Soc. Lond.* **6**, 163–284.
Chilton L. (1920) on a new isopodan genus (family Oniscidae) from Lake Corangamite, Victoria. *Proc. Linn. Soc. N.S.W.* **44**, 823–34.
Chilton C. (1922) A new isopod from central Australia belonging to the Phreatoicidae. *Trans. R. Soc. S. Aust.* **46**, 23–33.
Christianson A. G. and **Tichenor B. A.** (1968) *Industrial Waste Guide on Thermal Pollution*. U.S. Dept of Interior, Federal Water Pollution Control Administration.
Clarke F. W. (1924) The data of geochemistry (5th edn). *Bull. U.S. geol. Surv.* No. 770, 841 pp.
Clarke G. L., Edmondson W. T., and **Ricker W. E.** (1946) Mathematical formulation of biological productivity. *Ecol. Monogr.* **16**, 336–7.
Coaldrake J. E. (1961) The ecosystem of the coastal lowlands ('Wallum') of southern Queensland. *C.S.I.R.O. Aust. Bull.* No. 283.
Cole G. A. (1968) Desert limnology. In: *Desert Biology*, Vol. I (ed. G. W. Brown). Academic Press, New York.
Cole G. A. and **Brown R. J.** (1967) The chemistry of *Artemia* habitats. *Ecology* **48**, 858–61.
Coleman M. J. and **Hynes H. B. N.** (1970) The vertical distribution of the invertebrate fauna in the bed of a stream. *Limnol. Oceanogr.* **15**, 31–40.
Collins M. I. (1923) Studies in vegetation of arid and semi-arid New South Wales. Part I. The plant ecology of the Barrier District. *Proc. Linn. Soc. N.S.W.* **48**, 229–66.
Comita G. W. (1964) The energy budget of *Diaptomus siciloides*, Lilljeborg. *Verh. int. Verein. theor. angew. Limnol.* **15**, 646–53.
Commonwealth Department of Works (1961) The quality of sewage effluents for discharge to streams. In: *Report of the Proceedings of the Tenth Conference* [of professional officers representing the authorities controlling water supply and sewerage undertakings serving the cities and towns of Australia]. Canberra.
Conway E. J. (1942) Mean geochemical data in relation to oceanic evolution. *Proc. R. Ir. Acad.* (Ser. B) **48**, 119–59.
Cooke W. B. (1956) Colonization of artificial bare areas by microorganisms. *Bot. Rev.* **22**, 613–38.
Cooper W. E. (1965) Dynamics and productivity of a natural population of a fresh water amphipod *Hyalella azteca*. *Ecol. Monogr.* **35**, 377–94.
Cotton C. A. (1945) *Geomorphology. An Introduction to the Study of Landforms*. Whitcombe and Tombs, Wellington.

Cox B. G. (1965) Oscillations of Lake Wakatipu. *Trans. R. Soc. N.Z.* (Gen.) **1**, 183–90.
Croghan P. C. (1958) The survival of *Artemia salina* (L.) in various media. *J. exp. Biol.* **35**, 213–18.
Cummins K. W. (1967) [Compiler] Calorific equivalents for studies in ecological energetics (mimeographed, 2nd edn). Pymatuning Laboratory of Ecology, 52 pp.
Cunningham B. T., Moar N. T., Torrie A. W., and **Parr P. J.** (1953) A survey of the western coastal dune lakes of the North Island, New Zealand. *Aust. J. mar. Freshwat. Res.* **4**, 343–86.
Currey D. T. (1970) Lake systems. Western Victoria. *Aust. Soc. Limnol. Bull.* **3**, 1–13.
Cushing D. H. and **Nicholson H. F.** (1966) Methods of estimating algal production rates at sea. *Nature, Lond.* **212**, 310–11.
Dahl E. (1956) Ecological salinity boundaries in poikilosaline waters. *Oikos* **7**, 1–21.
Davis C. C. (1963) On questions of production and productivity in ecology. *Arch. Hydrobiol.* **59**, 145–61.
Davis J. J. (1965) Accumulation of radionuclides by aquatic insects. In: *Biological Problems in Water Pollution, Trans. 1962 Seminar.* Robert A. Taft Sanit. Eng. Cent., Cincinnati.
Davis J. J., Perkins R. W., Palmer R. F., Hanson W. C., and **Cline J. F.** (1958) Radioactive materials in aquatic and terrestrial organisms exposed to reactor water. *Proc. 2nd Intern. Conf. on the Peaceful Uses of Atomic Energy* (United Nations, Geneva, 1958) **18**, 423–8.
Deevey E. S. (1941) Limnological studies in Connecticut. VI. The quantity and composition of the bottom fauna of thirty-six Connecticut and New York lakes. *Ecol. Monogr.* **11**, 414–55.
Deevey E. S. (1964) Preliminary account of fossilization of zooplankton in Rogers Lake. *Verh. int. Verein. theor. angew. Limnol.* **15**, 981–92.
Delamare-Deboutteville C. (1960) *Biologie des Eaux Souterraines Littorales et Continentales.* Herman, Paris.
Dendy J. S. (1945) Predicting depth distribution of fish in three TVA [Tennessee Valley Authority] storage-type reservoirs. *Trans. Am. Fish. Soc.* **75**, 65–71.
Den Hartog C. (1960) Comments on the Venice System for the classification of brackish waters. *Int. Revue ges. Hydrobiol. Hydrogr.* **45**, 481–5.
Dew B. (1963) Animal life in caves. *Aust. Mus. Mag.* **15**, 158–61.
Dickinson P. (1952) Organic pollution of natural waters. *N.Z. Engng* **7**, 43–7.
Digby P. S. B. (1961) Mechanism of sensitivy to hydrostatic pressure in the prawn, *Palaemonetes varians* Leach. *Nature, Lond.* **191**, 366–8.
Dineen C. F. (1953) An ecological study of a Minnesota pond. *Am. Midl. Nat.* **50**, 349–76.
Doty M. S. (ed.) (1963) Proceedings of the conference on primary productivity measurement, marine and freshwater. *U.S. Atomic Energy Commission Report* TID 7633.
Drury G. H. (1966) *The Face of the Earth* (rev. edn). Penguin Books, Harmondsworth.
Dulhunty J. A. (1946) On glacial lakes in the Kosciusko region. *J. Proc. R. Soc. N.S.W.* **79**, 143–52.
Dumbleton L. J. (1969) A new species of *Ephydrella* Tonnoir and Malloch (Diptera: Ephydridae) from hot springs, and notes on other Diptera from mineralized waters. *N.Z. Ent.* **4**, 38–46.
Dumont H. J. (1969) A quantitative method for the study of periphyton. *Limnol. Oceanogr.* **14**, 303–7.
Dunk W. P. (1957) Weedicides. Their properties and application for the control of water weeds. *Aqua, Melb.* **9** (1), 3–9.
Dunstan B. (1921) Salt in Queensland. *Pub. geol. Surv. Qld* **268** (1).
Dussart B. (1966) *Limnologie. L'Étude des Eaux Continentales.* Gauthier-Villars, Paris.
Duthie H. C. and **Carter J. C. H.** (1970) The meromixis of Sunfish Lake, southern Ontario. *J. Fish. Res. Bd Can.* **27**, 847–56.
Dyce A. L. (1964) 'Tree holes' in our drier inland areas. *Aust. Soc. Limnol. Newsl.* **3** (1), 8–9.
Dyce A. L. (1970) Rot pockets in prickly pear in Australia—an example of an introduced regenerating habitat. *Aust. Soc. Limnol. Bull.* **2**, 20.
Eckstein Y. (1970) Physicochemical limnology and geology of a meromictic pond on the Red Sea shore. *Limnol. Oceanogr.* **15**, 363–72.
Edgar E. (1966) The male flowers of *Hydatella inconspicua* (Cheesem.) Cheesem. (Centrolepidaceae). *N.Z. Jl Bot.* **4**, 153–8.
Edmondson W. T. (1960) Reproductive rates of rotifers in natural populations. *Mem. Ist. Ital. Idrobiol.* **12**, 21–77.

Edmondson W. T. (1965) Reproductive rate of planktonic rotifers as related to food and temperature in nature. *Ecol. Monogr.* **35**, 61–111.
Edmondson W. T. (1968) A graphical model for evaluating the use of the egg ratio for measuring birth and death rates. *Oecologia* (Berl.) **1**, 1–37.
Edmondson W. T. (1969a) The present condition of the saline lakes in the Lower Grand Coulee, Washington. *Verh. int. Ver. theor. angew. Limnol.* **17**, 447–8.
Edmondson W. T. (1969b) Eutrophication in North America. In: *Eutrophication: Causes, Consequences, Correctives.* Nat. Acad. Sci., Washington, D.C.
Edward D. H. (1964) A cryptobiotic chironomid from south-western Australia. *Aust. Soc. Limnol. Newsl.* **3**, 29–30.
Edward D. H. (1968) Chironomidae in temporary freshwaters. *Aust. Soc. Limnol. Newsl.* **6** (1), 3–5.
Edwards R. W. and **Owens M.** (1960) The effects of plants on river conditions. I. Summer crops and estimates of net productivity of macrophytes in a chalk stream. *J. Ecol.* **48**, 151–60.
Eggleton F. E. (1956) Limnology of a meromictic, interglacial, plunge-basin lake. *Trans. Am. Microsc. Soc.* **75**, 334–78.
Ekman S. (1953) *Zoogeography of the Sea.* Sidgwick and Jackson, London.
Elliott J. M. (1967) Invertebrate drift in a Dartmoor stream. *Arch. Hydrobiol.* **63**, 202–37.
Elliott J. M. and **Minshall G. W.** (1968) The invertebrate drift in the River Duddon, English Lake District. *Oikos* **19**, 39–52.
Ellis P. and **Williams W. D.** (1970) The biology of *Haloniscus searlei* Chilton, an oniscoid isopod living in Australian salt lakes. *Aust. J. mar. Freshwat. Res.* **21**, 51–69.
Elster H. J. (1954a) Einige Gedanken zur Systematik, Terminologie und Zielsetzung der dynamischen Limnologie. *Arch. Hydrobiol* **20** (Suppl.), 487–523.
Elster H. J. (1954b) Über die Populationsdynamik von *Eudiaptomus gracilis* Sars und *Hetercope borealis* Fischer im Bodensee-Obersee. *Arch. Hydrobiol.* **20** (Suppl.), 546–614.
Eriksen C. H. (1966) Diurnal limnology of two highly turbid puddles. *Verh. int. Verein. theor. angew. Limnol.* **16**, 507–14.
Esterley C. O. (1912) The occurrence and vertical distribution of the Copepoda of the San Diego region. *Univ. Calif. Publs Zool.* **9**, 253–340.
Ettershank G., Fuller M., and **Brough E. J.** (1966) Hemiptera from saline waters in inland Australia. *Aust. J. Sci.* **29**, 144–5.
Evans A. J. (1970) Some aspects of the ecology of a calanoid copepod, *Pseudoboeckella brevicaudata* Brady 1875, on a subantarctic island. *A.N.A.R.E. Rep.* (Ser. B (11) Zool.) No. 110, 100 pp.
Fager E. W. (1969) Production of stream benthos: a critique of the method of assessment proposed by Hynes and Coleman (1968). *Limnol. Oceanogr.* **14**, 766–70.
Fair G. M. and **Geyer J. C.** (1954) *Water Supply and Waste-Water Disposal.* John Wiley, New York.
Fairbridge W. S. (1945) West Australian freshwater calanoids (Copepoda). I. Three new species of *Boeckella*, with a description of the developmental stages of *B. opaqua* n. sp., and a key to the genus. *J. Proc. R. Soc. West. Aust.* **29**, 25–65.
Fernando C. H. (1958) The colonization of small freshwater habitats by aquatic insects. 1. General discussion, methods and colonization in the aquatic Coleoptera. *Ceylon J. Sci. biol. Sci.* **1**, 117–54.
Fernando C. H. (1959) The colonization of small freshwater habitats by aquatic insects. 2. Hemiptera (The water-bugs). *Ceylon J. Sci. biol. Sci.* **2**, 5–32.
Fernie N. (1930) Water supplies from rock catchments in the Western Australian wheat belt. *J. Instn Engrs Aust.* **2**, 198–208.
Findenegg I. (1965) Factors controlling primary productivity, especially with regard to water replenishment, stratification, and mixing. *Mem. Ist. Ital. Idrobiol.* **18** (Suppl.), 105–19.
Fish G. R. (1963a) Observations on excessive weed growth in two lakes in New Zealand. *N.Z. Jl Bot.* **1**, 410–8.
Fish G. R. (1963b) Limnological conditions and growth of trout in three lakes near Rotorua. *Proc. N.Z. ecol. Soc.* **10**, 1–7.
Fish G. R. (1969) The oxygen content of some New Zealand lakes. *Verh. int. Verein. theor. angew. Limnol.* **17**, 392–403.
Fjerdingstad E. (1965) Some remarks on a new saprobic system. In: *Biological Problems in Water Pollution, Trans. 1962 Seminar.* Robert A. Taft Sanit. Eng. Cent., Cincinnati.

Flint E. A. (1938) A preliminary study of the phytoplankton in Lake Sarah (New Zealand). *J. Ecol.* **26**, 353–8.
Flint E. A. (1966) Toxic algae in some New Zealand freshwater ponds. *N.Z. vet. J.* **14**, 181–5.
Flowers S. and **Evans F. R.** (1966) The flora and fauna of the Great Salt Lake region, Utah. In: *Salinity and Aridity. New Approaches to Old Problems* (ed. H. Boyko). Junk, The Hague.
Fogg G. E. (1965) *Algal Cultures and Phytoplankton Ecology.* Univ. Wisconsin Press, Madison and Milwaukee.
Fogg G. E. (1968) *Photosynthesis.* The English Universities Press, London.
Fogg G. E. (1969) Extracellular products of phytoplankton. In: *A Manual on Methods for Measuring Primary Production in Aquatic Environments* (ed. R. A. Vollenweider). I.B.P. Handbook No. 12, Blackwell Scientific Publications, Oxford and Edinburgh.
Forbes S. A. (1887) The lake as a microcosm. *Peoria Sci. Ass. Bull.* **1887**, 77–87.
Forel F. A. (1892) Le Léman. *Monographie Limnologique.* Lausanne, Switzerland.
Foster R. F. and **McConnon D.** (1965) Relationships between the concentration of radionuclides in Columbia River Water and Fish. In: *Biological Problems in Water Pollution, Trans. 1962 Seminar.* Robert A. Taft Sanit. Eng. Cent., Cincinnati.
Francis G. (1878) Poisonous Australian lake. *Nature, Lond.* **18**, 11–2.
Frankenberg R. S. (1969) Studies on the evolution of galaxiid fishes with particular reference to the Australian fauna. Ph.D. Thesis, University of Melbourne.
Fredeen F. J. H. (1964) Bacteria as food for blackfly larvae (Diptera: Simuliidae) in laboratory cultures and in natural streams. *Can. J. Zool.* **42**, 527–48.
Freeman P. (1961) The Chironomidae (Diptera) of Australia. *Aust. J. Zool.* **9**, 611–737.
Frey D. G. (1960) The ecological significance of cladoceran remains in lake sediments. *Ecology* **41**, 684–99.
Frey D. G. (1969) The rationale of paleolimnology. *Mitt. int. Verein. theor. angew. Limnol.* **17**, 7–18.
Fromm P. O. and **Schiffman R. H.** (1958) Toxic action of hexavalent chromium on largemouth bass. *J. Wildl. Mgmt* **22**, 40–4.
Fryer G. (1957) The food of some freshwater cyclopoid copepods and its ecological significance. *J. Anim. Ecol.* **26**, 263–85.
Fryer G. (1968) Evolution and adaptive radiation in the Chydoridae (Crustacea: Cladocera): a study in comparative functional morphology and ecology. *Phil. Trans. R. Soc.* (Ser. B) **254**, 221–385.
Fryer G. (1969) Speciation and adaptive radiation in African lakes. *Verh. int. Verein. theor. angew. Limnol.* **17**, 303–22.
Gaarder T. and **Gran H. H.** (1927) Investigations of the production of plankton in the Oslo Fjord. *J. Cons. perm. int. Explor. Mer* **42**, 1–48.
Gage M. (1959) On the origin of some lakes in Canterbury. *N.Z. Geographer* **15**, 69–75.
Gage M. (1969) Rocks and landscape. In: *The Natural History of Canterbury* (ed. G. A. Knox). A. H. and A. W. Reed, Wellington.
Gauthier H. (1938) La vie dans les deserts subtropicaux. *Mem. Soc. Biogeogr.* **6**, 107–20. (Not seen.)
Geddes M. C. (1968) Studies on the limnology of the Marshall Reserve pond with special reference to *Boeckella triarticulata* (Thomson) (Copepoda, Calanoida). B.Sc. Honours Thesis, Monash University.
Gee J. M. (1969) A comparison of certain aspects of the biology of *Lates niloticus* (Linne) in endemic and introduced environment in East Africa. In: *Man-Made Lakes. The Accra Symposium* (ed. Letitia Obeng). Ghana Universities Press, Accra.
Gill E. D. (1970) *Rivers of History.* The Australian Broadcasting Commission, Sydney.
Gislén T. (1948). Aerial plankton and its conditions of life. *Biol. Rev.* **23**, 109–26.
Goede A. (1967) Tasmanian cave fauna: character and distribution. *Helictite, Sydney* **5**, 71–86.
Goldman C. R. (1961) Primary productivity and limiting factors in Brooks Lake, Alaska. *Verh. int. Verein. theor. angew. Limnol.* **14**, 120–4.
Goldman C. R. (1964) Primary productivity and micronutrient limiting factors in some North American and New Zealand lakes. *Verh. int. Verein. theor. angew. Limnol.* **15**, 365–74.
Goldman C. R. (1965) Micronutrient limiting factors and their detection in natural phytoplankton populations. *Mem. Ist. Ital. Idrobiol.* **18** (Suppl.), 121–35.

Goldman C. R. (1968) Aquatic primary production. *Am. Zool.* **8**, 31–42.
Goldman C. R., Mason D. T., and **Wood B. J. B.** (1963) Light injury and inhibition in Antarctic freshwater phytoplankton. *Limnol. Oceanogr.* **8**, 313–22.
Goldman C. R. and **Wetzel R. G.** (1963) A study of the primary productivity of Clear Lake, Lake County, California. *Ecology* **44**, 283–94.
Goode J. (1967) *Freshwater Tortoises of Australia and New Guinea (in the family Chelidae).* Lansdowne Press, Melbourne.
Gordon I. (1957) On *Spelaeogriphus*, a new cavernicolous crustacean from South Africa. *Bull. Br. Mus. nat. Hist. Zool.* **5**, 31–47.
Gorham E. (1961) Factors influencing supply of major ions to inland waters, with special reference to the atmosphere. *Bull. geol. Soc. Amer.* **72**, 795–840.
Gould J. (1863) *The Mammals of Australia*. Taylor & Francis, London.
Grange L. I. (1937) The Geology of the Rotorua-Taupo Subdivision, Rotorua and Kaimanawa Divisions. *Bull. geol. Surv. N.Z.* (N.S.) No. 37, 138 pp.
Green J. D. (1967) Studies on the zooplankton of Lake Pupuke. *Tane* **13**, 77–98.
Green J. D., Norrie P. H., and **Chapman M. A.** (1968) An internal seiche in Lake Rotoiti. *Tane* **14**, 3–11.
Gregory J. W. (1906) *The Dead Heart of Australia*. John Murray, London.
Griffin R. J. (1964) Quality classification of water. *Geol. Survey of N.S.W.* Report No. 17, 1–10.
Grigg G. C. (1965) Aspects of respiration in the Queensland lungfish, *Neoceratodus forsteri* (Krefft). *Aust. Soc. Limnol. Newsl.* **4** (1), 8–9.
Grindley J. R. (1964) Effect of low salinity water on the vertical migration of estuarine plankton. *Nature, Lond.* **203**, 781–2.
Hall D. J. (1964) An experimental approach to the dynamics of a natural population of *Daphnia galeata mendotae*. *Ecology* **45**, 94–112.
Hall R. E. (1961) On some aspects of the natural occurrence of *Chirocephalus diaphanus* Prévost. *Hydrobiologia* **17**, 205–17.
Hamilton A. L. (1969) On estimating annual production. *Limnol. Oceanogr.* **14**, 771–82.
Hardy A. C. (1956) *The Open Sea*. Collins, London.
Harmsworth R. V. and **Whiteside M. C.** (1968) Relation of cladoceran remains in lake sediments to primary productivity of lakes. *Ecology* **49**, 998–1000.
Harris J. E. (1953) Physical factors involved in the vertical migration of plankton. *Q. Jl Microsc. Sci.* **94**, 537–50.
Harris S. W. (1954) An ecological study of the waterfowl of the Pot-Holes area, Grant County, Washington. *Am. Midl. Nat.* **52**, 403–32.
Harrison A. D. (1965) River zonation in southern Africa. *Arch. Hydrobiol.* **61**, 380–6.
Hartland-Rowe R. (1966) The fauna and ecology of temporary pools in western Canada. *Verh. int. Verein. theor. angew. Limnol.* **16**, 577–84.
Hasler A. D. (1947) Eutrophication of lakes by domestic drainage. *Ecology* **28**, 383–95.
Hayes F. R. (1957) On the variation in bottom fauna and fish yield in relation to trophic level and lake dimensions. *J. Fish. Res. Bd Can.* **14**, 1–32.
Hayes F. R. (1963) Chemical characteristics of fresh water. *Great Lakes Res. Divn Univ. Michigan Publ.* No. 10.
Hayes F. R. and **Anthony E. H.** (1964) Productive capacity of North American lakes as related to the quantity and the trophic level of fish, the lake dimension, and the water chemistry. *Trans. Am. Fish. Soc.* **93**, 53–7.
Hayne D. W. and **Ball R. C.** (1956) Benthic productivity as influenced by fish predation. *Limnol. Oceanogr.* **1**, 162–75.
Healy J. (1963) Geology of the Rotorua district. *Proc. N.Z. ecol. Soc.* **10**, 53–8.
Hedgpeth J. W. (1959) Some preliminary considerations of the biology of inland mineral waters. *Archo Oceanogr. Limnol. Roma* **11** (Suppl.), 111–41.
Hedin S. A. (1904) *Scientific Results of a Journey in Central Asia, 1899 to 1902: Vol. 1, The Tarin River.* Stockholm, Lithographic Inst., Swedish Army, 523 pp.
Henderson C., Pickering Q. H., and **Tarzwell C. M.** (1960) The toxicity of organic phosphorus and chlorinated hydrocarbons to fish. In: *Biological Problems in Water Pollution, Trans. 1959 Seminar.* Robert A. Taft Sanit. Eng. Cent., Cincinnati.
Hickling C. F. (1961) *Tropical Inland Fisheries*. Longman, London.

Hickling C. F. (1968) *The Farming of Fish*. Academic Press, London.
Higginson F. R. (1966) The distribution of submerged aquatic angiosperms in the Tuggerah lakes system. *Proc. Linn. Soc. N.S.W.* **90**, 328–34.
Hillbricht-Ilkowska A., Gliwicz Z., and **Spodniewska I.** (1966) Zooplankton production and some trophic dependences in the pelagic zone of two Masurian lakes. *Verh. int. Verein. theor. angew. Limnol.* **16**, 432–40.
Hills E. S. (1940) *The Physiography of Victoria. An Introduction to Geomorphology.* Whitcombe and Tombs, Melbourne.
Hinton H. E. (1960) Cryptobiosis in the larva of *Polypedilum vanderplanki* Hint. (Chironomidae). *J. Insect Physiol.* **5**, 286–300.
Hirsch A. (1958) Biological evaluation of organic pollution of New Zealand streams. *N.Z. Jl Sci. Technol.* **1**, 500–53.
Hoare R. A. (1966) Problems of heat transfer in Lake Vanda, a density stratified Antarctic lake. *Nature, Lond.* **210**, 787–9.
Hobbie J. E. (1969) A method for studying heterotrophic bacteria. In: *A Manual on Methods for Measuring Primary Production in Aquatic Environments* (ed R. A. Vollenweider). I.B.P. Handbook No. 12, Blackwell Scientific Publications, Oxford and Edinburgh.
Hobbs D. F. (1948) Trout Fisheries in New Zealand. Their development and management. *Fish Bull. N.Z.* No. 9., 175 pp.
Holan G. (1969) New halocyclopropane insecticides and the mode of action of D.D.T. *Nature, Lond.* **221**, 1025–9.
Holthuis L. B. (1960) Two new species of atyid shrimps from subterranean waters of N.W. Australia (Decapoda Natantia). *Crustaceana* **1**, 47–57.
Holthuis L. B. (1968) On Hymenosomatidae (Crustacea Decapoda Brachyura) from fresh water, with the description of a new species. *Beaufortia* **15**, 109–21.
Hopkins C. L., Solly S. R. B., and **Ritchie A. E.** (1969) D.D.T. in trout and its possible effect on reproductive potential. *N.Z. Jl mar. Freshwat. Res.* **3**, 220–9.
Horne F. R. (1967) Effects of physical-chemical factors on the distribution and occurrence of some southeastern Wyoming phyllopods. *Ecology* **48**, 482–7.
Horsfall R. A. (1969) Administration and management of water supply in Victoria. *Proc. R. Soc. Vict.* **83**, 27–35.
Horton P. A. (1961) The bionomics of brown trout in a Dartmoor stream. *J. Anim. Ecol.* **30**, 311–38.
Hrbáček J. (1962) Species composition and the amount of zooplankton in relation to the fish stock. *Rozp. ČSAV, Řada Mat. a přír. Ved* **72** (10), 1–116.
Hubbert M. K. (1969) Energy resources. In: *Resources and Man.* National Academy of Science, Natural Research Council. W. H. Freeman, San Francisco.
Huet M. (1951) Nocivité des boisements en Epicéas (*Picea excelsa* Link.) pour certains cours d'eaux de l'Ardenne Belge. *Verh. int. Verein. theor. angew. Limnol.* **11**, 189–200.
Hughes J. C. and **Lund J. W. G.** (1962) The rate of growth of *Asterionella formosa* Hass. in relation to its ecology. *Arch. Mikrobiol.* **42**, 117–29.
Husmann S. (1966) Versuch einer ökologischen Gliederung des interstitiellen Grundwassers in Lebensbereiche einiger Prägung. *Arch. Hydrobiol.* **62**, 231–68.
Husmann S. (1970) Weitere Vorschläge für eine Klassifizierung subterraner Biotope und Biocoenosen der Süßwasserfauna. *Int. Revue ges. Hydrobiol. Hydrogr.* **55**, 115–29.
Hussainy S. U. (1969) Ecological studies on some microbiota of lakes in western Victoria. Ph.D. Thesis, Monash University.
Hutchinson G. E. (1932) Experimental studies in ecology. I. The magnesium tolerance of Daphniidae and its ecological significance. *Int. Revue ges. Hydrobiol. Hydrogr.* **28**, 90–108.
Hutchinson G. E. (1937) A contribution to the limnology of arid regions. *Trans. Conn. Acad. Arts Sci.* **33**, 47–132.
Hutchinson G. E. (1941) Ecological aspects of succession in natural populations. *Am. Nat.* **75**, 406–18.
Hutchinson G. E. (1951) Copepodology for the ornithologist. *Ecology* **32**, 571–7.
Hutchinson G. E. (1957) *A Treatise on Limnology* Vol. I. John Wiley & Sons, New York.
Hutchinson G. E. (1967) *A Treatise on Limnology* Vol. II. John Wiley & Sons, New York.
Hutchinson G. E. and **Löffler H.** (1956) The thermal classification of lakes. *Proc. natn. Acad. Sci. U.S.A.* **42**, 84–6.

Hutton J. T. and **Leslie T. I.** (1958) Accession of non-nitrogenous ions dissolved in rainwater to soils in Victoria. *Aust. J. agric. Res.* **9**, 492–507.
Hynes H. B. N. (1960) *The Biology of Polluted Waters.* Liverpool University Press, Liverpool.
Hynes H. B. N. (1961a) The invertebrate fauna of a Welsh mountain stream. *Arch. Hydrobiol.* **57**, 344–88.
Hynes H. B. N. (1961b) The effect of water-level fluctuations on littoral fauna. *Verh. int. Verein. theor. angew. Limnol.* **14**, 652–6.
Hynes H. B. N. (1963) Imported organic matter and secondary productivity in streams. *Int. Congr. Zool.* **16**, 271–3.
Hynes H. B. N. (1964) The use of biology in the study of water pollution. *Chemy Ind.*, March, 1964, 435–6.
Hynes H. B. N. (1969) The enrichment of streams. In: *Eutrophication: Causes, Consequences, Correctives.* Nat. Acad. Sci., Washington, D.C.
Hynes H. B. N. (1970a) The ecology of stream insects. *A. Rev. Ent.* **15**, 25–42.
Hynes H. B. N. (1970b) *The Ecology of Running Waters.* Liverpool University Press, Liverpool.
Hynes H. B. N. and **Coleman M. J.** (1968) A simple method of assessing the annual production of stream benthos. *Limnol. Oceanogr.* **13**, 569–73.
Illies J. (1961) Versuch einer allgemeinen biozönotischen Gliederung der Fließgewässer. *Int. Revue ges. Hydrobiol. Hydrogr.* **46**, 205–13.
Illies J. (1963) The Plecoptera of the Campbell and Auckland Islands. *Rec. Dom. Mus.*, Wellington **4**, 255–65.
Illies J. (1964) The invertebrate fauna of the Huallaga, a Peruvian tributary of the Amazon River, from the sources down to Tingo Maria. *Verh. int. Verein. theor. angew Limnol.* **15**, 1077–83.
Illies J. (1968) The first wingless stonefly from Australia. *Psyche, Cambr.* **75**, 328–33.
Illies J. and **Botosaneanu L.** (1963) Problèmes et méthodes de la classification et de la zonation écologique des eaux courantes, considerées surtout du point de vue faunistique. *Mitt. int. Verein. theor. angew. Limnol.* **12**, 1–57.
Ingram W. M., Mackenthun K. M., and **Bartsch A. F.** (1966) *Biological Field Investigative Data for Water Pollution Surveys.* U.S. Department of the Interior; Federal Water Pollution Control Administration, W.P.B. U.S. Government Printing Office.
Irwin J. (1968) Observations of temperatures in some Rotorua district lakes. *N.Z. Jl mar. Freshwat. Res.* **2**, 591–605.
Ivlev V. S. (1939) Transformation of energy by aquatic animals. *Int. Revue ges. Hydrobiol. Hydrogr.* **38**, 449–58.
Ivlev V. S. (1945) [The biological productivity of waters] (in Russian). *Uspekhi Sovremennoi Biologii* **19** (1), 98–120.
Ivlev V. S. (1966) The biological productivity of waters. *J. Fish. Res. Bd Can.* **23**, 1727–59. (Translation of Ivlev (1945) by W. E. Ricker.)
Jack R. L. (1921) Salt and gypsum resources in South Australia. *Bull. geol. Surv. S. Aust.* No. 8, 118 pp.
Jackson D. F. and **Dence W. A.** (1958) Primary productivity in a dichothermic lake. *Am. Midl. Nat.* **59**, 511–17.
Jenkin P. M. (1932) Report of the Percy Sladen expedition to some Rift Valley Lakes in Kenya in 1929. VII. Summary of the ecological results, with special reference to the alkaline lakes. *Ann. Mag. nat. Hist.* (ser. 10) **18**, 133–81.
Jennings J. N. (1957) Coastal dune lakes as exemplified from King Island, Tasmania. *Geogrl J.* **123**, 59–70.
Jennings J. N. and **Ahmad N.** (1957) The legacy of an ice cap. The lakes in the western part of the central plateau of Tasmania. *Aust. Geogr.* **7**, 62–75.
Jennings J. N., Noakes L. C., and **Burton G. M.** (1964) Notes on the Lake George and Lake Bathurst excursion. *A.N.Z.A.A.S. Canberra 1964* (Reprinted by Aust. Comm. Bureau of Mineral Resources, Geology and Geophysics, in 1967).
Jolly V. H. (1952) A preliminary study of the limnology of Lake Hayes. *Aust. J. mar. Freshwat. Res.* **3**, 74–91.
Jolly V. H. (1957a) Thermal stratification in some New Zealand lakes. *Proc. N.Z. ecol. Soc.* **4**, 43–4.

Jolly V. H. (1957b) A review of the copepod genus *Boeckella* in New Zealand. *Trans. R. Soc. N.Z.* **84**, 855–65.
Jolly V. H. (1958) A preliminary study of some New Zealand lakes. *Verh. int. Verein. theor. angew. Limnol.* **13**, 436–8.
Jolly V. H. (1965) Diurnal surface concentrations of zooplankton in Lake Taupo, New Zealand. *Hydrobiologia* **25**, 466–72.
Jolly V. H. (1966) The limnetic Crustacea of six reservoirs in the Sydney area of New South Wales. *Verh. int. Verein. theor. angew. Limnol.* **16**, 727–34.
Jolly V. H. (1968) The comparative limnology of some New Zealand lakes. I. Physical and chemical. *N.Z. Jl mar. Freshwat. Res.* **2**, 149–61.
Jolly V. H. and **Chapman M. A.** (1966) A preliminary biological study of the effects of pollution on Farmer's Creek and Cox's River, New South Wales. *Hydrobiologia* **27**, 160–92.
Jones J. R. E. (1964) *Fish and River Pollution*. Butterworths, London.
Journals of the House of Commons 1778–80. 37, col. 311. (Quoted in Turner I. (1968) *The Australian Dream*, p. 6. Sun Books, Melbourne.)
Juday C. (1922) Quantitative studies of the bottom fauna in the deeper waters of Lake Mendota. *Trans. Wis. Acad. Sci. Arts Lett.* **20**, 461–93.
Juday C. (1924) Summary of quantitative investigations on Green Lake, Wisconsin. *Int. Revue ges. Hydrobiol. Hydrogr.* **12**, 2–12.
Juday C. (1940) The annual energy budget of an inland lake. *Ecology* **21**, 438–50.
Jutson J. T. (1934) The physiography (geomorphology) of Western Australia. *Bull. geol. Surv. West. Aust.* No. 95, 366 pp.
Kajak Z. and **Ryback J. I.** (1966) Production and some trophic dependences in benthos against primary production and zooplankton production of several Masurian lakes. *Verh. int. Verein. theor. angew. Limnol.* **16**, 441–51.
Kaushik N. K. and **Hynes H. B. N.** (1968) Experimental study on the role of autumn-shed leaves in aquatic environments. *J. Ecol.* **56**, 229–43.
Kerekes J. and **Nursall J. R.** (1966) Eutrophication and senescence in a group of prairie-parkland lakes in Alberta, Canada. *Verh. int. Verein. theor. angew. Limnol.* **16**, 65–73.
Kerswill C. J. (1967) Studies on effects of forest spraying with insecticides 1952–63, on fish and aquatic invertebrates in New Brunswick streams. *J. Fish Res. Bd Can.* **24**, 701–8.
Keup L. E., Ingram W. M., and **Mackenthun K. M.** (1967) [Compilers] *Biology of Water Pollution*. United States Department of the Interior: Federal Water Pollution Control Administration.
Khlebovich V. V. (1968) Some peculiar features of the hydrochemical regime and the fauna of mesohaline waters. *Marine Biol.* **2**, 47–9.
Kiefer F. (1967) Cyclopiden aus salzhaltigen Binnengewässern Australiens (Copepoda). *Crustaceana* **12**, 292–302.
Kinne O. and **Kinne E. M.** (1962) Rates of development in embryos of a cyprinodont fish exposed to different temperature–salinity–oxygen combinations. *Can. J. Zool.* **40**, 231–53.
Knight-Jones E. W. and **Qasim S. Z.** (1955) Responses of some marine plankton animals to changes in hydrostatic pressure. *Nature, Lond.* **175**, 941–2.
Knight-Jones E. W. and **Qasim S. Z.** (1959) Effects of pressure pulses on the behaviour of some plankton animals. *Proc. Ind. Sci. Congr. Assoc.* (46th session).
Kolkwitz R. and **Marsson M.** (1908) Ökologie der pflanzlichen Saprobien. *Ber. dtsch. bot. Ges.* **26**, 505–19.
Kolkwitz R. and **Marsson M.** (1909) Ökologie der tierische Saprobien. Beiträge zur Lehre von der biologische Gewässerbeurteilung. *Int. Revue ges. Hydrobiol. Hydrogr.* **2**, 126–52.
Kronberger H. (1966) Developments in desalination and nuclear power. *Optima* **16**, 50–8.
Kuznetzov S. I. (1968) Recent studies on the role of microorganisms in the cycling of substances in lakes. *Limnol. Oceanogr.* **13**, 211–24.
Lagler K. F. (1968) Capture, sampling and examination of fish. In: *Methods for Assessment of Fish Production in Fresh Waters*. I.B.P. Handbook No. 3, Blackwell Scientific Publications, Oxford and Edinburgh.
Lake J. S. (1957) Trout populations and habitats in New South Wales. *Aust. J. mar. Freshwat. Res.* **8**, 414–50.
Lake J. S. (1959) The freshwater fishes of New South Wales. *N.S.W. State Fish. Res. Bull.* **5**, 1–19.
Lake J. S. (1967a) Rearing experiments with five species of Australian freshwater fishes. I. Inducement to spawning. *Aust. J. mar. Freshwat. Res.* **18**, 137–53.

Lake J. S. (1967b) Rearing experiments with five species of Australian freshwater fishes. II. Morphogenesis of ontogeny. *Aust. J. mar. Freshwat. Res.* **18**, 155–73.

Lake J. S. (1967c) Principal fishes of the Murray-Darling River system. In: *Australian Inland Waters and their Fauna: Eleven Studies* (ed. A. H. Weatherley). A.N.U. Press, Canberra.

Lake J. S. (1970) Reproductive adaptations of some Australian freshwater fishes (abstract only). *Aust. Soc. Limnol. Bull.* **3**, 22–3.

Lake J. S. (1971) *Freshwater Fishes and Rivers of Australia.* Thomas Nelson, Sydney.

Lake J. S. (undated) Freshwater fish of the Murray-Darling River system. *N.S.W. State Fish. Res. Bull.* **7**, 1–48.

Larkin P. A. and Northcote T. G. (1958) Factors in lake typology in British Columbia, Canada. *Verh. int. Verein. theor. angew. Limnol.* **13**, 252–63.

Le Cren D. (1969) Estimates of fish populations and production in small streams in England. In: *Symposium on Salmon and Trout in Streams* [H. R. MacMillan lectures in fisheries] (ed. T. G. Northcote). Inst. Fisheries, Univ. Brit. Columbia, Vancouver.

Lee L. C. (1969) Osmotic regulation and salinity tolerance of *Taeniomembras microstomus* (Gunter 1861) (Pisces: Mugiliformes: Atherinidae) from Australian salt lakes. *Aust. J. mar. Freshwat. Res.* **20**, 157–62.

Lee L. C. and Williams W. D. (1970) Meristic differences between two conspecific fish populations in Australian salt lakes. *J. Fish. Biol.* **2**, 55–6.

Lemmermann-Bremen E. (1899) Ergebnisse einer Reise nach dem Pacific (H. Schauinsland 1896–97). Planktonalgen. *Abh. naturv. Ver. Bremen* **16**, 313–98.

Lenz F. (1925) Chironomiden und Seetypenlehre. *Naturwissenschaften* **13**, 5–10.

Lenz F. (1927) Chironomiden aus norwegischen Hochgebirgsseen. Zugleich ein Beitrag zur Seetypenfrage. *Nyt Mag. Naturvid.* **66**, 111–92.

Leopold L. B. (1953) Downstream change of velocity in rivers. *Am. J. Sci.* **251**, 606–24.

Leopold L. B., Wolman M. G., and Miller J. P. (1964) *Fluvial Processes in Geomorphology.* Freeman, San Francisco and London.

Liebmann H. (1951) Handbuch der Frischwasser- und Abwasserbiologie. Bd. I. Oldenburg Vlg, Munich.

Liebmann H. (1962) Handbuch der Frischwasser- und Abwasserbiologie. Bd. II. Fischer Vlg, Munich.

Lim K. H. (1964) The biology of *Austrochiltonia subtenuis* (Sayce) 1902. M.Sc. Thesis, Monash University.

Lim K. M. and Williams W. D. (1971) Ecology of *Austrochiltonia subtenuis* (Sayce) (Amphipoda: Hyalellidae). *Crustaceana* **20**, 19–24.

Lindeman R. L. (1941) Seasonal food-cycle dynamics in a senescent lake. *Am. Midl. Nat.* **26**, 636–73.

Lindeman R. L. (1942) The trophic-dynamic aspect of ecology. *Ecology* **23**, 399–413.

Linden C. A. (1968) A theoretical investigation of internal waves in the Blue Lake. *Horace Lamb Centre for Oceanogr. Res. Research Paper* No. 27.

Little E. C. S. (1966) The invasion of man-made lakes by plants. In: *Man-Made Lakes* (ed. R. H. Lowe-McConnell). Institute of Biology and Academic Press, London.

Livingstone D. A. (1963) Chemical composition of rivers and lakes. *Prof. Pap. U.S. geol. Surv.* No. 440.

Livingstone D. A. and Boykin J. C. (1962) Vertical distribution of phosphorus in Linsley Pond mud. *Limnol. Oceanogr.* **7**, 57–62.

Llewellyn L. C. (1968) Adaptations of *Madigania unicolor* (spangled perch) to temporary waters. *Aust. Soc. Limnol. Newsl.* **6** (1), 6–7.

Lockwood A. P. M. (1959) The osmotic and ionic regulation of *Asellus aquaticus* (L.). *J. exp. Biol.* **36**, 546–55.

Löffler H. (1961) Beiträge zur Kenntnis der Iranischen Binnengewässer. II. Regionallimnologische Studie mit besonderer Berücksichtigung der Crustaceenfauna. *Int. Revue ges. Hydrobiol. Hydrogr.* **46**, 309–407.

Lowndes A. G. (1952) Hydrogen ion concentration and the distribution of freshwater Entomostraca. *Ann. Mag. nat. Hist.* (Ser. 12) **5**, 58–65.

Lund J. W. G. (1950) Studies on *Asterionella formosa* Hass. II. Nutrient depletion and the spring maximum. *J. Ecol.* **38**, 1–35.

Lund J. W. G. (1965) The ecology of the freshwater phytoplankton. *Biol. Rev.* **40**, 231–93.

Lund J. W. G. (1967) Planktonic algae and the ecology of lakes. *Sci. Prog., Lond.* **55**, 401–19.

Lundbeck J. (1926) Die Bodentierwelt norddeutscher Seen. *Arch. Hydrobiol.* **7** (Suppl.), 1–473.
Lundbeck J. (1936) Untersuchungen über die Bodenbesiedlung der Alpenrandseen. *Arch. Hydrobiol.* **10**, 207–358.
Luther H. and **Rzóska J.** (1971) *Project Aqua.* I.B.P. Handbook No. 21, Blackwell Scientific Publications, Oxford and Edinburgh.
Macan T. T. (1961) Factors that limit the range of freshwater animals. *Biol. Rev.* **36**, 151–98.
Macan T. T. (1963) *Freshwater Ecology.* Longmans Green, London.
Macfadyen A. (1948) The meaning of productivity in biological systems. *J. Anim. Ecol.* **17**, 75–80.
Macfadyen A. (1957) *Animal Ecology: Aims and Methods.* Sir Isaac Pitman & Sons, London.
Mackenthun K. M. (1965) *Nitrogen and Phosphorus in Water: an Annotated Selected Bibliography of their Biological Effects.* U.S. Dept Health, Education, and Welfare, Washington, D.C.
Mackenthun K. M., Ingram W. M., and **Porges R.** (1964) *Limnological Aspects of Recreational Lakes.* U.S. Public Health Service Publication No. 1167.
Mackereth F. J. H. (1966) Some chemical observations of post-glacial lake sediments. *Phil. Trans. R. Soc.* (Ser. B) **250**, 165–213.
Mackintosh N. A. (1937) The seasonal circulation of the Antarctic macroplankton. '*Discovery*' *Rep., Cambridge* **16**, 365–412.
Maguire B. (Jnr) (1963) The passive dispersal of small aquatic organisms and their colonization of isolated bodies of water. *Ecol. Monogr.* **33**, 161–85.
Main A. R. (1968) Ecology, systematics and evolution of Australian frogs. *Adv. Ecol. Res.* **5**, 37–86.
Mann K. H. (1965) Energy transformations by a population of fish in the River Thames. *J. Anim. Ecol.* **34**, 253–75.
Mann K. H. (1969) The dynamics of aquatic ecosystems. *Adv. Ecol. Res.* **6**, 1–81.
Margalef R. (1968) *Perspectives in Ecological Theory.* Univ. Chicago Press, Chicago.
Marples B. J. (1962) *An Introduction to Freshwater Life in New Zealand.* Whitcombe and Tombs, Wellington.
Martin A. A. (1967a) Australian anuran life histories: some evolutionary and ecological aspects. In: *Australian Inland Waters and Their Fauna: Eleven Studies* (ed. A. H. Weatherley). A.N.U. Press, Canberra.
Martin A. A. (1967b) The biology of tadpoles. *Aust. Mus. Mag.* **15**, 326–30.
McAllister C. D., Parsons T. R., Stephens K., and **Strickland J. D. H.** (1961) Measurements of primary production in coastal seawater using a large plastic sphere. *Limnol. Oceanogr.* **6**, 237–58.
McCall G. J. H. (1967) The progress of meteorites in Western Australia and its implications. In: *Yearbook of Astronomy 1968.* Eyre and Spottiswoode, London.
McConnell W. J. and **Sigler W. F.** (1959) Chlorophyll and productivity in a mountain river. *Limnol. Oceanogr.* **4**, 335–51.
McDowall R. M. (1964a) The affinities and derivation of the New Zealand freshwater fish fauna. *Tuatara* **12** (2), 59–67.
McDowall R. M. (1964b) A bibliography of the indigenous freshwater fishes of New Zealand. *Trans. R. Soc. N.Z.* (Zool.) **5** (1), 1–38.
McLaren I. A. (1963) Effects of temperature on growth of zooplankton and the adaptive value of vertical migration. *J. Fish. Res. Bd Can.* **20**, 685–727.
McLay C. L. (1968) A study of drift in the Kakanui River, New Zealand. *Aust. J. mar. Freshwat. Res.* **19**, 139–49.
McLean J. A. (1966) A comparative ecological study of three stream faunas in the Auckland area. *Tane* **12**, 97–102.
McLeod I. R. (1965) (ed.) *Australian Mineral Industry: The Mineral Deposits.* Dept of National Development, Bureau of Mineral Resources, Geology and Geophysics. Bull. No. 72.
McMichael D. F. (1952) An ecological study of Warrah Creek Warrah Sanctuary N.S.W. B.Sc. Honours Thesis, University of Sydney.
Mees G. F. (1962) 4. The subterranean freshwater fauna of Yardie Creek Station, North West Cape, Western Australia. *J. Proc. R. Soc. West. Aust.* **45**, 24–32.

Mellanby K. (1967) *Pesticides and Pollution.* (New Naturalist.) Collins, London.
Merilainen J. (1970) On the limnology of the meromictic Lake Valkiajarvi, in the Finnish Lake District. *Annls bot. fenn.* **7**, 29–51.
Milne A. (1961) Definition of competition among animals. *Symp. Soc. exp. Biol.* **15**, 40–61.
Minshall G. W. (1967) Role of allochthonous detritus in the trophic structure of a woodland springbrook community. *Ecology* **48**, 139–49.
Minshall G. W. (1968) Community dynamics of the benthic fauna in a woodland springbrook. *Hydrobiologia* **32**, 305–39.
Moore W. G. (1963) Some interspecies relationships in anostracan populations of certain Louisiana ponds. *Ecology* **44**, 131–9.
Morrissy N. M. (1967) The ecology of trout in South Australia. Ph.D. Thesis, University of Adelaide.
Morrissy N. M. (1970) Report on marron in farm dams (*Cherax tenuimanus*). Report V, Department of Fisheries and Fauna, Western Australia, Perth.
Mortimer C. H. (1941) The exchange of dissolved substances between mud and water in lakes [Sections I and II]. *J. Ecol.* **29**, 280–329.
Mortimer C. H. (1942) The exchange of dissolved substances between mud and water in lakes [Sections III and IV]. *J. Ecol.* **30**, 147–201.
Mortimer C. H. (1956) The oxygen content of air-saturated fresh waters, and aids in calculating percentage saturation. *Mitt. int. Verein. theor. angew. Limnol.* **6**, 1–20.
Mulvany J. B. (1951) Report on a mission to Europe in connection with disposal of gasification wastes, stream pollution and sewage treatment. *State Rivers and Water Supply Commission*, Techn. Bull. No. 4, 1–46 (Melbourne).
Muttkowski R. A. (1918) The fauna of Lake Mendota. A qualitative and quantitative survey with special reference to the insects. *Trans. Wis. Acad. Sci. Arts Lett.* **19**, 374–482.
National Academy of Sciences (1969) *Eutrophication: Causes, Consequences, Correctives.* [Proceedings of a symposium held in Madison 1967] Nat. Acad. Sci., Washington, D.C.
National Academy of Sciences (1970) *Recommended Procedures for Measuring the Productivity of Plankton Standing Stock and Related Oceanic Properties.* Nat. Acad. Sci., Washington, D.C.
Nauwerck A. (1963) Die Beziehungen zwischen Zooplankton und Phytoplankton im See Erken. *Symb. bot. upsal.* **17** (5), 1–163.
Nees J. and **Dugdale R. C.** (1959) Computation of production for populations of aquatic midge larvae. *Ecology* **40**, 425–30.
Negus C. L. (1966) A quantitative study of growth and production of unionid mussels in the River Thames at Reading. *J. Anim. Ecol.* **35**, 513–32.
Nelson D. J. and **Scott D. C.** (1962) Role of detritus in the productivity of a rock-outcrop community in a piedmont stream. *Limnol. Oceanogr.* **7**, 396–413.
Nicholls A. G. (1933) On the biology of *Calanus finmarchicus*. III. Distribution and diurnal migration in the Clyde sea-area. *J. mar. biol. Ass. U.K.* **19**, 139–64.
Nicholls A. G. (1946) Syncarida in relation to the interstitial habitat. *Nature, Lond.* **158**, 934.
Nicholls A. G. (1958) The population of a trout stream and the survival of released fish. *Aust. J. mar. Freshwat. Res.* **9**, 319–50.
Nixon S. W. (1969) A synthetic microcosm. *Limnol. Oceanogr.* **14**, 142–5.
Norman J. R. (1963) *A History of Fishes* (2nd edn rev. P. H. Greenwood). Ernest Benn, London.
Nygaard G. (1949) Hydrobiological studies of some Danish ponds and lakes. II. The quotient hypothesis and some new or little-known phytoplankton organisms. *Kgl. Danske Videnskab. Selskab, Biol. Skrifter* **7**, 293.
Nygaard G. (1955) On the productivity of five Danish waters. *Verh. int. Verein. theor. angew. Limnol.* **12**, 123–33.
Odum H. T. (1956) Primary production in flowing waters. *Limnol. Oceanogr.* **1**, 102–17.
Odum H. T. (1957) Trophic structure and productivity of Silver Springs, Florida. *Ecol. Monogr.* **27**, 55–112.
O'Farrell A. F. (1949) The Blue Hole Expedition, 1949. A short-term limnological study of the Gara River, New South Wales. *J. New England Univ. Coll. Sci.* **5**, 46–82.
Ohle W. (1956) Bioactivity, production and energy utilization of lakes. *Limnol. Oceanogr.* **1**, 139–49.

Økland J. (1964) The eutrophic lake Borrevann (Norway)—an ecological study on shore and bottom fauna with special reference to gastropods, including a hydrographic survey. *Folia limnol. scand.* **13**, 1–337.
Olive J. R. (1955) Some aspects of plankton associations in the high mountain lakes of Colorado (U.S.A.). *Verh. int. Verein. theor. angew. Limnol.* **12**, 425–35.
Ollier C. D. (1967) Maars: Their characteristics, varieties and definition. *Bull. Volcanol.* **31**, 45–73.
Ollier C. D. and **Joyce E. B.** (1964) Volcanic physiography of the western plains of Victoria. *Proc. R. Soc. Vict.* **77**, 357–76.
Olsen S. (1967) Recent trends in the determination of orthophosphate in water. In: *Chemical Environment in the Aquatic Habitat* (ed. H. L. Golterman and R. S. Clymo). N.V. Noord-Hollandsche Uitgevers Maatschappij, Amsterdam.
Owens M. (1965) Some factors involved in the use of dissolved-oxygen distributions in streams to determine productivity. *Mem. Ist. Ital. Idrobiol.* **18** (Suppl.), 209–24.
Packard A. (1955) Theories of vertical migration (unpublished manuscript of paper read to a meeting of the New Zealand Oceanographic Institute in Wellington on 31 March 1955).
Palmer C. M. (1962) *Algae in Water Supplies*. U.S. Public Health Service Publication No. 657.
Pantle R. and **Buck H.** (1955) Die biologische Überwachung der Gewässer und die Darstellung der Ergebnisse. *Gas- u. WassFach* **96**, 604.
Parsons W. T. (1963) Water hyacinth, a pest of world waterways. *J. Dep. Agric. Vict.* **61**, 23–7.
Pennak R. W. (1957) Species composition of limnetic zooplankton communities. *Limnol. Oceanogr.* **2**, 222–32.
Percival E. (1949) Summary of a report on a chemical survey of Lakes Lyndon and Pearson from March 1947 to October 1948. In: *N. Cant. Acclim. Soc. Ann. Report* No. 85.
Peterson D. F. (1966) Man and his water resource. Faculty Honor lecture, Utah State University, Logan, Utah, 1–44.
Peterson J. A. (1966) Glaciation of the Frenchman's Cap National Park. *Pap. Proc. R. Soc. Tasm.* **100**, 117–29.
Peterson J. A. (1968) Cirque morphology and pleistocene ice formation conditions in southeastern Australia. *Aust. Geograph. Studies* **6**, 67–83.
Phillips J. S. (1929) A report on the food of trout and other conditions affecting their well-being in the Wellington district. *Fish. Bull. N.Z.* **1**, 1–31.
Phillips J. S. (1931) A further report on conditions affecting the well-being of trout in New Zealand. *Fish. Bull. N.Z.* **3**, 1–27.
Pieczyńska E. and **Szczepańska W.** (1966) Primary production in the littoral of several Masurian lakes. *Verh. int. Verein. theor. angew. Limnol.* **16**, 372–9.
Playfair G. I. (1914) Contributions to a knowledge of the biology of the Richmond River. *Proc. Linn. Soc. N.S.W.* **39**, 93–151.
Pollard D. (1966) Landlocking in diadromus salmonoid fishes—with special reference to the common jollytail (*Galaxias attenuatus*). *Aust. Soc. Limnol. Newsl.* **5** (2), 13–16.
Por F. D. (1968) Solar lake on the shores of the Red Sea. *Nature, Lond.* **218**, 860–1.
Pora E. A. (1969) L'importance du facteur rhopique (équilibre ionique) pour la vie aquatique. *Verh. int. Verein. theor. angew. Limnol.* **17**, 970–86.
Potts W. T. W. and **Parry G.** (1964) *Osmotic and Ionic Regulation in Animals*. Pergamon Press, Oxford.
Powell A. W. G. (1946) Ecology of the fresh water fauna of Lake St Clair, particularly the Copepoda, with special reference to diurnal and seasonal variations in conditions. *Pap. Proc. R. Soc. Tasm.* **1945**, 63–127.
Pütter A. (1924) Der Umfang der Kohlensäurer-reduktion durch die Planktonalgen. *Pflügers Arch. ges. Physiol.* **205**, 293–312.
Ragotzkie R. A. and **Bryson R. A.** (1963) Correlation of currents with the distribution of adult *Daphnia* in Lake Mendota. *J. mar. Res.* **12**, 157–72.
Ravera O. and **Tonolli V.** (1956) Body size and number of eggs in diaptomids, as related to water renewal in mountain lakes. *Limnol. Oceanogr.* **1**, 118–22.
Rawson D. S. (1951) The total mineral content of lake waters. *Ecology* **32**, 669–72.

Rawson D. S. (1953) The bottom fauna of Great Slave Lake. *J. Fish. Res. Bd Can.* **10**, 486–520.
Rawson D. S. (1955) Morphometry as a dominant factor in the productivity of large lakes. *Verh. int. Verein. theor. angew. Limnol.* **12**, 164–75.
Rawson D. S. (1956) Algal indicators of trophic lake types. *Limnol. Oceanogr.* **1**, 18–25.
Rawson D. S. and **Moore J. E.** (1944) The saline lakes of Saskatchewan. *Can. J. Res.* (Ser. D) **22**, 141–201.
Reid L. W. (1966) Wastewater pollution and general eutrophication of a hydroelectric impoundment. *J. Wat. Pollut. Control Fed.* **38**, 165–74.
Report from the Senate Select Committee on water pollution (1970) *Water Pollution in Australia.* Aust. Comm. Govt Print. Office, Canberra.
Report of the first, second, third, fourth, and fifth Interstate Conferences on Artesian Water (1913, 1914, 1921, 1925, 1928). Australian capital cities, Govt Printers.
Report of the State Development Committee on the Introduction of European Carp into Victorian waters. Govt Printer, Melbourne.
Richter C. P. and **Maclean A.** (1939) Salt taste thresholds of humans. *Am. J. Physiol.* **126**, 1–6.
Ricker W. E. (1946) Production and utilization of fish populations. *Ecol. Monogr.* **16**, 374–91.
Ricker W. E. (1968) (ed.) *Methods for Assessment of Fish Production in Fresh Waters.* I.B.P. Handbook No. 3, Blackwell Scientific Publications, Oxford and Edinburgh.
Ricker W. E. and **Foerster R. E.** (1948) Computation of fish production. *Bull. Bingham oceanogr. Coll.* **11**, 173–211.
Riek E. F. (1953) The Australian freshwater prawns of the family Atyidae. *Rec. Aust. Mus.* **23**, 111–21.
Riek E. F. (1959) The Australian freshwater Crustacea. In: *Biogeography and Ecology in Australia* (eds A. Keast, R. L. Crocker, C. S. Christian). Junk, The Hague.
Ringer S. (1883) *J. Physiol* **4**, 29–42. (Not seen.)
River Murray Commission (1957) Report on the River Murray flood problem (with particular reference to the 1956 flood). Dept of Defence Production, Melbourne.
Rodhe W. (1949) The ionic composition of lake waters. *Verh. int. Verein. theor. angew. Limnol.* **10**, 377–86.
Rodhe W. (1955) Can plankton production proceed during winter darkness in subarctic lakes? *Verh. int. Verein. theor. angew. Limnol.* **12**, 117–22.
Rodhe W., Vollenweider R. A., and **Nauwerck A.** (1958) The primary production and standing crop of phytoplankton. In: *Perspectives in Marine Biology* (ed. A. A. Buzzati-Traverso). Univ. California Press, Los Angeles.
Roos T. (1957) Studies on upstream migration in adult stream-dwelling insects. I. *Rep. Inst. Freshwat. Res. Drottningholm* **38**, 167–93.
Round F. E. and **Brook A. J.** (1959) The phytoplankton of some Irish loughs and an assessment of their trophic status. *Proc. R. Ir. Acad.* (Ser. B) **60**, 167–91.
Round F. E. (1965) *The Biology of the Algae.* Edward Arnold, London.
Rounsefell G. A. (1946) Fish production in lakes as a guide for estimating production in proposed reservoirs. *Copeia* **1946** (1), 29–40.
Rudd R. L. (1964) *Pesticides and the Living Landscape.* University of Wisconsin Press, Madison and Milwaukee.
Rudjakov J. A. (1970) The possible causes of diel vertical migrations of planktonic animals. *Marine Biol.* **6**, 98–105.
Russell H. C. (1886) Anniversary address [to the Royal Society of N.S.W. in 1885]. *Proc. R. Soc. N.S.W.* **19**, 1–27.
Russell H. C. (1887) Notes upon floods in Lake George. *Proc. R. Soc. N.S.W.* **20**, 241–60.
Russell I. C. (1893) Geological reconnaissance in central Washington. *Bull. U.S. geol. Surv.* No. 108, 168 pp.
Ryckman R. E. and **Ames C. T.** (1953) Insects reared from cacti in Arizona. *Pan-Pacif. Ent.* **29**, 163–4.
Rylov V. M. (1963) Freshwater Cyclopoida. In: *Fauna of U.S.S.R.* Vol. III, No. 3. Crustacea (ed. E. N. Pavlovskii). (Trans. A. Mercado.) Israel Program for Scientific Translations, Jerusalem.

Rzóska J. (1961) Observations on tropical rainpools and general remarks on temporary waters. *Hydrobiologia* **17**, 265–86.
Samuel L. W. (1951) The salinity and hardness of some rivers and streams in the southwest of Western Australia. *Annual Report of the Government Chemical Laboratories for 1949, Agriculture, Forestry, and Water Supply Division,* Appendix, 159–63.
Sars G. O. (1889) On some freshwater Ostracoda and Copepoda raised from dried Australian mud. *Vid. Selsk. Forh. Christiania* **8**, 1–79.
Sars G. O. (1914) *Daphnia carinata* King and its remarkable varieties. *Arch. Math. Naturv.* **34** (1), 1–14.
Sattler W. (1963) Über den Körperbau und Ethologie der Larve und Puppe von *Macronema* Pict. (Hydropsychidae), ein als Larve sich von 'Mikro-Drift' ernährendes Trichopter aus dem Amazongebiet. *Arch. Hydrobiol.* **59**, 26–60.
Sauberer F. (1950) Die spektrale Strahlungsdurchlässigkeit des Eises. *Wett. Leben* **2** (9/10), 193–7.
Saunders G. W. (1969) Some aspects of feeding in zooplankton. In: *Eutrophication: Causes, Consequences, Correctives.* Nat. Acad. Sci., Washington, D.C.
Sawyer C. N. (1966) Basic concepts of eutrophication. *J. Wat. Pollut. Control Fed.* **38**, 737–44.
Schindler D. W. (1968) Feeding, assimilation and respiration rates of *Daphnia magna* under various environmental conditions and their relation to production estimates. *J. Anim. Ecol.* **37**, 369–86.
Schminke H. K. and **Noodt W.** (1968) Discovery of Bathynellacea, Stygiocaridacea and other interstitial Crustacea in New Zealand. *Naturwissenschaften* **4**, 184–5.
Schmitz W. (1961) Fließgewässerforschung-Hydrographie und Botanik. *Verh. int. Verein. theor. angew. Limnol.* **14**, 541–86.
Schwoerbel J. (1961) Über die Lebensbedingungen und die Besiedlung des hyporheischen Lebensraumes. *Arch. Hydrobiol.* **25** (Suppl.), 182–214.
Schwoerbel J. (1966) *Methoden der Hydrobiologie.* Franckh'sche Verlagshandlung, Stuttgart.
Scott W. W. (1939) *Standard Methods of Chemical Analysis* (5th edn, ed. N. H. Furman). Van Nostrand, Princeton.
Sculthorpe C. D. (1967) *The Biology of Aquatic Vascular Plants.* Edward Arnold, London.
Serventy D. L. (1938) *Palaemonetes australis* Dakin in south-western Australia. *J. R. Soc. West. Aust.* **24**, 51–7.
Shapiro J. (1957) Chemical and biological studies on the yellow organic acids of lake water. *Limnol. Oceanogr.* **2**, 161–79.
Sheldon R. W. and **Parsons T. R.** (1967) *A Practical Manual on the Use of the Coulter Counter in Marine Science.* Coulter Electronic Sales Co., Canada.
Simmonds M. A. (1962) Notes on pollution of ground-water supplies in Queensland, Australia. *Proc. Soc. Wat. Treat. Exam.* **11**, 76–83.
Simpson E. S. (1928) Problems of water supply in Western Australia. *Rep. Australas. Ass. Advmt Sci.* **18**, 634–74.
Sládeček V. (1965) The future of the saprobity system. *Hydrobiologia* **25**, 518–37.
Sládeček V. (1969) The measures of saprobity. *Verh. int. Verein. theor. angew. Limnol.* **17**, 546–59.
Sládečková A. (1962) Limnological investigation methods for the periphyton ('Aufwuchs') community. *Bot. Rev.* **28**, 286–350.
Slobodkin L. B. (1962) Energy in animal ecology. *Adv. Ecol. Res.* **1**, 69–101.
Smith F. E. MS (1955) A restudy of the Lindeman approach to community dynamics. Pp. 27–32 in 'Lectures on population dynamics', multigraphed by Scripps Institution of Oceanography, La Jolla, California. (Quoted in part in Ricker's translation of Ivlev's 1945 paper, p. 1746).
Smith J. B., Tatsumoko M., and **Hood D. W.** (1960) Carbamino carboxylic acids in photosynthesis. *Limnol. Oceanogr.* **5**, 425–31.
Sorokin Yu I. (1965) On the trophic role of chemosynthesis and bacterial biosynthesis in water bodies. *Mem. Ist. Ital. Idrobiol.* **18** (Suppl.), 187–205.
Sorokin Yu I. (1969) Part I. General methods [bacterial production]. In: *A Manual on*

Methods for Measuring Primary Production in Aquatic Environments (ed. R. A. Vollenweider). I.B.P. Handbook No. 12, Blackwell Scientific Publications, Oxford and Edinburgh.
Stahl B. (1959) The developmental history of the chironomid and *Chaoborus* faunas of Myers Lake. *Invest. Indiana Lakes Streams* **5**, 47–102.
State Rivers and Water Supply Commission (1954) River Gaugings. Vol. VII. 1943–48. Govt Printer, Melbourne.
Steemann Nielsen E. (1952) The use of radioactive carbon (C^{14}) for measuring organic production in the sea. *J. Cons. perm. int. Explor. Mer* **18**, 117–40.
Steemann Nielsen E. (1965) On the terminology concerning production in aquatic ecology with a note about excess production. *Arch. Hydrobiol.* **61**, 184–9.
Steenis C. G. G. J. van (1949) Podostemaceae. *Flora Malesiana Bull.* (Ser. 1) **4**, 65–8.
Steenis C. G. G. J. van (1952) Rheophytes. *Proc. R. Soc. Qld* **62**, 61–8.
Steinböck O. (1958) Grundsätzliches zum 'Kryoeutrophen' See. *Verh. int. Verein. theor. angew. Limnol.* **13**, 181–90.
Stokell G. (1955) *Fresh Water Fishes of New Zealand.* Simpson and Williams, Christchurch.
Stoner D. (1923) Insects taken at hot springs, New Zealand. *Entopath. News* **34**, 88–90.
Stout V. M. (1964) Studies on temporary ponds in Canterbury, New Zealand. *Verh. int. Verein. theor. angew. Limnol.* **15**, 209–14.
Stout V. M. (1969a) Lakes in the mountain region of Canterbury, New Zealand. *Verh. int. Verein. theor. angew. Limnol.* **17**, 404–13.
Stout V. M. (1969b) Life in lakes and ponds. In: *The Natural History of Canterbury* (ed. G. A. Knox). A. H. and A. W. Reed, Wellington.
Stout V. M. et al. (1969) The invertebrate fauna of the rivers and streams. In: *The Natural History of Canterbury* (ed. G. A. Knox). A. H. and A. W. Reed, Wellington.
Straškraba M. (1968) Der Anteil der höheren Pflanzen an der Produktion der stehenden Gewässer. *Mitt. int. Verein. theor. angew. Limnol.* **14**, 212–30.
Strickland J. D. H. (1960) Measuring the production of marine phytoplankton. *Bull. Fish. Res. Bd Can.* **122**, 1–172.
Strickland J. D. H. (1965) Production of organic matter in the primary stages of the marine food chain. In: *Chemical Oceanography*, Vol. I (eds J. P. Riley and G. Skirrow). Academic Press, New York.
Strøm K. Münster (1932) Tyrifjord, a limnological study. *Skrift. ut. Norske Vidensk.-Akad. Oslo. I. Math.-Naturw.Klasse*, **1** (3), 1–84.
Sutcliffe W. H., Baylor, E. R., and Menzel D. W. (1963) Sea surface chemistry and Langmuir circulation. *Deep Sea Res.* **10**, 233–43.
Swain R., Wilson I. S., Hickman J. L., and Ong J. E. (1970) *Allanspides helonomus* gen. et sp. nov. (Crustacea: Syncarida) from Tasmania. *Rec. Queen Vict. Mus.*, No. 35, 1–13.
Symposium über den Einfluss der Strömungsgeschwindigkeit auf die Organismen des Wassers (1962). *Schweiz. Z. Hydrol.* **24**, 353–484.
Tabor H. Z. (1966) Solar ponds. *Sci. J.* **2** (6), 66–71.
Talling J. F. (1951) The element of chance in pond populations. *Naturalist, Hull* **1951**, 157–70.
Tansley A. G. (1935) The use and abuse of vegetational concepts and terms. *Ecology* **16**, 284–307.
Tansley A. G. (1939) *The British Islands and their Vegetation.* Cambridge University Press, Cambridge.
Tarzwell C. M. (1965) In: The value and use of water quality criteria: discussion. In: *Biological Problems in water Pollution, Trans. 1962 Seminar* Robert A. Taft Sanit. Eng. Cent., Cincinnati.
Taub F. B. (1969a) A biological model of a freshwater community: a gnotobiotic ecosystem. *Limnol. Oceanogr.* **14**, 136–42.
Taub F. B. (1969b) Gnotobiotic models of freshwater communities. *Verh. int. Verein. theor. angew. Limnol.* **17**, 485–96.
Teal J. M. (1957) Community metabolism in a temperate cold spring. *Ecol. Monogr.* **27**, 283–302.
Thienemann A. (1920) Biologische Seetypen und die Gründung einer hydrobiologischen Anstalt am Bodensee. *Arch. Hydrobiol.* **13**, 347–70.

Thienemann A. (1931) Der Produktionsbegriff in der Biologie. *Arch. Hydrobiol.* **22**, 606–22.
Thomas E. A. (1965) The eutrophication of lakes and rivers, cause and prevention. In: *Biological Problems in Water Pollution*, Trans. 1962 Seminar Robert A. Taft Sanit. Eng. Cent., Cincinnati.
Tilly L. J. (1968) The structure and dynamics of Cone Spring. *Ecol. Monogr.* **38**, 169–97.
Timms B. V. (1967) Ecological studies on the Entomostraca of a Queensland pond with special reference to *Boeckella minuta* Sars (Copepoda: Calanoida). *Proc. R. Soc. Qd* **79**, 41–70.
Timms B. V. (1968a) Comparative species composition of limnetic planktonic crustacean communities in south-east Queensland, Australia. *Hydrobiologia* **31**, 474–80.
Timms B. V. (1968b) Water renewal in reservoirs, and its relation to body size and egg numbers in copepods. *Hydrobiologia* **31**, 481–91.
Timms B. V. (1969) A preliminary limnological survey of the Wooli Lakes, New South Wales. *Proc. Linn. Soc. N.S.W.* **94**, 105–12.
Timms B. V. (1970a) Chemical and zooplankton studies on lentic habitats of north-eastern New South Wales. *Aust. J. mar. Freshwat. Res.* **21**, 11–33.
Timms B. V. (1970b) Variations in the water chemistry of four small lentic localities in the Hunter Valley, New South Wales, Australia. *Aust. Soc. Limnol. Bull.* **3**, 36–9.
Timms B. V. (1970c) Aspects of the limnology of five small reservoirs in New South Wales. *Proc. Linn. Soc. N.S.W.* **95**, 46–59.
Timms B. V. and **Midgley S. H.** (1969) The limnology of Borumba Dam, Queensland. *Proc. R. Soc. Qd* **81**, 27–42.
Truesdale G. A., Downing A. L., and **Lowden G. F.** (1955) The solubility of oxygen in pure water and sea water. *A. appl. Chem. Lond.* **5**, 53–62.
Twidale C. R. (1968) *Geomorphology with Special Reference to Australia*. Nelson, Melbourne.
Twomey S. (1953) The identification of individual hygroscopic particles in the atmosphere by a phase-transition method. *J. appl. Physics, Lancaster* **24**, 1099–102.
Tyler J. E. (1965) Colour of 'pure' water. *Nature, Lond.* **208**, 549–50.
Tyler J. E. (1968) The Secchi disc. *Limnol. Oceanogr.* **13**, 1–6.
Tyler P. A. (1970) Taxonomy of Australian desmids. I. The genus *Micrasterias* in Tasmania and south-east Australia. *Br. phycol. J.* **5**, 211–34.
Tyler P. A. and **Marshall K. C.** (1967a) Form and function in manganese-oxidizing bacteria. *Arch. Mikrobiol.* **56**, 344–53.
Tyler P. A. and **Marshall K. C.** (1967b) *Hyphomicrobia*—a significant factor in manganese problems. *J. Am. Wat. Wks Ass.* **59**, 1043–8.
Tyler P. A. and **Marshall K. C.** (1967c) Microbial oxidation of manganese in hydro-electric pipelines. *Antonie van Leeuwenhoek* **33**, 171–83.
United States Dept of the Interior (1957) Preliminary toxicity studies with hexadecanol reservoir evaporation reduction. Bureau of Reclamation Chemical Engineering Lab. Rep., No. 51-10. (Not seen.)
Vandel A. (1965) *Biospeleology. The Biology of Cavernicolous Animals*. Pergamon Press, London and New York.
Venables J. R. C. (1970) Factors affecting salinity in Western District streams. *Aqua, Melb.* Sept., 20–1.
Verduin J. (1956a) Primary production in lakes. *Limnol. Oceanogr.* **1**, 85–91.
Verduin J. (1956b) Energy fixation and utilization by natural communities in western Lake Erie. *Ecology* **37**, 40–9.
Victorian Yearbook (1968) Commonwealth Bureau of Census and Statistics, Melbourne.
Voigt G. K. (1960) Alteration of the composition of rainwater by trees. *Am. Midl. Nat.* **63**, 321–6.
Vollenweider R. A. (1960) Beiträge zur Kenntnis optischer Eigenschaften der Gewässer und Primärproduktion. *Mem. Ist. Ital. Idrobiol.* **12**, 201–44.
Vollenweider R. A. (1968) Scientific fundamentals of the eutrophication of lakes and flowing waters, with particular reference to nitrogen and phosphorus as factors in eutrophication. *Organization for Economic Co-operation and Development*, Paris. DAS/CSI/68.27, 159 pp.
Vollenweider R. A. (1969) (ed.) *A Manual on Methods for Measuring Primary Production in*

Aquatic Environments. I.B.P. Handbook No. 12, Blackwell Scientific Publications, Oxford and Edinburgh.
Walker K. F. (1969) The ecology and distribution of *Halicarcinus lacustris* (Brachyura: Hymenosomatidae) in Australian inland waters. *Aust. J. mar. Freshwat. Res.* **20**, 163–73.
Walker K. F., Williams W. D., and **Hammer U. T.** (1970) The Miller method for oxygen determination applied to saline lakes. *Limnol. Oceanogr.* **15**, 814–5.
Waring G. A. (1965) Thermal springs of the United States and other countries of the world. A summary. *Prof. Pap. U.S. geol. Surv.* No. 492.
Waters T. F. (1965) Interpretation of invertebrate drift in streams. *Ecology* **46**, 327–34.
Waters T. F. (1966) Production rate, population density, and drift of a stream invertebrate. *Ecology* **47**, 595–604.
Waters T. F. (1969) The turnover ratio in production ecology of freshwater invertebrates. *Amer. Nat.* **103**, 173–85.
Watson J. A. L. (1968) Australian dragonflies. *Aust. Mus. Mag.* **16**, 33–8.
Watson J. A. L. (1969) Taxonomy, ecology, and zoogeography of dragonflies (Odonata) from the north-west of Western Australia. *Aust. J. Zool.* **17**, 65–112.
Weatherley A. H. (1958a) Tasmanian farm dams in relation to fish culture. *C.S.I.R.O. Div. Fish. Oceanogr. Tech. Paper* No. 4, 24 pp.
Weatherley A. H. (1958b) Growth, production, and survival of brown trout in a large farm dam. *Aust. J. mar. Freshwat. Res.* **9**, 159–66.
Weatherley A. H. (1962) Notes on distribution, taxonomy and behaviour of tench *Tinca tinca* (L.) in Tasmania. *Ann. Mag. nat. Hist.* **4** (13), 713–19.
Weatherley A. H., Beevers J. R., and **Lake P. S.** (1967) The ecology of a zinc polluted river. In: *Australian Inland Waters and their Fauna: Eleven Studies* (ed. A. H. Weatherley). A.N.U. Press, Canberra.
Weatherley A. H. and **Lake J. S.** (1967) Introduced fish species in Australian inland waters. In: *Australian Inland Waters and their Fauna: Eleven Studies* (ed. A. H. Weatherley). A.N.U. Press, Canberra.
Weatherley A. H. and **Nicholls A. G.** (1955) The effects of artificial enrichment of a lake. *Aust. J. mar. Freshwat. Res.* **6**, 443–68.
Webster R. G. (undated) The salinity of surface waters in Victoria, Australia. State Rivers and Water Supply Commission, Melbourne. Mimeographed. B7–32.
Weigmann G. and **Schminke H. K.** (1970) *Wandesia glareosa* n. sp., eine Grundwassermilbe aus Australien (Acari, Hydrachnellae). *Arch. Hydrobiol.* **67**, 268–75.
Welch P. S. (1935) *Limnology*. McGraw-Hill, New York and London.
Welch P. S. (1948) *Limnological Methods*. Blakiston, Philadelphia.
Wells F. B. and **Taylor M. E. U.** (1970) The ecology of Waikoropupu Springs—a preliminary report. *N.Z. Limnol. Soc. Newsl.* **6**, 18–9.
West G. S. (1909) The algae of the Yan Yean Reservoir, Victoria: a biological and oecological study. *J. Linn. Soc.* (Bot) **39**, 1–88.
Westlake D. F. (1965) Some basic data for investigations of the productivity of aquatic macrophytes. *Mem. Ist. Ital. Idrobiol.* **18** (Suppl.), 229–48.
Wetzel R. G. (1964) A comparative study of the primary productivity of higher aquatic plants, periphyton, and phytoplankton in a large, shallow, lake. *Int. Revue ges. Hydrobiol. Hydrogr.* **49**, 1–61.
Wetzel R. G. (1965) Techniques and problems of primary productivity measurements in higher aquatic plants and periphyton. *Mem. Ist. Ital. Idrobiol.* **18** (Suppl.), 249–67.
Wetzel R. G. (1966) Variations in productivity of Goose and hypereutrophic Sylvan Lakes, Indiana. *Invest. Indiana Lakes Streams* **7**, 147–84.
Wetzel R. G. (1968) Dissolved organic matter and phytoplankton productivity in marl lakes. *Mitt. int. Verein. theor. angew. Limnol.* **14**, 261–70.
Wetzel R. G. (1969a) Excretion of dissolved organic compounds by aquatic macrophytes. *BioScience* **19**, 539–40.
Wetzel R. G. (1969b) Factors influencing photosynthesis and excretion of dissolved organic matter by aquatic macrophytes in hard-water lakes. *Verh. int. Verein. theor. angew. Limnol.* **17**, 72–85.
Whipple G. C. (1948) *The Microscopy of Drinking Water*. (4th edn rev. Fair G. M. and Whipple M. C.). John Wiley and Sons, New York.

Whitford L. A. (1960) The current effect and growth of fresh-water algae. *Trans. Am. microsc. Soc.* **79**, 302–9.
Whitley G. P. (1945) New sharks and fishes from Western Australia. Part 2. *Aust. Zool.* **11**, 1–42.
Williams G. E. (1969) Flow conditions and estimated velocities of some central Australian stream floods. *Aust. J. Sci.* **31**, 367–9.
Williams W. D. (1964a) Subterranean freshwater prawns (Crustacea: Decapoda: Atyidae) in Australia. *Aust. J. mar. Freshwat. Res.* **15**, 93–106.
Williams W. D. (1964b) Some chemical features of Tasmanian inland waters. *Aust. J. mar. Freshwat. Res.* **15**, 107–22.
Williams W. D. (1965a) Ecological notes on Tasmanian Syncarida (Crustacea: Malacostraca), with a description of a new species of *Anaspides*. *Int. Revue ges. Hydrobiol. Hydrogr.* **50**, 95–126.
Williams W. D. (1965b) Subterranean occurrence of *Anaspides tasmaniae* (Thomson) (Crustacea, Syncarida). *Int. J. Spel.* **1**, 333–7.
Williams W. D. (1966) Conductivity and the concentration of total dissolved solids in Australian lakes. *Aust. J. mar. Freshwat. Res.* **17**, 169–76.
Williams W. D. (1967a) The chemical characteristics of lentic surface waters in Australia. In: *Australian Inland Waters and their Fauna: Eleven Studies* (ed. A. H. Weatherley). A.N.U. Press, Canberra.
Williams W. D. (1967b) The changing limnological scene in Victoria. In: *Australian Inland Waters and their Fauna: Eleven Studies* (ed. A. H. Weatherley). A.N.U. Press, Canberra.
Williams W. D. (1968a) *Australian Freshwater Life. The Invertebrates of Australian Inland Waters.* Sun Books, Melbourne.
Williams W. D. (1968b) The distribution of *Triops* and *Lepidurus* (Branchiopoda) in Australia. *Crustaceana* **14**, 119–26.
Williams W. D. (1969a) Eutrophication of lakes. *Proc. R. Soc. Vict.* **83**, 17–26.
Williams W. D. (1969b) Freshwater ecology as an applied science. *Proc. ecol. Soc. Aust.* **4**, 32–8.
Williams W. D. (1969c) Energy transformations in salt lakes. *Aust. Soc. Limnol. Bull.* **1**, 9–12.
Williams W. D. (1970) Salt lake ecosystems. *Aust. Soc. Limnol. Bull.* **3**, 18–19.
Williams W. D. (1972) The uniqueness of salt lake ecosystems. *Proc. UNESCO/IBP Symp. on Productivity Problems of Freshwaters.* Warsaw.
Williams W. D. and **Siebert B. D.** (1963) The chemical composition of some surface waters in central Australia. *Aust. J. mar. Freshwat. Res.* **14**, 166–75.
Williams W. D., Walker K. F., and **Brand G. W.** (1970) Chemical composition of some inland surface waters and lake deposits of New South Wales, Australia. *Aust. J. mar. Freshwat. Res.* **21**, 103–16.
Willington C. M. (1959) The mineral industry on Yorke Peninsula. *Min. Rev., Adelaide* **110**, 8–9.
Willis J. H. (1964) Vegetation of the basalt plains in western Victoria. *Proc. R. Soc. Vict.* **77**, 397–418.
Wilson A. T. and **Wellman H. W.** (1962) Lake Vanda: An Antarctic lake. *Nature, Lond.* **196**, 1171–3.
Wilson S. (1879) *Salmon at the Antipodes.* Edward Stanford, London.
Winberg G. G. (1936) [Some general questions of productivity of lakes] (in Russian). *Zool. zh.* **15**, 587–603.
Winterbourn M. J. (1968) The faunas of thermal waters in New Zealand. *Tuatara* **16**, 111–22.
Winterbourn M. J. (1969) The distribution of algae and insects in hot spring thermal gradients at Waimangu, New Zealand. *N.Z. Jl mar. Freshwat. Res.* **3**, 459–65.
Winterbourn M. J. and **Brown T. J.** (1967) Observations on the faunas of two warm streams in the Taupo thermal region. *N.Z. Jl mar. Freshwat. Res.* **1**, 38–50.
Wisely B. (1953) Two wingless alpine stoneflies (Order Plecoptera) from southern New Zealand. *Rec. Canterbury Mus.* **6**, 219–31.
Wofner H. and **Twidale C. R.** (1967) Geomorphological history of the Lake Eyre basin. In: *Landform Studies from Australia and New Guinea* (eds J. N. Jennings and J. A. Mabbutt). A.N.U. Press, Canberra.

Wood R. B. (in press) The production of *Spirulina* in open lakes. [Paper delivered at conference on 'Preparing nutritional protein from *Spirulina*', Stockholm, June 1968.] Swedish Council for Applied Research, Stockholm.

Wood W. E. (1924) Increase of salt in soil and streams following the destruction of native vegetation. *J. R. Soc. West. Aust.* **10**, 35–47.

Woodwell G. M. (1970) Effects of pollution on the structure and physiology of ecosystems. *Science, N.Y.* **168**, 429–33.

World Health Organization (1963) *International Standards for Drinking Water*. Geneva.

Worthington E. B. and Ricardo C. K. (1936) Scientific results of the Cambridge expedition to the East African lakes 1930–1, No. 17. The vertical distribution and movements of the plankton in Lakes Rudolf, Naivasha, Edward and Bunyoni. *J. Linn. Soc.* (Zool.) **40**, 33–69.

Wright J. C. and Mills I. K. (1967) Productivity studies on the Madison River, Yellowstone National Park. *Limnol. Oceanogr.* **12**, 568–77.

Wynne-Edwards V. C. (1962) *Animal Dispersion in Relation to Social Behaviour*. Oliver and Boyd, London.

Yaldwyn J. C. (1959) Notes on Crustacea Decapoda Natantia from subterranean waters in New Zealand. *Vie Milieu* **9**, 334–9.

Yaron Z. (1964) Notes on the ecology and entomostracan fauna of temporary rainpools of (sic) in Israel. *Hydrobiologia* **24**, 489–513.

Zelinka M. and Marvan P. (1961) Zur Präzisierung der biologischen Klassifikation der Reinheit fliessender Gewässer. *Arch. Hydrobiol.* **57**, 389–407.

Index
LAKES AND RESERVOIRS

The following abbreviations are used for the Australian states: N.S.W. (New South Wales), Qld (Queensland), S.A. (South Australia), Tas. (Tasmania), Vic. (Victoria), and W.A. (Western Australia). The North Island of New Zealand is abbreviated to N.I.N.Z. and the South Island to S.I.N.Z.

Other abbreviations used are L. (Lake) and Res. (Reservoir). Where 'L.' should precede the entry it has been placed after it but is preceded by a comma. Where the comma has been omitted 'L.' is correctly placed.

Ackland, L. (S.I.N.Z.) (43°38′ S, 171°07′ E), 81
Ada, L. (Tas.) (41°52′ S, 146°28′ E), Table 12:1
Albacutya, L. (Vic.) (35°45′ S, 141°58′ E), 232
Ainsworth, L. (N.S.W.) (28°47′ S, 153°36′ E), 58
Albert, L. (east Africa) (1°44′ N, 31°00′ E), 259
Albina, L. (N.S.W.) (36°25′ S, 148°16′ E), 56
Alexandrina, L. (S.A.) (35°26′ S, 139°10′ E), 257
Amy, L. (S.A.) (37°13′ S, 139°47′ E), Fig. 10:6
Aral Sea (U.S.S.R.) (45°00′ N, 60°00′ E), 15
Aroarotamahine, L. Mayor Island (N.I.N.Z.) (37°17′ S, 176°16′ E), Fig. 5:3, Table 5:1, 4, 16, 52, 62, 74, 83, 116
Arrow, L. (Ireland) (54°03′ N, 8°20′ W), 103
Arthurs Lakes (Tas.) (42°00′ S, 146°56′ E), 79, 101
Aswan Dam (Egypt) (23°58′ N, 32°52′ E), 232

Babadag lakes (Rumania) (44°53′ N, 28°46′ E), 16
Baikal, L. (U.S.S.R.) (53°00′ N, 108°00′ E), 50, 63
Barrine, L. (Qld) (17°15′ S, 145°38′ E), Table 12:1, 51

Basin lakes (Vic.) (38°20′ S, 143°27′ E); see West Basin Lake
Beeac, L. (Vic.) (38°12′ S, 143°37′ E), Fig. 10:5, 197, 204
Beloie, L. (U.S.S.R.) (57°00′ N, 37°30′ E), Table 4:1, 37, 41, 43, 68
Black Sea (Europe) (44°00′ N, 35°00′ E), 15
Blue L. (N.S.W.) (36°24′ S, 148°19′ E), 56
Blue L. (S.A.) (37° 51′ S, 140° 47′ E), Plate 3:2, Figs 1:1 and 1:2, Table 1:4, 51, 52, 95
Boga, L. (Vic.) (35°28′ S, 143°38′ E), 231
Boemingen, L. (Qld) (25°33′ S, 153°04′ E), Fig. 1:2, Table 1:4, 59
Bonney, L. (Antarctica) (77°43′ S, 162°23′ E) 87
Borax L. (Calif., U.S.A.) (38°59′ N, 122°39′ W), 30, 34, 35
Borumba Dam (Qld) (26°31′ S, 152°35′ E), 113
Browne L. (S.A.) (37°51′ S, 140°46′ E), Plate 3:2
Brunner, L. (S.I.N.Z.) (42°37′ S, 171°27′ E), 130
Buchanan, L. (Qld) (21°35′ S, 145°52′ E), 50, 210, 231
Bullenmerri, L. (Vic.) (38°15′ S, 143°07′ E), 51, 197, 205
Bunga, L. (Vic.) (37°52′ S, 148°03′ E), 60

INDEX OF LAKES AND RESERVOIRS

Bunyoni, L. (east Africa) (1°10′ S, 29°50′ E), 74, 79
Burley Griffin, L. (Aust. Capital Terr.) [artificial] (35°18′ S, 149°07′ E), 237
Burragorang, L. (N.S.W.) (34°00′ S, 150°26′ E); see Warragamba Dam (synonym)

Canobolas, L. (N.S.W.) (33°12′ S, 148°58′ E), 83
Clarence Lagoon (Tas.) (42°06′ S, 146°19′ E), 56
Caspian Sea (U.S.S.R.–Iran) (42°00′ N, 50°00′ E), 15, 56
Castle L. (Calif., U.S.A.) (ca. 41°18′ N, 122°22′ W), 72
Christabel, L. (S.I.N.Z.) (42°25′ S, 172°15′ E), Table 12:1
Clear L. (Calif., U.S.A.) (39°06′ N, 122°50′ W), 33, 238
Club L. (N.S.W.) (36°25′ S, 148°17′ E), 56
Coalstoun Lakes (Qld) (25°37′ S, 151°54′ E), 52
Cockajemmy Lakes (Vic.) (37°30′ S, 142°30′ E), 53
Coleridge, L. (S.I.N.Z.) (43°19′ S, 171°31′ E), 72
Condah, L. (Vic.) [Now largely drained] (38°03′ S, 141°51′ E), 53
Cooking L. (Alberta, Canada) (53°25′ N, 113°00′ W), 28
Cootapatamba, L. (N.S.W.) (36°28′ S, 148°16′ E), Fig. 6:5, 56
Coradgill, L. (Vic.) (38°07′ S, 143°21′ E), Fig. 10:5
Corangamite, L. (Vic.) (38°11′ S, 143°25′ E), Plate 10:3, Figs 1:1, 1:2, 10:2, 10:5, 10:7, Tables 1:4, 10:1, 12:1, 52, 92, 97, 197, 198, 199, 207, 232, 233
Crater L. (Oregon, U.S.A.) (42°52′ N, 122°05′ W), 90
Crescent, L. (Tas) (42°11′ S, 147°10′ E), Table 12:1, 79, 100, 101, 113
Cultus L. (Brit. Col., Canada) (49°04′ N, 122°00′ W), 39
Cundare, L. (Vic.) (38°09′ S, 143°37′ E), 204

Dead Sea (Israel) (31°30′ N, 35°00′ E), 202
Derryclare L. (Ireland) (53°28′ N, 9°47′ W), 103
Disappointment, L. (W.A.) (23°20′ S, 122°40′ E), 50
Dobson, L. (Tas.) (42°42′ S, 146°34′ E), 79
Dutton, L. (S.A.) (31°48′ S, 137°09′ E), 231

Eacham, L. (Qld) (17°17′ S, 145°38′ E), Table 12:1, 51
Edgar, L. (Tas.) (43°00′ S, 146°19′ E), 50

Edward, L. (east Africa) (0°20′ S, 29°30′ E), 74, 79
Edward, L. (S.A.) (37°38′ S, 140°36′ E), Figs 1:1, 1:2, Table 1:4, 259
Eildon, L. [Res.] (Vic.) [artificial] (37°07′ S, 145°57′ E), 258
Elingamite, L. (Vic.) (38°21′ S, 143°01′ E), 51, 113
Eliza, L. (S.A.) (37°15′ S, 139°52′ E), Fig. 10:6, 197
Ellesmere, L. (S.I.N.Z.) (43°46′ S, 172°28′ E), 61
Erie, L. (North America) (42°10′ N, 81°20′ W), 30, 257
Erken, L. (Sweden) (59°25′ N, 18°15′ E), 29, 30, 107
Esrom, L. (Denmark) (56°00′ N, 12°22′ E), 126
Eucumbene, L. (N.S.W.) [artificial] (36°03′ S, 148°43′ E), Plate 11:1
Euramoo, L. (Qld) (17°10′ S, 145°38′ E), Table 12:1, 51
Eyre, L. (S.A.) (28°30′ S, 137°15′ E), Plates 10:1, 10:2, Figs 10:1, 10:3, Table 12:1, 9, 10, 50, 193, 195, 197

Fayetteville Green L. (New York, U.S.A.) (43°01′ N, 76°00′ W), 75, 86, 87
Fellmongery, L. (S.A.) (37°10′ S, 139°46′ E), Fig. 10:6
Fergus, L. (Tas.) (41°57′ S, 146°03′ E), Table 12:1
Fowler, L. (S.A.) (35°05′ S, 137°37′ E), 231

Galilee, L. (Qld) (22°22′ S, 145°48′ E), 50, 210
George, L. (N.S.W.) (35°06′ S, 149°26′ E), Plate 3:1, 50, 94
Gippsland Lakes (Vic.) (Centred on 38°00′ S, 147°37′ E), 60; see also L. Bunga
Gnarpurt, L. (Vic.) (38°04′ S, 143°23′ E), Fig. 10:5, 198
Gnotuk, L. (Vic.) (38°13′ S, 143°06′ E), Figs 10:4, 10:5, Tables 10:4, 12:1, 51, 197, 198, 207, 208
Goose L. (Indiana, U.S.A.) (41°12′ N, 85°52′ W), Fig. 2:3, 32
Grassmere, L. (S.I.N.Z.) (41°43′ S, 174°10′ E), 211
Great L. (Tas.) (41°53′ S, 146°45′ E), 50, 79, 101, 233
Great Salt L. (Utah, U.S.A.) (41°15′ N, 112°30′ W), 16, 202, 204
Great Slave L. (Northwest Terr., Canada) (61°30′ N, 114°20′ W), Table 4:1
Green L. (N.I.N.Z.) (ca. 38°20′ S, 176°22′ E); see L. Rotowhero

Green L. (Wisconsin, U.S.A.) (43°48′ N, 89°00′ W), 67, 68

Hattah L. (Vic.) (34°44′ S, 142°21′ E), Table 12:1, 234
Hayes, L. (S.I.N.Z.) (44°58′ S, 168°50′ E), 113, 116
Heaton, L. (N.I.N.Z.) (40°07′ S, 175°17′ E), 58
Hiawatha, L. (N.S.W.) (29°50′ S, 153°17′ E), 79
Hindmarsh, L. (Vic.) (36°04′ S, 141°55′ E), 232, 233
Hot L. (Washington, U.S.A.) (*ca.* 47°23′ N, 119°30′ W), 86, 87
Hypipamee Crater L. (Qld) (*ca.* 17°19′ S, 145°28′ E), 53

Ida, L. (S.I.N.Z.) (43°13′ S, 171°33′ E), 81

Judd, L. (Tas.) (42°58′ S, 146°25′ E), Table 12:1

Kariah, L. (Vic.) (38°10′ S, 143°13′ E), Fig. 10:5
Keilambete, L. (Vic.) (38°13′ S, 142°53′ E), 51, 197
Kutubu, L. (New Guinea) (6°24′ S, 143°18′ E), 79
Kylemore L. (Ireland) (53°33′ N, 9°51′ W), 103

Lagoon of Islands (Tas.) (42°06′ S, 146°57′ E), Table 12:1
Lahontan Lakes (Nevada, U.S.A.), 4
Laurentian Great Lakes (North America), 56, 67, 104; *see also* North American Great Lakes
Leake, L. (S.A.) (37°37′ S, 140°35′ E), Figs 1:1, 1:2, Table 1:4, 259
Leake, L. (Tas.) (42°01′ S, 147°49′ E), 101
Lefroy, L. (W.A.) (31°15′ S, 121°43′ E), 231
Leg of Mutton L. (S.A.) (37°51′ S, 140°46′ E), Plate 3:2
Leven, L. (Scotland) (56°13′ N, 3°24′ W), 30
Linsley Pond (Conn., U.S.A.) (41°19′ N, 72°46′ W), Table 4:1, 70
Little Manitou L. (Saskatchewan, Canada) (49°25′ N, 92°50′ W), 202
Lochiel, L. (S.A.) [Also known as Bumbunga L.] (33°55′ S, 138°11′ E), 231
Lyndon, L. (S.I.N.Z.) (43°19′ S, 171°42′ E), 58, 72, 73, 81, 100, 129

MacDonnell, L. (S.A.) (32°02′ S, 133°02′ E), 231

Manapouri, L. (S.I.N.Z.) (45°32′ S, 167°32′ E), Plate 3:7, 56, 100
Maraetai, L. (N.I.N.Z.) [artificial] (38°20′ S, 175°44′ E), 254
Marion, L. (S.I.N.Z.) (42°41′ S, 172°14′ E), Table 12:1
Mariut, L. (Egypt) (31°10′ N, 29°55′ E), 33
Marymere, L. (S.I.N.Z.) (43°08′ S, 171°51′ E), 55
Mendota, L. (Wisconsin, U.S.A.) (43°07′ N, 88°25′ W), 47, 66, 68, 75
Miers, L. (Antarctica) (78°07′ S, 163°54′ E), 87
Miko*ł*ajskie, L. (Poland) (*ca.* 53°50′ N, 21°35′ E), 34
Minchin, L. (S.I.N.Z.) (42°51′ S, 171°49′ E), 54
Minnie Water, L. (N.S.W.) (29°48′ S, 153°17′ E), Table 12:1
Miraflores Third Locks L. (Panama) [artificial] (*ca.* 9°00′ N, 79°30′ W), 86
Modewarre, L. (Vic.) (38°15′ S, 144°07′ E), 205
Mono L. (Calif., U.S.A.) (38°00′ N, 119°00′ W), 16
Moses L. (Washington, U.S.A.) (47°08′ N, 119°21′ W), 59
Mt Eccles Crater L. (Vic.) (38°02′ S, 141°55′ E), 52
Mt Gambier Lakes (S.A.) (37°50′ S, 140°47′ E), Plate 3:2; *see also* Blue L. (S.A.) and Valley L.
Mt Kosciusko lakes (N.S.W.) (36°27′ S, 148°16′ E), Table 12:1; *see also* L. Albina, Blue L. (N.S.W.), Club L., and L. Cootapatamba
Mt Le Brun lakes (Qld) (25°37′ S, 151°54′ E), 52
Mt Quincan Crater L. (Qld) (17°18′ S, 145°35′ E), 52
Mt Ruapehu Crater L. (N.I.N.Z.) (39°17′ S, 175°33′ E), Plate 1:1
Mueller, L. (S.I.N.Z.) (43°25′ S, 170°02′ E), Table 12:1
Mystic Park lakes (Vic.) (Centred on 35°35′ S, 143°43′ E), 231

Naivashá, L. (east Africa) (0°45′ S, 36°24′ E), 74
Nakura, L. (east Africa) (0°12′ S, 36°08′ E), 17
Ngapouri, L. (N.I.N.Z.) (38°17′ S, 176°22′ E), 52
Nicholls, L. (Tas.) (42°40′ S, 146°40′ E), 81
Nipigon, L. (Ontario, Canada) (49°50′ N, 88°30′ W), Table 4:1, 68
Nkugute, L. (east Africa) (*ca.* 0°25′ S, 30°10′ E), 79

INDEX OF LAKES AND RESERVOIRS

North American Great Lakes (North America), 234, 257, 259; see also Laurentian Great Lakes

Ohakuri, L. (N.I.N.Z.) [artificial] (38°26′ S, 176°08′ E), 37, 107
Okaro, L. (N.I.N.Z.) (38°17′ S, 176°22′ E), 52
Okataina, L. (N.I.N.Z.) (38°08′ S, 176°25′ E), Table 12:1, 52
Omapere, L. (N.I.N.Z.) (35°21′ S, 173°48′ E), 53
Omeo, L. (Vic.) (36°58′ S, 147°41′ E), 51
Ontario, L. (North America) (43°40′ N, 77°30′ W), 68, 257

Pearson, L. (S.I.N.Z.) (43°06′ S, 171°47′ E), 58, 100
Pedder, L. (Tas.) (42°56′ S, 146°08′ E), Plate 3:4, Table 12:1, 55, 65, 130, 131, 232, 315
Pernatty Lagoon (S.A.) (31°32′ S, 137°14′ E), 231
Pink Lakes (Vic.) [Also known as salt lakes of Linga] (35°03′ S, 141°44′ E), Table 10:1, 92, 231
Pink L. (W.A.) (33°50′ S, 121°50′ E), 231
Prion, L. (Macquarie Island) (ca. 54°30′ S, 158°57′ E), 79
Pukaki, L. (S.I.N.Z.) (44°00′ S, 170°11′ E), 93
Pupuke, L. (N.I.N.Z.) (36°53′ S, 174°52′ E), Fig. 5:4, 52, 83, 117
Purrumbete, L. (Vic.) (38°17′ S, 143°14′ E), 51, 109, 113, 126

Quill, L. (S.I.N.Z.) (44°40′ S, 167°40′ E), 56

Razelm lakes (Rumania) (44°51′ N, 28°59′ E), 16
Rea, L. (Ireland) (53°11′ N, 8°35′ W), 103
Red Rock lakes (Vic.) (38°15′ S, 143°30′ E), Plate 10:3, Table 12:1; see also Red Rock 'Tarn' and L. Werowrap
Red Rock 'Tarn' (Vic.) (38°15′ S, 143°30′ E), Plate 10:3, Figs 1:1, 1:2, Table 1:4
Robe, L. (S.A.) (37°13′ S, 139°48′ E), Fig. 10:6
Rotoehu, L. (N.I.N.Z.) (38°01′ S, 176°32′ E), 52, 79
Rotoiti, L. (N.I.N.Z.) (38°01′ S, 176°24′ E), Fig. 5:10, 36, 52, 95
Rotokawau, L. (i) (N.I.N.Z.) (35°02′ S, 173°11′ E), Table 12:1
Rotokawau, L. (ii) (N.I.N.Z.) (38°05′ S, 176°26′ E), 52

Rotomohana, L. (N.I.N.Z.) (38°16′ S, 176°26′ E), Table 12:1, 52, 88
Rotongaio, L. (N.I.N.Z.) (38°49′ S, 176°04′ E), 52
Rotopounamu, L. (N.I.N.Z.) (39°02′ S, 175°44′ E), Table 12:1
Rotorua, L. (N.I.N.Z.) (38°06′ S, 176°17′ E), 53, 79, 101, 103
Rotowhero, L. (N.I.N.Z.) [Also known as Green L.] (ca. 38°20′ S, 176°22′ E), Table 12:1
Roundabout, L. (S.I.N.Z.) (43°37′ S, 171°05′ E), 81
Rudolf, L. (east Africa) (4°00′ N, 36°00′ E), 4, 74, 116, 259

St Clair, L. (S.A.) (37°20′ S, 139°55′ E), Fig. 10:6, 197
St Clair, L. (Tas.) (42°04′ S, 146°10′ E), Plate 3:6, Table 12:1, 56, 115, 116
Salt L. (Beachport, S.A.) (37°13′ S, 140°00′ E), Fig. 10:6
Salt L. (Sutton, S.I.N.Z.) (45°35′ S, 170°05′ E), Plate 10:4, Figs 1:1, 1:2, Tables 1:4, 10:1, 210
Salt lakes of Linga (Vic.) [Also known as the Pink Lakes] (35°03′ S, 141°44′ E), Table 10:1
Salton Sea (Calif., U.S.A.) (33°25′ N, 115°50′ W), 210
Sarah, L. (S.I.N.Z.) (43°03′ S, 171°47′ E), 100, 102
Seal, L. (Tas.) (42°38′ S, 141°37′ E), Table 12:1
Severson L. (Minn., U.S.A.) (ca. 46°55′ N, 95°30′ W), 44
Soap L. (Wash., U.S.A.) (47°23′ N, 119°30′ W), Fig. 5:5, 33, 85, 86, 87
Sodon L. (Mich., U.S.A.) (42°19′ N, 83°17′ W), 75
Sorell, L. (Tas.) (42°07′ S, 147°10′ E), Table 12:1, 79, 98, 100, 101, 102, 113
Swan L. (N.I.N.Z.) (36°20′ S, 174°08′ E), 58
Sylvan L. (Indiana, U.S.A.) (41°29′ N, 85°22′ W), 34

Takapuna, L. (N.I.N.Z.) (36°53′ S, 174°52′ E); see Pupuke, L. (synonym)
Tanganyika, L. (east Africa) (7°00′ S, 30°00′ E), 15
Tarawera, L. (N.I.N.Z.) (38°12′ S, 176°26′ E), 53, 88
Tarli Karng, L. (Vic.) (37°37′ S, 146°45′ E), Table 12:1, 54
Taupo, L. (N.I.N.Z.) (38°48′ S, 175°56′ E), Plate 3:3, 52, 53, 116

Te Anau, L. (S.I.N.Z.) (45°14′ S, 167°46′ E), 56, 100, 130
Tegid, L. (Wales) (52°53′ N, 3°38′ W), 234
Tekapo, L. (S.I.N.Z.) (43°53′ S, 170°32′ E), 93
Tikitapu, L. (N.I.N.Z.) (38°10′ S, 176°25′ E), Table 5:1, 52
Toghraklik-köl (Tibet) (*ca.* 40°50′ N, 87°00′ E), 59
Tooms L. (Tas.) (42°14′ S, 146°48′ E), 101
Toorourrong Res. (Vic.) (37°28′ S, 145°08′ E), 100
Torrens, L. (S.A.) (31°00′ S, 137°50′ E), 50
Tyrrell, L. (Vic.) (35°21′ S, 142°50′ E), 56, 231

Valley L. (S.A.) (37°51′ S, 140°46′ E), Plate 3:2
Vanda, L. (Antarctica) (77°35′ S, 161°39′ E), Fig. 5:6, 87
Victoria, L. (east Africa) (1°00′ S, 33°00′ E), 259

Waikare, L. (N.I.N.Z.) (37°27′ S, 175°13′ E), 58
Waikaremoana, L. (N.I.N.Z.) (38°46′ S, 177°06′ E), 54
Wairarapa, L. (N.I.N.Z.) (41°12′ S, 175°15′ E), 58

Wakatipu, L. (S.I.N.Z.) (45°05′ S, 168°33′ E), Plate 3:5, Fig. 5:9, 56, 94, 100, 111
Wanaka, L. (S.I.N.Z.) (44°30′ S, 169°08′ E), 56
Warragamba Dam (N.S.W.) 34°00′ S, 150°26′ E), 107, 111
Washington L. (Wash., U.S.A.) (46°36′ N, 122°22′ W), 257
Werowrap, L. (Vic.) (38°15′ S, 143°30′ E), Plate 10:3, 30, 203, 208
West Basin L. (Vic.) (38°20′ S, 143°27′ E), 87, 197
Windermere (U.K.) (54°20′ N, 2°57′ W), 105, 257
Woods L. (Tas.) (42°05′ S, 147°00′ E), 101
Wooli lakes (N.S.W.) (29°50′ S, 153°17′ E), *see* L. Hiawatha and L. Minnie Water

Yamma Yamma, L. (Qld) [Also known as L. McKillop] (26°19′ S, 141°25′ E), Plate 9:2
Yan Yean Res. (Vic.) (37°34′ S, 145°09′ E), 98, 100, 101, 102
Yaninee, L. (S.A.) (32°59′ S, 135°15′ E), 231

Zürichsee (Switzerland) (47°14′ N, 8°41′ E), 257

Index
ORGANISMS

All supra-generic taxa are listed in the general index.

Abramis, 166
Acanthocyclops vernalis Sars, 187
Acartia clausii Giesbrecht, 209
Acmea, 210
Acnanthes, Table 12:5
Adelotus brevis (Gunther), 178
Aedes (Halaedes) australis (Erich.),
  Table 10:3
A. natronius Edwards, 16
A. (Ochlerotatus) vigilax (Sk.),
  Table 10:3
Agraptocorixa, Table 6:7
Aldia, 159
Alisma plantago-aquatica L., Table 6:6
Allanaspides, 173
Allotrissocladius, 178
A. amphibius Freeman, 178
Alona, Fig. 9:4, 187
A. abbreviata Sars, Table 12:6
A. affinis (Leydig), Table 12:6
A. kendallensis Henry, Fig. 9:4
A. rectangula Sars, 187
Alosa pseudoharengus (Wilson), 109
Amphipleura, 208
Amphora, Table 12:5, 208
Anabaena, 70, 93, 102, 219
A. circinalis (Kuetz.), 208
A. spiroides Klebahn, 208
A. cylindrica Lemmermann, 72
Anacystis, 102, 219
Anaspides tasmaniae (Thomson), 153, 172, 260
Aneura, 152
Anguilla, 205

A. australis Schmidt, Table 11:3
Angianthus preissianus (Steetz) Benth., 208
Anisops, Tables 6:7, 9:2
A. assimilis White, 211
A. gratus Hale, Table 10:3
A. thienemanni Lundblad, Table 10:3
A. wakefieldi White, 211
Ankistrodesmus, Table 12:5
Anodonta, 40
Anommatophasma candidum Mees, 172
Antocha, Table 12:6
Anuraeopsis, 111
Aphanizomenon, 102, 219
Apocyclops dengizicus (Lepeschkin), 210
A. dimorphus Kiefer, 210
Apodya lactea (Ag.), 242
Aponogeton, 129
A. distachyos L.f., Table 6:6
Apteryoperla longicauda Illies, 160
Arcella, 131
Archichauliodes, Table 8:1
A. guttiferus Walk., Table 12:6
Artemia, 15, 16
A. salina L., Fig. 1:5, 14, 15, 211
Ascomorpha saltans Bartsch, 111
Asellus, 244
A. aquaticus (L.), Fig. 1:5, Table 2:2, 13, 244
Asplanchna, 111, 186–7
Astacopsis gouldi Clark, 229
Astacus fluviatilis L., 74
Asterionella, 100–2, 105, 219
A. gracillima (Hantzsch), 100
Atalophlebioides, 159

Australomedusa baylii Russell, Table 10:2, 204
Austrochiltonia, Table 6:7, 206–7
  A. subtenuis (Sayce), Tables 6:5, 10:3
  A. australis (Sayce), Table 10:3
Austrolestes, Table 9:2
Austrosimulium, Table 8:5
Azolla, 129
  A. rubra L., 185

Baetis vagans McDunnough, Table 2:2, 40–1
Bagous, Table 10:3
Barbus, 166
Batillariella estuarina Tate, 207
Batrachospermum, Table 12:5, 152
Beggiatoa alba (Vaucher), 242
Bidyanus bidyanus (Mitchell) Table 11:3
Bodo, 242
Boeckella, Figs 6:1, 6:2, Tables 9:2, 12:6, 107, 109, 111, 116–17, 189
  B. delicata Percival, Table 6:3
  B. dilatata Sars, Fig. 6:2, 111, 116
  B. fluvialis Henry, Table 6:3, Figs 6:2, 9:4, 107, 187
  B. geniculata Bayly, Table 6:3
  B. hamata Brehm, Table 6:3, Fig. 6:2
  B. major Searle, Table 6:3
  B. minuta Sars, Tables 6:3, 6:4, Figs 6:1, 9:4, 107, 116, 187, 189
  B. montana Bayly, Table 6:3, 189
  B. opaqua Fairbridge, 178
  B. propinqua Sars, Table 6:3, Figs 6:2, 6:6, 116
  B. propinqua longisetosa Smith, 115–16
  B. pseudochelae Searle, Table 6:3, Fig. 6:2
  B. robusta Sars, Table 6:3, 109
  B. rubra Smith, Table 6:3
  B. symmetrica Sars, Table 6:3, Fig. 6:1, 109, 185
  B. triarticulata (Thomson), Tables 6:3, 10:2, Fig. 6:1, 16, 107, 117, 185, 186–7, 189, 204
Bosmina, 70, 71, 189
  B. meridionalis Sars, Fig. 9:4, 113, 116–17
Botryococcus, 101
  B. braunii Kuetz, 100–1
Brachionus, 16, 122
  B. plicatilis Müller, Table 10:2, 16, 92, 203, 210
  B. novae-zelandiae Morris, 16
  B. caudatus Barrois and Daday, 16
  B. calyciflorus Pallas, 16
Branchinella, Table 6:7, 178
Branchiura sowerbyi Beddard, Table 6:5
Bufo marinus (L.), 260

Cactoblastus cactorum (Berg), 176
Calamoecia, Figs 6:3, 6:4, 107, 111, 189, 204, 209
  C. ampulla (Searle), Table 6:3, Fig. 6:3, 107
  C. attenuata (Fairbridge), Table 6:3, Fig. 6:4
  C. australica Sars, Fig. 6:3
  C. canberra Bayly, Table 6:3, Fig. 6:4
  C. clitellata Bayly, Table 10:2, 14, 16, 92, 203–4, 209
  C. expansa (Sars), Fig. 6:3
  C. gibbosa (Brehm), Table 6:3, Fig. 6:4
  C. lucasi Brady, Table 6:3, Figs 6:3, 9:4, 107, 117, 187
  C. salina (Nicholls), Table 10:2, 14, 22, 92, 201, 203–4, 209
  C. tasmanica (Smith), Table 6:3, Fig. 6:3, 109, 185, 187
  C. trifida Bayly, Table 6:3, Fig. 6:4
  C. ultima (Brehm), Fig. 6:4
Calanus, 117–18
Callitriche intermedia Hoffmann, Table 6:6
  C. stagnalis Scop., Table 6:6, 185
  C. heterophylla Pursh emend. Darby, Table 6:6
Calypogeia, 152
Camptocercus similis Sars, Fig. 9:4
Campylopus, 152
Candonocypris, Tables 6:5, 9:2
  C. candonoides (King), Table 12:6
Capitella capitata (Fabricius), Table 10:3
Carassius auratus (Bloch), 229
  C. carassius (L.), 229
Carchesium, 242
Carex, 36
Caridina, Table 6:7, 74
  C. thermophila Riek, 173
Centrolepis, 130
Ceratium, 100–1
  C. hirundinella (Müller) Schrank, 123
Ceratonereis erythraeensis Fauvel, Table 10:3, 209
Ceratophyllum, 129
  C. demersum L., 37
Ceriodaphnia, 113, 116, 189
  C. cornuta Sars, Fig. 9:4, 187
  C. dubia (Herrick), 116
  C. quadrangula (Müller), Table 12:6
Chaetoceros, 208
Chaoborus, Tables 2:2, 4:1, 6:7, 43, 83, 238
Chara, 129, 152
Chelodina, 155
  C. longicollis (Shaw), 155
  C. expansa Gray, 155
Cherax, 74
  C. tenuimanus (Smith), 189, 229
Chimarrha, 166

INDEX OF ORGANISMS

Chironomus, Tables 2:2, 4:1, 9:2, 43, 126, 211, 244
  C. bathophilus (=*anthracinus*) Kieffer, 126
  C. occidentalis Skuse, Table 6:5
  C. plumosus L., 126
  C. zealandicus Hudson, 129
Chlamydomanas, 189
Chlorella, 111
Chlorohydra, 187
Chydorus, Fig. 9:4, 187, 189
  C. eurynotus Sars, Fig. 9:4
  C. sphaericus Müller, 70, 189
Cladophora, Table 12:5, 152, 208, 242, 257
Claucoma, 242
Closterium, Table 12:5
Cocconeis, Table 12:5, 208
Coleochaete, Table 12:5
Coleps, 242
Coloburiscus, Table 8:3
  C. humeralis (Walker), Table 8:5
Colpidium, 242
Conochiloides coenobasis Skorikov, 189
Cordylophora, Table 10:3
Cosmarium, Table 12:5, 102
Cotulus vulgaris Levyns, 208
Coxiella, 207, 210
  C. striata (Reeve), Table 10:3, 207, 209
  C. striatula (Menke), 207
Coxiellada gilesi (Angas), 211
Craterocephalus, 204
Crinia pseudinsignifera Main, 178
Crocodilus johnstoni Krefft, 155, 229
  C. porosus Schneider, 155
Cruregens fontanus Chilton, 172
Cryptochironomus, Table 6:5
Cryptomonas, 111
Ctenopharyngodon, 227, 258
  C. idellus (Cuvier and Valenciennes), 227
Culex, Table 9:2, 244
  C. rotoruae Belkin, 211
Culicoides loughnani Edwards, 176
Cyclops sens. str., 111
Cyclorana, 178
Cyclotella, 101–2, 208
  C. stelligera Cleve and Grunow, 111
Cycnogeton, 129
Cymbella, Table 12:5
Cypretta, Table 12:6, 187
Cypricercus, 187
  C. sanguineus Chapman, 187
Cypridopsis, Table 6:5
  C. australis Henry, Fig. 9:4
  C. inaequivalva Klie, 16
Cyprinodon macularius Baird and Girard, 22
Cyprinus, 227
  C. carpio L., 227, 229, 259
Cypris, Table 9:2, 185, 187
  C. kaiapoiensis Chapman, 186–7

Cyzicus, 178

Daphnia, Table 9:2, 70, 75, 109, 113, 115–16, 120, 185, 189
  D. carinata King, Figs 6:5, 9:4, Table 12:6, 111, 113, 116, 121, 186–7, 191
  D. galeata mendotae Sars, 42
  D. 'longispina', 120
  D. lumholtzi Sars, Fig. 9:4, 111, 113, 116, 122, 187
  D. 'magna', Table 12:2, 40, 120, 238
  D. 'pulex', 15
Daphniopsis pusilla Serventy, Table 10:2, 203
Dasyhelea, 178
Deleatidium, Tables 8:3, 8:5, 159
Diacypris, Table 10:2, 22, 203, 211
Diaphanosoma, 111
  D. excisum Sars, Fig. 9:4, 113
  D. brachyurum Liéven, 113
Diaprepocoris, Table 6:7
Diaptomus coeruleus (Fischer), 15, 120
  D. graciloides (Lilljeborg), 107
  D. lumholtzi Sars, Table 6:3, 107
  D. siciloides Lilljeborg, Table 2:3, 44
  D. spinosa Daday, 16
  D. transvaalensis Methuen, 16
Dinobryon, Table 12:5, 100–1, 105, 219
Dinotoperla, Table 12:6, 253
Diploneis, Table 12:5
Dorylaimis, 178
  D. stagnalis Dujardin, Table 6:5
Draparnaldia, Table 12:5
Dugesia, Table 6:5
Dunaliella, 208
  D. salina (Dunal) Teodoresco, 92, 97
Dunhevedia crassa King, Fig. 9:4

Ecnomus, Table 6:5
Ectocyclops phaleratus (Koch), Table 12:6
Egeria densa Planch., Table 6:6
Eichhornia, 129
  E. crassipes (Mart.) Solms, Table 6:6, 152, 259
Eleocharis, 129
Elodea, 129
  E. canadensis Michx, Table 6:6, 259
Elseya, 155
Emydura, 155
  E. macquari (Cuvier), 155
Enithares, Table 6:7
Enochrus (Methydrus) andersoni Blackb., Table 10:3
Enteromorpha, 207
Ephydrella, Table 10:3, 211
  E. novaezealandiae Tonnoir and Mallock, 211
  E. thermarum Dumbleton, 211

Epistylis, 242
Epithemia, Table 12:5
Eristalis, 244
Escherichia coli I (Migula) Castellani, Table 12:7, 219, 254
Euastrum, 99
Eucyclops, 111, 187–8
  E. agilis (Koch) (=*E. serrulatus* (Fischer)), Table 12:6
  E. serrulatus (Fischer), Fig. 9:4
Eucypris, Table 6:8, 187
Euglena, 102, 242
Eulimnadia, 178
Euphausia frigida Hans, 118
  E. triacantha Holt and Tattersall, 118
Euastacus armatus (Von Martens), 229

Fluvialosa richardsoni (Castelnau), 109
Fragilaria, 101, 105, 219
Frustularia, Table 12:5
Fusarium aqueductum Lager, 242

Gadopsis marmoratus Richardson, Table 11:3
Gaimardia, 130
Galaxias, 109
  G. attenuatus Jenyns, 205
Gambusia affinis (Baird and Girard), 174, 229
Geotridium, 242
Gladioferens spinosus Henry, Fig. 9:4
Glenodinium, 101
Glossamia aprion (Richardson), Table 8:4
Glyceria, 129
  G. maxima (Hartm.) Holmb., Table 6:6
Gobiomorphus, 172
  G. coxii (Krefft), 159
Gomphocythere australica Hussainy, Table 6:5
Gomphonema, Table 12:5, 242
Gomphosphaeria, 103
Gymnodinium aeruginosum Stein, 30, 208

Halicarcinus lacustris (Chilton), Table 10:3, 207
Halicyclops ambiguus Kiefer, Table 10:2, 203, 209
Haloniscus searlei Chilton, Fig. 1:5, Table 10:3, 14, 22, 200, 207, 209
Halophila, 208
Haloragis, 129
Haplochromis, 259
Helicopsyche, Table 8:3
  H. iltona Mosely and Kimmins, Table 8:5
Hemianax papuensis (Burmeister), 185
Hemiboeckella searlei Sars, 185
Heterias, Table 6:5
Heterolaophonte, Table 10:2, 204

Hexanematichthys leptaspis Bleeker, Table 8:4
Hexarthra fennica (Levander), Table 10:2, 16, 203
  H. jenkinae (de Beauchamp), Table 10:2, 16, 203
Homo sapiens, 215
Hydatella, 130
  H. inconspicua (Cheesem.) Cheesem., 130
Hydra, Tables 8:1, 12:6
Hydrachna, Table 9:2
Hydrocleys nymphoides (Humb. and Bonpl. ex Willd.) Buchen., Table 6:6
Hydrodictyon recticulatum (L.) Lagerh., 257
Hydrometra, 133
Hydromys, 155
  H. chrysogaster Geoffroy, 155
Hydropsyche, Table 8:3
  H. cockerelli Banks, Table 12:4, 240
Hyla, Table 9:2, 175, 187
  H. lesueri Dumeril and Bibron, Fig. 8:2, 159
Hyphomicrobium, 223
Hypseleotris compressus (Krefft), Table 8:4
Hyridella menziesi (Gray), 129

Ilyocryptus, Table 6:5
  I. sordidus (Liéven), Table 12:6
  I. spinifer Herrick, Fig. 9:4
Ilyodromus, Tables 6:8, 12:6
Iris pseudacorus L., Table 6:6
Ischnura aurora (Brauer), Table 10:3

Juncus, 129

Keratella, 111, 122, 186
Kurtus gulliveri Castelnau, Table 8:4

Lagarosiphon, 129
  L. major (Ridl.) Moss, Table 6:6, 36
Lates calcarifer (Bloch), Table 11:3
  L. niloticus (L.), 259
Lechriodus fletcheri (Boulenger), 185
Leiopelma, 155
Lemna, 129
  L. minor L., 185
Lepidozia, 152
Lepidurus, Tables 6:7, 9:2, 185, 187
  L. apus viridis Baird, 178
Lepilaena bilocularis T. Kirk, 129
  L. preissi (Lehm.) F. Muell., 207
Leptomitus (=*Apodya*) *lacteus* (ag.), 242
Leydigia acanthocercoides (Fischer), Fig. 9:4
Limnadia, 178
  L. stanleyana King, 178
Limnodopsis, 178
Limnodrilus, 244
  L. hoffmeisteri Claparede, Table 6:5

INDEX OF ORGANISMS

Limnodynastes tasmaniensis Günther, 185
Limnoxenus, Table 9:2
Lovenula africana Daday, 16
Lovettia sealii Johnson, Table 11:3
Ludwigia palustris (L.) Ell., Table 6:6
Lumbriculus, Table 12:6
L. variegatus (Müller), Table 6:5
Lymnaea tomentosa Pfeiffer, Table 12:6
Lynceus, 178

Maccullochella macquariensis (Cuv. and Val.), Table 11:3, 234
Macrobrachium, Table 6:7
Macrocyclops albidus (Jurine), Fig. 9:4, Table 12:6
Macrogyrus, 133
Macrothrix spinosa King, Table 12:6, 189
Macquaria australasica (Cuv. and Val.), Table 11:3
Madigania unicolor (Gunther), Table 11:3, 162
Maundia, 129
Melanotaenia nigrans (Richardson), Table 8:4
Melosira, Table 12:5, 101, 105
M. granulata (Ehr.) Ralfs, 101
Merismopedia, Table 12:5
Merragata, 133
Mesochra baylyi Hamond, Table 10:2, 204
Mesocyclops leuckarti (Claus), Fig. 9:4, Table 12:6, 19, 111, 113, 187, 189, 191
M. decipiens (Kiefer), 111
Mesostoma, Table 10:3
Mesovelia, 133
Metacyclops, 211
Metridia, 118
M. gerlachei Giesbrecht, 118
Micrasterias, 99
M. hardyi West, Plate 6:1, Table 6:1, 98
Microcoleus lyngbyaceus (Kuetz.) Crouan, 208
Microcyclops, 111, 211
M. arnaudi Sars, Table 10:2, 203, 209, 211
M. monacanthus (Kiefer), 211
Micronecta, Table 6:7
M. gracilis Hale, Table 10:3
Microvelia, Table 9:2
Milligania, 130
Milyeringa veritas Whitley, 172
Moina, 189
M. micrura Kurz, Fig. 9:4, 113
Mougeotia, Table 12:5
Mucor, 242
Myriophyllum, 207
M. brasiliense Cambess., Table 6:6
M. verrucosum Lindl., 129
Mytilocypris, Table 10:2, 204

Naeogeus, 133
Navicula, Table 12:5, 208, 242
Necterosoma, Table 10:3
Neidium, Table 12:5
Nematocentris maculata (Weber), Table 8:4
Neobatrachus, 178
Neoceratodus, 162
N. forsteri (Krefft), Table 8:4, 155
Neoscatella vittithorax Malloch, 211
Newnhamia fenestrata King, Table 6:5, Fig. 9:4
Nitella, Table 12:5, 129
Nitzschia, Table 12:5, 208, 242
Nodularia, 102
N. spumigena Mertens (in Jurgens), 92, 97, 208, 257
Notholca, 111
Nuphar, 129
Nymphea, 129
Nymphoides, 129

Oedogonium, 185
Olinga, Table 8:3
O. feredayi (McLachlan), Table 8:5
Oncorhynchus nerka Walbaum, 39
O. tschawytscha Walbaum, 205, 229
Onychocamptus bengalensis Sewell, Table 10:2, 204
Oocystis, 101
Opuntia inermis (=*O. stricta* Haw.), 176
Ornithorhynchus anatinus Shaw and Nodder, 155
Orthocladius, 126
Oscillatoria, Table 12:5, 102, 103, 152, 208, 219, 242
O. rubescens De Cand., 102, 257
O. linosa Ag., 257
Oxyethira albiceps (McLachlan), Table 8:5

Palaemonetes australis Dakin, 207
Papyrus, 259
Paraborniella tonnoiri Freeman, 178
Paracalliope, Table 6:7
Paracyclops fimbriatus (Fischer), Table 12:6
Paranephrops planifrons White, 174
Parartemia, 210
P. zietziana Sayce, Table 10:2, 14, 22, 200, 203
Paratelphusa, 189
P. transversa (von Martens), 189
Paratya, Table 6:7
P. australiensis Kemp, Table 12:6, 162, 172
P. (Paratyà) curvirostris (Heller), 172
Parisia, 172
P. gracilis Williams, 172
P. unguis Williams, 172
Pediastrum, 102, 189

Perca fluviatilis L., 205, 229
Percalates colonorum (Gunther), Table 11:3
Peridinium, 101
Petromyzon marinus L., 234
Pettancylus, Table 12:6
Phaenocora, Table 6:5
Philypnodon grandiceps Krefft, 205
Phormidium, 152, 242
Phragmites, 129
Phreatomerus latipes (Chilton), 173–4
Physa, 187
Physastra, Table 9:2
 P. gibbosa (Gould), Table 12:6
Pimephales promelas Rafinesque, Table 12:2, 238
Pinus radiata D. Don, 259
Pisidium novaezealandiae (Deshayes), 129
Pistia stratiotes L., 259
Planorbis, Table 9:2
 P. corunna Gray, Table 8:5
Platycypris, Table 10:2, 22, 203
Plectroplites ambiguus (Richardson), Tables 8:4, 11:3, 153, 234
Pleurotaenium, Table 12:5
Plumatella repens (L.), Table 12:6
Polyarthra, 83, 111
 P. euryptera (Wierzejski), 83
 P. longiremis Carlin, 83
 P. major (Burckhardt), 83
 P. remata (Skorikow), 83
 P. vulgaris Carlin, 83
Polypedilum vanderplanki Hinton, 178
Polyphemus pediculus (L.), 15
Potamogeton, 129, 207
 P. crispus L., Table 6:6
 P. pectinatus L., 129, 207–8
Potamopyrgus, Table 8:3
 P. antipodum (Gray), Table 8:5, 174
 P. spelaeus (Franenfeld), 172
Potamothrix bavaricus (Oschmann), Table 6:5
Pratia platycalyx (F. Muell) Benth., 208
Prionocypris, Table 9:2, 185
Procladius, Table 10:3
Procorixa, Table 10:3
Prosopium williamsoni, Fig. 12:1
Prototroctes oxyrhynchus Gunther, 260
Pseudemydura, 155
Pseudochydorus, Table 6:5
Psychoda, 244
Puntius, 227
Pycnocentrodes, Tables 8:3, 8:5

Ranunculus aquatilis L., Table 6:6
 R. fluitans Lam., Table 6:6
Retropinna, 109
 R. victoriae Stokell, 205
Rhantus, Table 9:2

Rhizosolenia, Table 12:5, 101
Rhoicosphenia, Table 12:5
Rhopolodia, Table 12:5
Rivularia, Table 12:5
Robertsonia propinqua (T. Scott), Table 10:2, 204
Rorippa nasturtium-aquaticum (L.) Hayek, Table 6:6
Ruppia maritima L., 207
 R. spiralis L. ex Dumort, 129
Rutilus rutilus (L.), 229

Sagittaria trifolia L., Table 6:6
Salicornia australasica (Moq.) Hj. Eichler, 208
Salmo, 166
 S. gairdneri Richardson, 83, 205, 227, 229
 S. salar L., 229
 S. trutta L., 83, 172, 205, 229, 259, 260
Salvelinus manaycush Wahlb., 234
Salvinia, 129
 S. auriculata Aubl., Table 6:6, 259
Saycia, 178
Scapholeberis, 133
Scatella, 211
Scenedesmus, Table 12:5, 102, 105, 111, 189
Schizothrix friessii (Ag.) Gomont, 208
Scirpus, 36, 129
Scleropages leichhardti (Gunther), Table 8:4, 155
Selliera radicans Cav., 208
Sergentia, 126
Sigara, Tables 6:7, 9:2
 S. arguta (White), 211
Simocephalus, Table 9:2, 187
 S. acutirostratus King, Fig. 9:4
 S. elizabethae (King), Fig. 9:4
 S. exspinosus (Koch), 15
 S. vetulus (Müller), Table 12:6
Simulium ornatipes Sk., Table 12:6
Sphaerotilus natans Kutzing, 242
Spirogyra, Table 12:5, 152, 185, 242, 257
Spirostomum ambiguum Ehrb., 19
Spirulina, 74, 102
 S. subsalsa Oersted, 208
 S. platensis (Nordst.) Gomont, 16
Spongilla, Table 12:6
Staurastrum, Table 12:5, 99, 100, 101, 102
Staurodesmus, 101
Stephanodiscus, 101
Stictochironomus, 126
Stigeoclonium, Table 12:5, 242, 245
Stizostedion canadense (Smith), 83
Stygiocaris, 172
 S. lancifera Holthuis, 172
 S. stylifera Holthuis, 172
Suaeda australis (R.Br.) Moq., 208
 S. maritima Dumort, 208

INDEX OF ORGANISMS

Surirella, Table 12:5, 242
Symphitoneuria wheeleri Banks, 207
Synchaeta, 111
S. pectinata Ehrenberg, 189
Synechococcus, 30
Synedra, Table 12:5, 101, 219
Synura, 219

Tabellaria, Table 12:5, 101, 105, 219
T. fenestrata (Lyngbye) Kützing, 257
Taeniomembras, Table 10:3, 204
T. microstomus (Günther), 200
Tandanus tandanus Mitchell, Tables 8:4, 11:3
Tanypus, 43
Tanytarsus, 126
T. (Tanytarsus) barbitarsus Freeman, Table 10:3, 207
Tenagomysis, Table 6:7
Thiobacteria, 29
Thymallus, 166
Tilapia, 227, 258
Tinca tinca (L.), 227, 229
Torrenticola queenslandica Domin., 152
Trapa natans L., Table 6:6
Tribonema, Table 12:5
Trichechus, 258

Trintoperla, Table 12:6
Triops australiensis australiensis (Spencer and Hall), Plate 9:3, 178
Triploceras, 99
Tropocyclops, 111
Tubifera, 244
Tubifex, Table 12:6, 244
T. tubifex (Müller), Table 6:5
Typha, 36

Ulothrix, 152, 242, 257
Unio, 40

Vallisneria, 129
V. spiralis L., Table 6:6
Vaucheria, 152
Volvox, 189
Vorticella, 242

Wilsonia rotundifolia Hook, 208
Wolffia, 129

Xuthotricha, Table 6:5

Zannichellia palustris L., 129
Zostera, 208
Zygnema, 257

Index
GENERAL

In addition to subject matter, included in this index are all references to rivers (and other lotic environments) and to all taxa higher than those of species and genus. With regard to the names of such taxa, when the only text reference is to an anglicized or common name it is so indexed, but when reference is to both anglicized or common *and* scientific names, only the latter is fully indexed, the non-scientific name being merely cross-referenced. For convenience of reference, rivers have been indexed as proper nouns followed by the term river, irrespective of the normal word order in general use.

abrasion, 138
Acheron River, 148
active uptake of ions, 13
Adelaide, 163, 257
adventive aquatic macrophytes, 129
afforestation, 258
Afon Hirnant, 45
Africa, 79, 116, 136, 152, 167, 178, 225, 259
African
  lakes in the Rift Valley, 74
  Rift Valley, 18
Agnes River, 148
agricultural
  fertilizers, 256
  malpractices, 258
  supplies of water, 220–2
air pollution, 10
Alberta, 28
aldrin, 238
alewife, 109
algae, 27, 109, 111, 131, 151, 219, 242, 244, 245, 246, 257, 258
algal
  blooms, 102, 189
  chemo-organotrophy, 29
  growth, 257
  standing crop, 30
  succession, 105
  toxicity, 219

Alice
  Springs, 225
  River, 136
alien aquatic macrophytes of Australia and New Zealand, 130
alkaline, 4, 97
  carbonate lakes, 3
  lakes, 19, 74
  waters, 4, 15, 16, 75
alkalinity, 4
alkyl phosphates, 237
Allen curve, 39, 40, 41, 43
allochthonous
  energy input for streams, 169
  organic matter, 17, 44, 71
allometry, 123
allotrophic lakes, 71
alluvial fans, 58
Alvie, 16, 52
Amadeus Basin, 216
Amazon River, 136
amictic, 79, 81, 87
amoebae, 204
Amphibia, 172, 175
Amphipoda, 42, 69, 74, 132, 153, 172, 174, 190, 206, 209
amphipod(s), *see* Amphipoda
anabranches, 58, 151
anaerobic bacteria, 82

GENERAL INDEX

analytical methods, 2
Ancylidae, *see* Ferrissiidae
anemotrophic, 71
angiosperms, 129, 152, 158, 207, 208
angling, 228
Anguillidae, 226
Anguilliformes, 226
anionic diagrams, 6
Anisoptera, 154, 247
Annelida, 206, 253
annual turnover ratio, 68
Anostraca, 132, 177, 178, 185, 187, 200, 203, 205
anostracans, *see* Anostraca
antagonism, 252
Antarctic, 79, 87, 88, 118, 120, 194
   convergence, 118
antecedent drainage, 137
anthurid, 172
Aphelocheiridae, 162
Aplochitonidae, 226
applied
   aspects of limnology, 215
   ecology, 215
apterism, 160, 172
aqualin, 258
aquarium fish, 229
aqueducts, 215, 234
Aramac, 210
Arctic, 79
argillites, 99, 100
arheic
   drainage, 137, 195
   regions, 193
   system, 197
artesian
   basins, 171, 216
   waters, 216
Arthurs Range, 55
Arthur Valley, 56
asellid, *see* Asellidae
Asellidae, 153, 244
Asellota, 132
ash-free dry weight, 27
Asia, 129, 227
astatic lakes, 94
athalassic
   saline waters, 20, 204, 209, 210
   waters, 1, 8, 9, 10, 12, 15, 16
atherinid, *see* Atherinidae
Atherinidae, 109, 200, 204
Atherton Tableland, 51, 52, 53
Atlantic, 118
   salmon, 229
atmospheric
   pollutants, 146
   supply of oceanic salts, 1

atomic
   absorption spectrophotometry, 2
   waste disposal, 240
attenuation coefficients, 92
Atyidae, 154, 162, 172, 173, 190
atyids, *see* Atyidae
Auckland, 52, 83, 117, 186
Aufwuchs, 65, 123, 130–1
Australasia, 23, 56, 93, 94, 95, 98, 99, 109
Australasian lakes, 111
Australia, 11, 12, 40, 58, 59, 60, 61, 79, 81, 93, 101, 102, 106, 107, 108, 110, 111, 112, 114, 117, 118, 121, 126, 130, 131, 132, 135, 136, 139, 140, 144, 146, 148, 151, 152, 153, 154, 155, 156, 160, 162, 165, 166, 167, 169, 171, 172, 173, 174, 177, 179, 180, 181, 182, 184, 185, 187, 191, 193, 194, 195, 202, 204, 205, 206, 207, 209, 210, 215, 216, 217, 218, 220, 221, 222, 224, 225, 227, 228, 229, 232, 233, 234, 235, 239, 240, 244, 248, 253, 256, 257, 259, 260
Australian
   Alps, 221
   Atomic Energy Commission, 240
   bass, 226
   rivers, discharge and drainage areas, 136
austral seablight, 208
Austroperlidae, 153
autochthonous organic matter, 17, 44
autotrophic level, 25
autotrophs, 26, 28
Avoca River, 148

Babinda Creek, 152
bacillariophyceans, *see* diatoms
Bacteria, 44, 75, 91, 107, 109, 131, 169, 173, 204, 219, 236, 242, 244, 258
bacterial
   biosynthesis, 29
   growth determination, 44
   plankton, 102
   pollution, 254
   production, 44
bacteriological culture methods, 255
Baetidae, 154, 247
barbel, 166
Barkly Basin, 216
barnacles, 16
Barramundi, 226
Barwon River, 52, 148
base-level profile, 138
Bass Strait, 58
bathing, 230
Bathurst, 83
bathymetric map, 61
Bathynellacea, 171

Bay of Plenty, 52
Beachport, 60, 199, 200, 202
bed-load, 138
beetle(s), see Coleoptera
behavioural adaptations to running waters, 161
Benmore, 222
benthic
  biomass, 126
  fauna of Lake Purrumbete, 127
  invertebrate fauna of running waters, 153, 154
  invertebrates, effect of flooding on, 156
  productivity, 70
  productivity estimates, 68
  regions, 64
  species and depth in lake, 128
benthos, 41, 45, 65, 123
  of lakes, 123–9
  mean standing crop, 67
billabong, 58
binodal seiches, 93, 94
biochemical oxygen demand, 235, 254, 255
biogeochemical cycles, 169
biological
  assessment of pollution, 256
  effects of direct changes by man upon inland aquatic environment, 234
  magnification, 238
  significance of ionic proportions, 14–17
biomass, 26, 44
biosphere, 24
biotic
  indices of pollution, 256
  transference, 259
bivalve molluscs, 13, 124, 129, 132
blackfish, 226
Blepharoceridae, 153, 158, 159, 162, 166
blepharocerids, see Blepharoceridae
blood worms, see Chironomidae
blooms, see algal blooms
blowholes, 173
blue-green algae, see Cyanophyta
Blue Mountains, 230
B.O.D., see biochemical oxygen demand
body fluids, 13
boeckellids, 122, 123
bog lakes, 18
bogs, 18
boiling mud pools, 173
bony bream, 109
bores, 173
bottom deposits of beds of running waters, 145
boundary layer, 159
Bourke, 144
Bowen Basin, 216
brackish, 167
  lakes, 208

brackish (continued)
  waters, 16
bradyauxesis, 123
branchiopods, 16, 184, 185, 187, 190
bream, 166
brine wastes, 239
Brisbane, 181, 189
  River, 239
Britain, 103, 179, 229, 239
British
  Columbia, 39
  Isles, see Britain
  Royal Commission on Sewage Disposal, 255
Broken Hill, 60
brown
  humified lakes, 64
  trout, 40, 146, 229
browsers, 69
bryophytes, 152
bugs, see Hemiptera
Bulimba, 239
Burdekin River, 136
burning-off, 258
burrowing moth, 176

cactus, 176
caddis, see Trichoptera
Caenidae, 154, 167, 247
Calamoceratidae, 237
calanoid, see Calanoida
Calanoida, 14, 15, 16, 76, 92, 109, 111, 113, 116, 178, 185, 186, 189, 191, 201, 203, 204, 209
calanoid
  copepod associations, 106
  distributions, 108, 110, 112, 114
  species composition and size differentiation, 107
calcium
  and the availability of carbon in one or other of several different forms, 73–5
  as a factor in lake productivity, 73–4
caldera, 51, 52
  lakes, 68
California, 16, 30, 33, 72, 210, 238
calories, 27
calorific
  equivalents for some aquatic biota, 27
  values, 44
calorimetry, 27
Campaspe River, 148
Canada, 102, 224
canals, 225, 234
Canberra, 50, 237, 251, 252
cane toad, 260
Canning Basin, 216
Canterbury, 54, 55, 79, 81, 99, 100, 129, 172
capacity of river, 138

GENERAL INDEX

Cape Campbell, 211
Cape Flattery, 59
Cape York, 155
  Peninsula, 59
Captain's Flat, 237
carbamates, 238
carbamino carboxylates, 75
carbon, 74
carbon dioxide, 150
  in inland waters, 20–3
carina, 121
Carnarvon Basin, 216
carp, 227, 259, 260
catchment grazing, 217
catfish, 226
cattle, 220
cavernicolous, 172
caves, 171, 172
  and wells, fauna of, 172
cellular tolerance, 14, 201
cellulase, 169
Central Plateau of Tasmania, 55, 56, 60, 79
centric diatoms, 102
Centrolepidaceae, 130
Centropomidae, 109
Ceratopogonidae, 129, 133, 175, 176, 178, 203, 205, 209, 247
ceratopogonid(s), see Ceratopogonidae
Channel country, 142
chaoborid larvae, 124, 126
charophyceans, 152
Charters Towers, 210
cheimomictic lakes, 80, 81, 82, 84, 119
Chelonia, 155
chemical
  coagulation, 219
  composition of some Australian rivers, 149
  composition of some natural waters, 7
  composition of spring water, 173
  correction, 219
chemocline, 75, 85, 86, 87
chemosynthesis, 29
China, 136, 225
china-clay, 237
Chironomidae, 40, 124, 126, 127, 129, 132, 154, 157, 160, 163, 164, 167, 174, 175, 178, 191, 206, 211, 244, 245, 246, 247, 248, 252
chironomids, see Chironomidae
  of lake muds, 126
Chlamydobacteriales, 223
chlordane, 237
chloride in rain water, 9
chlorinated
  hydrocarbons, 237
  naphthalenes, 238
chlorination, 219
chlorite schists, 100
Chlorococcales, 102, 189

Chlorophyceae, 151, 152, 208, 242
chlorophycean index, 102, 103
chlorophycean(s), see Chlorophyceae
chlorophyll, 73
  a, 208
  estimates, 27
cholera, 254
Christchurch, 61, 186
chrysophycean plankton, 101
chrysophycean(s), 131
chrysophytes, 152
chydorid(s), 70, 71, 185, 186
chytrid fungi, 105
ciliates, 174, 175, 204, 242
circadian rhythm, 120
circulation patterns of some lakes, 80
cirque lakes, 55, 62
Cladocera, 15, 107, 109, 113, 116, 120, 123, 127, 133, 178, 185, 186, 187, 189, 190, 191, 203, 205, 238, 247
cladoceran(s), see Cladocera
  remains, 70
Clarence
  Basin, 216
  River, 136
Clarke–Bumpus sampler, 27, 116
clay-pans, 60
clinograde, 23
closed lake, 49
Clupeiformes, 226
Clutha River, 222
^{14}C methodology, 31, 35
CMU, 238
Cnidaria, see Coelenterata
coarse fish, 257
coastal
  athalassic saline waters, 204
  ponds, 181, 185
  salt lakes, 9
cobalt, 72, 73
Coelenterata, 154, 206, 248
cold
  monomictic lakes, 79, 81
  thereimictic lakes, 81
Coleoptera, 131, 132, 154, 157, 159, 162, 164, 172, 174–6, 178, 190, 206, 207, 247
coleopterans, see Coleoptera
coliform
  bacteria, 219, 254
  standards, 254
Collie River, 147
Colorado, 102
colorimetry, 92
coloured
  dissolved organic matter, 92
  sulphur bacteria, 102
colourless sulphur bacteria, 29
colour of lakes, 92–3
Columbia River, 136, 239, 241

common salt, 231
competence of river, 138
competitive exclusion, 191
completely submerged, rooted macrophytes, 129
complexometric titration, 2
compound index for phytoplankton, 102, 103
Conchostraca, 177, 178, 185, 187
conchostracan(s), see Conchostraca
conductivity, 12
Cone Spring, 47
Connecticut, 126
consequent streams, 137
Conservation Bill, 235
conservation, list of aquatic sites in Australia and New Zealand provisionally proposed for, 233
consumption, 26
contact water, 96
conversion factors, 27
Coolamon Creek, 152
Cooper's Creek, 140, 142
Coorong, 58
Copepoda, 14, 15, 16, 19, 44, 70, 92, 107, 109, 117, 118, 122, 131, 178, 186, 187, 189, 190, 191, 201, 203, 204, 205, 209, 210, 211, 247
copepodites, 113, 116, 117
copepods, see Copepoda
copper sulphate, 252, 258
coregonid fish, 257
Corixidae, 167, 190, 207
corixids, see Corixidae
corrasion, 138
correction factors for altitude for percentage saturation of oxygen, 21
corrosion, 138, 146
Coulter counter, 31
Cox's gudgeon, 159
Cox's River, 153, 154, 245, 246
crab(s), 13, 189, 207
Cranbourne, 99
crayfish, 74, 172, 173, 189, 229
creeks, 49
crocodiles, 155, 229
crop, 29, 69
crops grown under irrigation, 222
Crustacea, 27, 43, 65, 70, 73, 115, 119, 127, 131, 132, 152, 153, 154, 171, 172, 175, 184, 185, 190, 238, 253
crustaceans, see Crustacea
crustacean plankton, 41
cryptobiotic, 178
cryptorheic
    drainage, 137
    system, 193
Culicidae, 132, 133, 167, 175, 178, 206, 229, 257

culicids, see Culicidae
Cullarin Range, 50
cultural eutrophication, 97, 241, 256
currents, 96
cyanides, 249
cyanophyceans, see Cyanophyta
Cyanophyta, 16, 19, 30, 70, 72, 92, 93, 97, 102, 131, 152, 157, 174, 190, 208, 219, 242, 257
cyclic salts, 146, 193
cyclomorphosis, 113, 120–3
    in *Boeckella minuta*, 124
    in *Boeckella propinqua*, 122
    in *Daphnia carinata*, 121
    in dinoflagellates, 123
    in rotifers, 122
Cyclopoida, 111, 116, 117, 178, 187, 191, 203, 209, 211
cyclopoids, see Cyclopoida
    food of, 111
Cyprinae, 205, 209
Cypriniformes, 226

2,4-D, 238, 258
dalapon, 338
Daly River, 136
dams, 232
Dandenong Creek, 225
Danube River, 16, 135, 136
Darling River, 58, 144, 146, 147, 149
daschis, 59
dawn rise, 115, 120
DDD, 238
DDT, 237, 238, 252
dead water, zones of, 142
Decapoda, 132, 206
decomposers, 26
defeated drainage, 137
definition
    of pond, 179
    of production and productivity, 29
deflation basins, 60
deforestation, 258
dendritic drainage patterns, 142
Denmark, 70, 126
deoxygenation, 242
Derwent River, 136, 223
desalination, 217
desert pup fish, 22
Desmideae, 97, 98, 99, 100, 102, 103
desmids, see Desmideae
detergents, 236, 248, 256
detergent swans, 248
detritovores, 169
detritus, 31
    chain, 47
diapause, 162, 178
diaptomids, 122

GENERAL INDEX

diatoms, 27, 65, 98, 100, 102, 105, 111, 131, 151, 208, 219, 242, 245, 257
diatreme, 54
2,4-dichlorophenoxyacetic acid, 238
dichloro-diphenyl-trichloro-ethane, 237
dichothermic, 87
dieldrin, 238
differential accumulation of radionuclides, 240
diffuse radiation, 89
diffusion, 13
dimictic lakes, 79, 80, 81, 82, 84, 85, 119
dinoflagellates, 111, 123
dinitro compounds, 238
dinitro-ortho-cresol, 238
Diptera, 16, 131, 132, 154, 157, 159, 161, 164, 190, 206, 211, 244, 247
dipterans, *see* Diptera
diquat, 258
direct
 effects of man on inland waters, 232–4
 radiation, 89
discharge of rivers, 136, 140, 141
dispersal, 160
 mechanisms, 191
 of biota of small bodies of water, 190–1
dissolved
 organic matter, 17, 75
 oxygen, 20
distillation treatment, 258
diurnal
 fluctuations in drift, 165
 vertical migration of zooplankton, 113–20
DNOC, 238
domestic
 refuse, 236
 supplies of water, 215–19
Douglas, 197, 199
dragonflies, *see* Odonata
drainage, 232
 patterns or systems, 137, 193–7, 233
dredging, 233
drift, 163–6
 amount of, 164
 and bottom fauna, 164
 significance of, 165
drinking water, 215
 quality, 218
 standards, 218
 upper salinity limits for, 220
dry
 fallout, 8
 weight, 27
Drysdale River, 147
duck-billed platypus, 155, 229
duck-shooting, 230
dune barrage lakes, 59
Dunedin, 189

dune lakes, 58, 74
dysentery, 254
Dytiscidae, 132, 167, 207
dytiscid, *see* Dytiscidae

eastern water-rat, 155
East Gippsland Basin, 216
E. coli I, *see Escherichia coli* I
ecological efficiencies, 44
ecosystem
 concept, 24
 dynamics, 47
ecosystems, 169, 211
Edmondson's method of determining secondary production, 41–2
EDTA, 2, 8
eels, 172
effects
 of pollutants on inland waters, 240–53
 of thermal pollution in rivers, 253
 on rivers of radioactive waste, 253
egesta, 26
egestion, 44
egg diapause, 162
Egypt, 33, 210, 232
Ekman-Birge grab, 28
Elat, 87
electronic particle counter, 31
Eleotridae, 172
Elmidae, 162, 166
Elminthidae, *see* Elmidae
emergent macrophytes, 36, 129
emission flame photometry, 2
empidids, 162
empneuston, 71
Emschergenossenschaft, 235
Emscher River Commission, 235
endorheic drainage, 137, 193, 195, 197, 211
endrin, 238
energy
 budget, 25, 44, 46–7
 content, 24
 pathways, 169
 pathways in Cone Spring, 46
 relationships in an ecosystem, 25
 transfer, 24, 44
 units, 27
England, 36, 40, 45, 215, 225, 235, 257
English Lake District, 105, 230
enteric virus diseases, 254
Entomostraca, 19, 188, 191
entomostracans, *see* Entomostraca
Ephemerellidae, 166
Ephemeroptera, 41, 45, 127, 153, 154, 157, 161, 164, 246, 247, 253
ephemeropterans, *see* Ephemeroptera
Ephydridae, 174, 206, 207, 211
ephydrids, *see* Ephydridae

epilimnion, 19, 23, 78, 79, 80, 82, 83
epilithic, 131
  algae, 35, 151, 158
epineuston, 131
epipelic, 131
  algae, 151
epiphytic, 131
  algae, 151, 158
epipleuston, 131
epipotamon, 167
epirhithron, 167
episodic rivers and streams, 139, 140
epizoic, 131
  algae, 151
EPN, 238
epsomite, 86
eriochrome black T, 2
erosion, 138
Escherichia coli I
  densities, 255
  standards, 254
Esperance Bay, 231
estuaries, 14, 118, 155, 205
ethyl p-nitrophenyl benzenephosphonothionate, 238
Eucla Basin, 216
eucrenon, 166
euglenoids, *see* Euglenophyta
Euglenophyta, 102, 151
euglenophyte plankton, 102
euphausiaceans, 118, 120
euphotic zone, 64, 67, 118
Europe, 40, 102, 126, 136, 166, 172, 191, 216, 225, 227, 228, 235
Eustheniidae, 153, 154
eutrophication, 256–8
eutrophic
  chlorococcal plankton, 102
  diatom plankton, 101
  dinoflagellate plankton, 101
  waters, 19, 33, 64, 66, 70, 102, 103, 124, 126, 256
eutrophy, 102
excreta, 26
exorheic drainage, 137, 193
explosion craters, 52
extension of valleys, 139
extinction coefficient, 90
extracellular products, 75

faecal
  contamination, 219, 254
  matter, 236
faeces, 26
Falmouth, 229
Far East, 225
farm dams, 49, 178–89
Farmer's Creek, 245, 246, 249, 250
fathead minnow, 238

fatty alcohols, 237
fault-scarp lake, 50
feeding, 44
ferric iron, 84
Ferrissiidae, 154, 166
Fiery Creek, 148
filtration of sewage, 219
Fiordland lakes, 100
fires, 258
firmly-bound CO_2, 11
fish, 24, 43, 65, 111, 131, 155, 159, 162, 166, 167, 172, 187, 189, 190, 200, 204, 205, 206, 217, 225, 237, 248–50, 252, 256–9
  farming, 225, 227
  hatcheries, 228
  introductions, effects of, 260
  of commercial or recreational importance, 226
  ponds, 227
  predation, 109
  production, 28, 37, 43, 157, 226, 227
  productivity, 77
  productivity and mean depth of lakes, 67
  zonations, 166
Fisheries Research Board of Canada, 237
Fisheries Research Division, New Zealand Marine Department, 61
fishing, 225–30
  methods, 225
Fitzroy
  Basin, 216
  River, 136
fjord lake, 56
flagellate(s), 65, 92, 204
Fletcher's frog, 185
floating-leafed, rooted macrophytes, 129
Florida, 47, 99
flowering plants, *see* angiosperms
flow rates, 138, 140–3
fluoride, 220
fluviatile lakes, 58
Fortescue River, 136, 216
Frankland Range, 54, 55
Fraser Island, 59, 101
free-floating macrophytes, 129
Frenchman's Cap National Park, 55
fresh athalassic waters, 12
fresh water, ionic composition of, 2
freshwater
  algae, *see* algae
  crab, *see* crab
  crayfish, *see* crayfish
  mussel, *see* mussel
frog, 155, 159, 178, 185
full supply level, 233
fungi, 169, 242, 258

Gadopsidae, 226
galaxiid(s), 155, 166

GENERAL INDEX

gammarid, 172
Ganges River, 136
Gas Clauses Act, 235
gas maar, 54
Gastropoda, 127, 172, 187, 206, 207, 209, 247
gastropods, *see* Gastropoda
Gause Principle, 103
Gayndah, 52
genetic introgression, 234
Georgina Basin, 216
geostrophic effects, 96
geothermal heat, 87, 88
Germany, 235
Gerridae, 133
geysers, 173
giant Tasmanian crayfish, 229
Gibson Desert, 193
glacial lakes, 54–6, 97, 111
Glenorchy, 55
Glossiphoniidae, 154, 247
Glossosomatinae, 166
gneisses, 100
gnotobiotic study, 211
goby, 159
golden perch, 153, 226
Gordon River Scheme, 223
Goulburn River, 58, 148
Goyder Channel, 196
graben lakes, 62
Grand Coulee, 85, 202
graphical method of estimating fish production, 38
grass carp, 227
grayling, 166, 260
grazing food chain, 47
Great Australian Artesian Basin, 171, 173, 216, 217
Great Dividing Range, 97, 98
Great Lakes, 234
Great Sandy Island, 59
grebes, 238
green alga(e), 97, 131, 207, 219, 242
green sulphur bacteria, 29, 75, 93
greywackes, 99, 100
gripopterygid, *see* Gripopterygidae
Gripopterygidae, 153, 154, 157, 159, 166
gross primary production, 28
ground water(s), 171, 216
Gulf of Carpentaria, 155
Gyrinidae, 133

hairback herring, 109
Haiti, 210
half-bound CO_2, 11
half-life of some radionuclides of biological importance, 240
Halplidae, 167
halite, 231
halobiont, 16
halocline, 118
halophyte species, 208
Hamersley, 216
Haplopoda, 107
Harappā civilization, 235
Hardy's theory, 118
harmonics, 93
Haroharo Caldera, 52
harpacticoid, 204, 209
harvest, 29, 67
harvesting, 36
Hebridae, 133
heleoplankton, 184
helminth eggs, 236
helocrenes, 173
Helodidae, 166
Hemiptera, 131, 132, 154, 162, 174, 206, 211, 247, 253
hemipterans, *see* Hemiptera
heptachlor, 237, 238
herbicides, 237
heterauxesis, 123
heterotrophic algae, 85
heterotrophs, 26, 28, 236
Hirudinea, 154, 244, 247
Holarctic, 113
holdfasts, 158
holomictic
  lakes, 85, 86
  saline lakes, 102
holomixis, 81, 119
homoiosaline, 12
Horokiwi Stream, 38, 40, 156, 157
horses, 220
hot water as pollutant, 239
humic acids, 18
Hungary, 87
Hydracarina, 127, 129, 164, 175, 190
hydracarines, *see* Hydracarina
Hydraenidae, 162, 166
hydraulic suckers, 159
Hydrobiidae, 154
Hydro Electric Commission of Tasmania, 54, 223
hydro-electricity, 222–4
hydrogen ion concentration, 17–19
hydrogen sulphide, 124
hydrological
  cycle, 193
  regions, 137
  regions of Australia, 194
  regions of Victoria, 194
Hydrometridae, 133
hydrophilid, *see* Hydrophilidae
Hydrophilidae, 154, 207
Hydropsychidae, 154, 161, 247
Hydroptilidae, 154, 167, 247
Hydrostychaceae, 152
hydroxy naphthol blue, 2

Hydrozoa, 204, 205
hydrozoan, see Hydrozoa
hylid, 159
Hymenosomatidae, 206, 207
hyperosmotic
  regulation, 13
  regulators, 13
hypersaline lagoons, 14
hyphomycetes, 169
hypocrenon, 166
hypolimnetic
  currents, 96
  oxygen depletion, 124, 257
hypolimnion, 23, 78, 79, 80, 82, 83, 84, 124, 198, 233
hyponeuston, 131
hypo-osmotic
  regulation, 14
  urine, 13
hypopotamon, 167
hyporheic biotope, 171
hyporhithron, 167
Hyriidae, 153

ice, 85
immature ecosystems, 211
impoundments, 217, 222, 232, 233, 234
India, 136, 210
Indiana, 32, 33, 34, 70
indicator species, 246, 256
indirect effects of man on inland waters, 258–60
Indo-China, 225
Indus River, 136, 235
industrial supplies of water, 219–20
inert material as pollutant, 237
infra-red radiation, 89
inherited drainage, 137
initial distinctions between standing water bodies, 49
inland saline waters, see salt lakes, saline streams
Innisfail, 152
Innot Creek, 173
inorganic
  ions of running waters, 148
  poisons, 237
Insecta, 27, 43, 127, 131, 132, 153, 157, 159, 160, 165, 172, 174, 190, 205, 206, 207, 209, 237, 238, 258
insectan behavioural patterns, 161
insecticides, 237, 238
insect(s), see Insecta
insequent streams, 137
interaction of factors influencing productivity, 77
intercellular lacunae of macrophytes, 37

intermittent
  rivers, 139, 161, 162
  streams, 139, 161, 162
internal seiches, 94, 95, 96
International Biological Programme, 26, 47
interstitial
  fauna, 171, 172
  ground waters, 172
  waters, 171
intestinal bacterium, 254
introduced fish, 259
invertebrates, 27
ion antagonism, 14
ion-exchange procedure, 2
ionic
  composition of Australian athalassic waters, 201
  composition of sea water, 1–2
  composition of world average fresh water, 1–2
  equilibrium, 14
  regulation, 14, 15
ion
  supply by action of hydrogen ions, 8
  supply by complex formation, 8
  supply by ion exchange, 8
  supply by oxidation and reduction, 8
  supply by solution, 8
Iowa, 41
Iraq, 210
Irish loughs, 103
iron, 73, 84
  bacteria, 29, 223
irrigation, 220, 221, 222
Island Stream, 153, 154, 164
Isopoda, 14, 45, 132, 153, 172, 173, 190, 200, 206, 207, 209, 244
isopod(s), see Isopoda
isotopes, 240
Israel, 87, 227

Java, 227
jet flows, 143
Johnstone's crocodile, 229
Johnstone River, 152

Kakanui River, 153, 154, 163, 164
Kalgoorlie, 60, 193
Kareeya, 223
Katherine, 172
kettle lakes, 55
Kielce, 227
Kiewa
  River, 149
  scheme, 222
kilocalories, 27
King Island, 59, 99, 181

GENERAL INDEX

Kosciusko plateau, 55
Kuratau Spit, 53

Lachlan River, 146
lacustrine sediments, 11
lagoon, 60
Lahontan Basin, 212
laminar flow, 143
land-clearing, 258
landslide lakes, 54
lateral lake, 58
lake(s), 45, 49, 217
  area, 61
  as a microcosm, 71
  associated with coastlines, 60
  associated with volcanic activity, 51
  behind sand dunes, 58
  bottoms, 26
  formed by collapse of lava flows, 53
  formed by volcanic damming, 53
  formed by wind action, 58
  held by morainic dams, 56
  in rock basins produced by deflation, 58
  in rock or clay basins produced by deflation, 60
  productivity, 66–77, 258
  typology, 126
  volume, 61
lake trout, 234
lateral migration of streams, 139
Latidae, 226
Latrobe River, 239
leaching requirement, 222
leech, see Hirudinea
lentic environments, 49, 135
Lepidoptera, 154, 206, 207, 247
lepidopteran, see Lepidoptera
leptocerid, see Leptoceridae
Leptoceridae, 153, 154, 167, 206, 207, 247
leptophlebiid, see Leptophlebiidae
Leptophlebiidae, 153, 154, 159, 166, 247
level of compensation, 64
lichens, 152
light and dark bottles, 30, 35
light, differential transmission of, 91
lightning, 19
limnetic region, 63
limnetic zooplankton, 187
limnocrenes, 173
limnogram, 93, 94
limno-humic acid, 18
limpets, 246
Linga, 92, 202
Lithgow, 245
Little River, 147, 223
littoral, 64
  benthic fauna, 124

littoral (continued)
  benthos, 123
  shelf, 124
littoriprofundal, 64
liverworts, 152, 158
load, 138
load of saltation, 138
lochans, 179
Lofu River, 15
London, 235
longicorn grubs, 175
longitudinal
  profile of rivers, 138
  seiche, 94
  zonation of lotic environments, 166–7
loricate rotifers, 122
loss on ignition, 11
lotic biota, factors controlling the, 156
lotic
  ecosystems, 169
  environments, 45, 49, 135–69
Lucas Heights, 240
lungfish, 155, 162

maars, 51, 52
Macarthur, 52
Macdonnell Ranges, 225
McLaren's theory, 119
McMurdo Sound, 87
Macquarie
  Island, 79
  perch, 226
Macquariidae, 226
macrobenthos, 45, 69
macro-invertebrates from Farmer's Creek and Cox's River, 247
macrophyte biomass, 36
macrophytes, 26, 27, 29, 30, 36–7, 64, 123, 152, 208, 257, 258
  adventive, 129
  in lakes, 129–30
  in ponds, 184
  of running water, 157
macrophytic growth, 257
Madagascar, 152, 172
Madison, 257
magnesium, 73
major cations, 2
major ions in sea water and in some athalassic saline waters, 202
Malacca, 227
Malacostraca, 172, 247
malacostracan, see Malacostraca
malathion, 238
Malaysia, 129, 227
Mallee, 56
mammals, 155, 258

manatee, 258
manganese, 73
manganese deposits, 223
Maribyrnong River (Creek), 149, 255
marine-brackish, 209
marine salt beds, 8
Maritime Services Board, 240
marron, 189, 229
marsupial, 155
Massachusetts, 47
material budgets, 26
mature ecosystems, 211
Maucha's field diagrams, 4, 5, 6
maximum
 depth of lakes, 62
 length of lakes, 62
 width of lakes, 62
mayflies, see Ephemeroptera
Mayor Island, 52, 62, 82, 116
mean
 area of lakes, 68
 depth of lakes, 62, 66, 68
 flow rates, 143
 width of lakes, 62
mecopteran, 153
medusae, 204
Megaloptera, 154, 158, 159, 161, 247, 253
megalopteran larvae, see Megaloptera
Melbourne, 98, 117, 178, 179, 220, 229, 230, 239, 248, 254, 255
mercury, 238
meres, 179
meromictic lakes, 75, 81, 82, 85–7, 93, 96, 102
meroplanktonic species, 104
meropleuston, 133
mesosaprobic zone, 244
mesotrophic
 desmid plankton, 102
 dinoflagellate plankton, 101
 lakes, 126
mesotrophy, 102
Mesoveliidae, 133
metalimnion, 78, 80, 83, 95
metapotamon, 167
metarhithron, 167
meteorites, 11
methaemoglobinaemia, 220
method
 for determining phytoplankton production, 30
 of determining the volume of a lake, 63
 for estimating fish production, 37
 of obtaining samples of biota, 27
 of presenting analytical results, 2–8
methoxychlor, 238
methyl mercury, 238
methyl-orange end-point, 4
Michigan, 41, 42

microbial
 decomposition, 169
 processes in lakes, 44
micro-ecosystems, 211, 213
micro-nutrients, 20, 72–3
Middle East, 217, 225, 227
midges, 129, 238
midnight sinking, 115, 120
Miller method of oxygen determination, 20
milliequivalents as units, 3
Millstream group of springs, 216
mirabilite, 86
Minnesota, 40, 44, 45, 47
Mississippi–Missouri river system, 136
Missouri River, 164
Mitchell River, 136
mites, see Hydracarina
Mitta Mitta River, 148
mixolimnion, 85, 86, 87
modified caldera lake, 52
Mollusca, 27, 69, 73, 126, 127, 154, 157, 164, 190, 206, 244, 246, 253
Molonglo River, 251, 252, 253
molybdenum, 72–3
monimolimnion, 85, 86, 87
monomictic lakes, 79, 81
Monotremata, 155
morphological
 adaptations to flowing water, 158–61
 features of Australian salt lakes, 197
morphology
 of lake basins, 61–3
 influence of on lake productivity, 66–71
morphometric parameters, 61–3, 84
Moscow, 68
mosquitoes, see Culicidae
mosquito fish, 229
mosses, 152, 158
moth-fly, 244
mound springs, 173, 174
Mount
 Abrupt, 53
 Disappointment, 98
 Donna Buang, 160
 Eccles, 52, 53
 Field National Park, 81
 Gambier, 95
 Hypipamee, 53
 Kosciusko, 81, 121
 Le Brun, 52
 Ngauruhoe, 223
 Quincan, 52
 Ruapehu, 18, 223
 Tongariro, 223
 Wellington, 54
mud, 84
mud-water interface, 126
Muganskaya steppe, 210

GENERAL INDEX

Murray
  Basin, 216
  cod, 226, 234
Murray–Darling river system, 58, 109, 136, 152, 155, 224
Murray–Murrumbidgee river system, 146
Murray River, 58, 136, 140, 141, 146, 147, 148, 155, 234
  lobster, 229
Murrumbidgee River, 136, 149, 155
mussels, 40, 129, 153, 191
Mysidacea, 132
Myxophyceae, *see* Cyanophyta
myxophycean plankton, 102

naid worms, *see* Naididae
Naididae, 246, 248
Nannochoristidae, 153
nannoplankton, 30, 65
natant shrimps, 65
natronophils, 16
Naurcoridae, 162
nauplii, 117
navigation, 224–5
Nebraska, 60
nekton, 64, 65, 123, 131
nektonic
  invertebrates, 132
  productivity, 70
nektoplanktonic, 65
Nelson, 58
Nematoda, 127, 131, 132, 154, 174, 178, 206, 248
nematodes, *see* Nematoda
net-phytoplankton, 30, 97
net-plankton, 65
net primary production, 28
neuston, 131
Nevada, 4, 202
New Brunswick, 237
New Guinea, 79, 114, 152, 229
New South Wales, 50, 58, 59, 60, 79, 83, 94, 99, 106, 107, 113, 136, 144, 146, 147, 149, 152, 153, 154, 175, 182, 183, 187, 189, 190, 191, 206, 217, 218, 222, 224, 231, 233, 245, 249, 250, 254, 259
New South Wales Department of Health, 218
New York State, 86, 87
New Zealand, 4, 16, 17, 18, 36, 37, 38, 40, 49, 52, 53, 54, 55, 56, 58, 59, 61, 62, 70, 72, 74, 79, 81, 82, 83, 93, 95, 97, 99, 101, 102, 106, 107, 108, 110, 111, 112, 113, 116, 117, 118, 121, 126, 129, 130, 131, 132, 133, 135, 136, 139, 140, 144, 146, 150, 151, 152, 153, 154, 155, 156, 157, 159, 160, 163, 166, 167, 169, 171, 172, 173, 174, 179, 180, 181, 184, 185, 187, 189, 191, 202, 210, 211, 213, 222, 224, 225,

New Zealand *(continued)*
  227, 228, 229, 231, 232, 233, 235, 237, 244, 246, 253, 254, 257, 260
Niagara Falls, 234
niche, 103
nicotine alkaloids, 237
Nile River, 136
nitrate, 105
  bacteria, 29
  in inland waters, 19–20
nitrite bacteria, 29
nitrogen, 83
nitrogen-assimilating bacteria, 19
nitrogen-fixing bacteria, 19
Nive River, 223
non-flagellate phytoplankton, 65
non-marine saline waters, 209; *see also* salt lakes
non-marine saline waters in New Zealand, 210–11
non-periodic water movements, 96
non-planktonic biota of lakes, 123–33
non-poisonous salts, 239
non-sulphur bacteria, 29
non-thermal springs, 172
non-volatile solids, 11
North America, 30, 56, 83, 109, 176, 179, 220, 221, 225, 235, 238, 244, 257
Northern Territory, 106, 136, 147, 149, 162, 172, 210, 229, 231
North Island (New Zealand), 18, 53, 58, 59, 106, 157, 173, 189, 211, 233
North Johnstone River, 152
Northland, 130
North Wales, 45
North West Cape, 172
Norwood Warp, 50
Notonectidae, 65, 167
notonectids, *see* Notonectidae
Notostraca, 132, 177, 178, 185
notostracans, *see* Notostraca
Nullarbor Plain, 50
numerical method of estimating fish production, 37
nutrient pathways in an ecosystem, 25
nutrients to a lake, supply of, 70
nymphal diapause, 162
Nymphulinae, 154, 247

obligate cavernicoles, 172
obsequent streams, 137
Odonata, 154, 161, 185, 190, 206, 207, 246, 247
odonatan nymphs, *see* Odonata
Odontoceridae, 154, 166, 247
odours from water, 219
Officer Basin, 216
Ohio River, 235

Ohio River Valley Water Sanitation Commission, 235
Oligochaeta, 43, 45, 69, 124, 126, 127, 132, 154, 157, 161, 164, 172, 175, 187, 244, 245, 248
oligochaetes, *see* Oligochaeta
oligomictic lakes, 79, 81, 82, 85, 96
oligosaprobic zone, 245
oligotrophic
  chlorococcal plankton, 101
  desmid plankton, 101
  diatom-desmid plankton, 101
  diatom plankton, 101
  dinoflagellate plankton, 101
  lakes, 33, 64, 66, 102, 103, 126, 217, 256
  lakes, plankton of, 101
oligotrophy, 217, 218
  causative factors of, 217
oniscoid, *see* Oniscoidea
Oniscoidea, 14, 207, 209
Ontario, 45, 68
open lake, 49
optical phenomena in lakes, 87–93
Ord–Victoria Region Basins, 216
Oregon, 90
organically combined carbon, 27
organic
  debris in air, 10
  phosphates, 237, 249
  poisons, 237, 238, 249, 250, 252
  trade residues, 236
  waste levels in rivers, 254
  weedicides, 258
organomass, 26
origin of lake basins, 49–61
Orthocladiinae, 132, 247
orthograde, 23
orthophosphate, 20
osmo-conformers, 14
osmoregulation, 12, 15, 200, 201
osmosis, 13
osmotic regulation, *see* osmoregulation
Ostracoda, 14, 16, 127, 132, 164, 178, 187, 189, 203, 204, 205, 209, 210, 211, 247
ostracod(s), *see* Ostracoda
Otago, 99, 202, 210
ox-bow lakes, 58, 139
Oxley Basin, 216
oxygen, 82, 124, 150
oxygen-depth curve, 23
oxygen
  in hypolimnion, 84
  in inland waters, 20–3
  in shallow ponds, 181
  sag curve, 242
  saturation values in salt waters, 22
over-grazing, 258
overturn, 84

Palaearctic, 210
palaemonid shrimps, 207, 229
palaeolimnological studies, 71
palaeolimnology, 11, 70
Panama Canal, 86
paraquat, 258
parathion, 238
Parliamentary Senate Select Committee, 235
Parnidae, 154, 157, 164
Paroo River, 147, 149
Patearoa, 210, 211
pathogenic bacteria, 219, 236
pathogens, 258
Pauhara, 58
3-(p-chlorophenyl)-1,1 dionethylurea, 238
pea mussels, 129
peat bogs, 26
peats, 71
pelagic, 63
pelagic eggs, 162
penetration through water of solar radiation, 87–91
percentage saturation of oxygen, 20
Perciformes, 226
periphyton, 29, 30, 34–6, 131
  biomass, 208
  net production, 35
permanent
  ponds, 124, 179, 184, 185, 187, 189
  streams or rivers, 139
Perth Basin, 216
pesticides, 237
pH
  as a factor in zooplankton vertical migration, 116
  ecological significance of, 19
  range in natural waters, 17–18
phenolphthalein end-point, 4
Philopotaminae, 166
phosphate, 105
  in inland waters, 19–20
phosphatic rocks, 20
phosphine, 20
phosphorus, 20, 83
  as a factor in lake productivity, 71–2
photo-inhibition, 32
photosynthesis, 29
photosynthesis-depth curves, 32
photosynthetic
  bacteria, 29, 64
  protistans, 64
  quotients, 31
phreatobious fauna, 171
Phreatoicidea, 132, 153, 172, 173
phreatoicids, *see* Phreatoicidea
Phryganeidae, 153
physiological adaptations to running waters, 161–2

phytobenthos, 65
phytoplankton, 27, 28, 29, 30-4, 64, 65, 74, 75, 97-105, 119, 151, 257, 258
  associations, 101
  controlling factors, 104-5
  diversity and its basis, 103-4
  seasonal aspects, 104-5
  seasons, 104
  some general aspects of distribution and broad geochemical correlations of, 97-101
  types and indices, 101-3
piedmont lakes, 56, 62
pigment analyses, 27
pigs, 220
pine, 259
Pine River, 223
Pirie-Torrens Basin, 216
planimetry, 61
plankton, 27, 64, 85
  in lakes, 97-123
  of ponds, 187
planktonic
  crustaceans, 19, 74
  productivity, 70
planorbid, see Planorbidae
Planorbidae, 127, 154
plastron, 162
platinum units, 92
platypus, 155, 229
Plecoptera, 153, 154, 157, 159, 160, 161, 162, 190, 246, 247, 253
plecopterans, see Plecoptera
Pleistocene, 50
Plenty, 229
Pleurodira, 155
pleuston, 65, 123, 131-3
Plotosidae, 163, 226
Poatina, 222
Podonomidae, 166
Podostemaceae, 152
poikilosaline, 12, 200
poison
  concentration and dosage, 250
  exposure time of, 250
  selective toxicity of, 252
  threshold dose of, 250
Poland, 34, 45, 227
pollen, 71
pollutants, classification of, 236
polluted air, 8
pollution, 234-56
  assessment of, 253-6
  definition, 234
  effects on inland aquatic ecosystems, 253
  recovery from, 244
Polycentropidae, 154, 161, 247
Polychaeta, 206, 209

polychaete, see Polychaeta
polymictic lakes, 79, 81, 119
Polyphemidae, 15
Polyphemoidea, 107
polyphosphates, 20
polysaprobic zone, 244
Polyzoa, 154, 248
pond(s), 49, 109, 116, 117, 124, 154, 178-89, 190, 232
  biological features of, 184
  biota of, 124
  chemical nature of, 181-2
  communities of, 182
  pH in, 181
  turbidity in, 181
  zooplankton of, 186
pools, 49, 109
Porifera, 127, 154, 248
Potamanthidae, 247, 248
potamon, 167, 168, 169
Potamonidae, 189
potamphytoplankton, 151
potamoplankton, 163
Potamopyrgidae, see Hydrobiidae
poultry, 220
powered boating, 230
p,p^1-dichlorodiphenyldichloroethane, 238
prawns, 162, 172, 190
pre-blooming, 258
predators, 69
Prescott Point, 196
pressure, 119
prickly pear, 176
primary production, 28, 29-37
  variations in rates of, 34
primary treatment, 236
production, 28, 44, 66
  by invertebrates, 40, 41, 45
  by stream insects, 43
  by stream invertebrates, 42
  estimates for fish, 39
productivity, 28, 29, 66, 69
  in unfertilized fish ponds, 227
  of lacustrine communities, 66
profundal, 64
  benthos, 123, 126, 129
  fauna, 124, 257
Project Aqua, 232, 233
protective nocturnalism, 119
Protista, see Protozoa
Protozoa, 124, 131, 172, 174, 178, 190, 204, 245
protozoans, see Protozoa
psammon, 65, 123, 131, 132
psephenid, see Psephenidae
Psephenidae, 154, 158, 159, 166, 247
Psychodidae, 166, 175
psychodids, see Psychodidae

puddles and rock-pools, 176–8
purple sulphur bacteria, 29, 75, 93
Pütter's hypothesis, 75
pyralid, see Pyralidae
Pyralidae, 127, 206
pyrethrum, 237

Queensland, 50, 51, 52, 53, 59, 74, 97, 99, 101, 106, 107, 113, 116, 124, 136, 140, 142, 152, 155, 162, 173, 176, 177, 210, 216, 217, 218, 223, 226, 229, 231, 233, 235, 239, 259, 260
quinnat salmon, 205, 229

radioactive wastes, 239, 240
  effect of on rivers, 253
Radiological Advisory Council of New South Wales, 240
radionuclide, 239, 240
  accumulation factors for *Hydropsyche*, 241
rainbow trout, 189, 227
rain-pools, 177, 178, 190
rates
  of fish production, 40
  of phytoplankton production, 33
  of production in macrophytes, 36
rat-tailed maggot, 244
raw sewage, 236
Razelm complex of lakes, 16
reception and reflection of solar radiation, 87–91
recreational and aesthetic uses of water, 230
redox potential, 84
Red Rock, 202, 203
replacement quotient, 76
reproductive adaptations of some Australian fish, 163
  to running waters, 162–3
reptiles, 155
repurified zone, 245
resequent streams, 137
reservoirs, 217, 232
residual phosphates, 20
respiration, 26, 44
respiratory rates, 25
resting eggs, 191
Rhabdocoela, 178, 248
rhabdocoels, see Rhabdocoela
rheocrenes, 173
rheophilous, 151
  algae, 152
  periphyton, 36
rhithron, 166, 168, 169
rhizomes, 37
rhodophytes, 152
Rhyacophilidae, 154, 157, 164, 166, 247
Richmond, 239
riffle beetles, 246
Rift Valley of Africa, 4, 79, 202

river
  discharge, 138
  meanderings, 139
  salinities, 146
  zooplankton, 152
rivers, 49, 135–69
  effects of inert suspended materials upon, 248
  effects of organic wastes upon, 248
  effects of poisons upon, 249
  effects of pollution upon, 240–53
rivulets, 49
roadside ditches, 178
Robe, 60, 199, 200, 202
Rock-and-Pillar Range, 210, 213
rock salt, 231
Romans, 235
Rome, 215
roots, 37
Rotatoria, see Rotifera
rotenone, 237
rot fluids of vegetation, 176
Rotifera, 16, 83, 92, 107, 109, 131, 152, 174, 178, 186, 187, 189, 190, 203, 205, 209, 210
rotifer(s), see Rotifera
  food of, 111
  production, 42
Rotorua, 52, 97, 173, 210
Rottnest Island, 231
rowing, 230
Roxburgh, 222
Royal Botanic Gardens, 230
Royal Commission on Sewage Disposal, 235
Ruapehu Peak, 120
Rudjakov theory, 120
running waters
  as ecosystems, 30, 169
  biological features of, 151–69
  biota of, 151–5
  biota, adaptations of, 157–63
  biota, controlling factors for, 155–7
  characteristics of, 135
  chemical features of, 146–50
  dissolved gases in, 150
  non-biological features of, 135–50
  physical features of, 139–46
  physiographical background, 135–9
run-off, 258
Russell River, 152
Russia, 37, 41, 45, 210, 225

Sacramento–San Joaquin river system, 136
sailing, 230
Saint Laurence Seaway, 225
salina, 9, 197
Salinaland, 60, 193
saline, 16
  athalassic waters, 11

GENERAL INDEX

eutrophic lakes, 103
lake(s), 28, 60, 92, 93, 195, 204
  mud, 208
  ponds, 211
  rivers, 148
  springs, 173
  streams, 146, 148
  water(s), 1, 12, 15, 97, 102, 103, 114, 193, 198, 200, 202, 207
  waters, lower limit of, 204
salinities of some Australian rivers and streams, 147
salinity, 1, 11, 12, 76, 85, 86
  as factor in lake productivity, 76
  change, 200
  discharge correlation, 148
  of inland saline waters, 198
  records for nektonic and macrobenthic fauna, 206
  stratification in rivers, 148
  tolerance, 13, 201, 205
  tolerances of zooplankton and microbenthic fauna, 205
salmon, 225
salmonid
  fish, 257, 260
  production in small streams, 40
salt, human taste threshold of, 218
salt-lake organisms, 198
salt lakes, 14, 30, 33, 71, 193, 199, 207, 211, 213
  artificial, 211
  as ecosystems, 211–13
  benthic microfauna of, 202–4
  biota of, 211
  derivation and affinities of fauna, 209–10
  flora of, 207–9
  macrobenthic fauna of Australian, 207
  macrophyte flora of, 207
  nekton and benthic macrofauna of, 204–7
  of northern Australia, 210
  periphytic algae in, 208
  physical and chemical features of, 197–202
  species diversity in, 211
  zooplankton of, 202–4
salt pans, 9, 197
  pools, 59
  resources, 231
  supplies, 231
salts
  of alkali and alkaline earth metals, 249
  of heavy metals, 249
Saprobiensystem, 245
saprobity indices, 254
Saskatchewan, 28, 202
saturation values for dissolved oxygen, 21
sauger, 83
Scalloped Bay, 196
Scotland, 30

sea lamprey, 234, 259
seasonal
  aspects of phytoplankton, 104–5
  aspects of zooplankton, 111–13
  succession of phytoplankton populations, 105
Seattle, 257
sea-urchin, 14
sea water, 6, 14
  ionic composition of, 2, 202
Secchi disc, 64, 91
  transparency, 91, 146
secondary production, 26, 28, 37–44
  indirect methods of determining, 43
  'predation' method of determining, 43
secondary treatment of sewage, 236
secular changes in salinity, 199
sedimentation, 219
seiches, 93–6
self-absorption, 37
Sericostomatidae, 154, 157, 247
Serpentine River, 65
Serranidae, 226
seston, 27, 31, 91, 92
sevin, 238
sewage, 236
  effluents, 256
  fungus, 242, 245
  ponds, 179
  treatment, 236
Shannon River, 147
sheathed bacterium, 242
sheep, 220
shield shrimp, see Notostraca
shoot flows, 143
shoreline development, 62
shore terrace, 124
short-tailed eel, 226
shrimps, 74, 207, 229
Siberia, 50, 227
silica, 105
silk, 159, 161
silver perch, 226
Silver Springs, 47
Simuliidae, 154, 157, 160, 161, 162, 166, 169
simuliid larvae, see Simuliidae
Sinai Peninsula, 87
Siphlonuridae, 154, 167
Sixth Creek, 147, 149, 165
skin-diving, 230
sloughs, 179
snails, 129, 245, 246, 258
Snob's Creek, 253
snow, 8, 85
Snowy
  Mountains Scheme, 221, 222, 224
  River, 136, 147
sodium
  arsenite, 258

sodium (continued)
  versenate, see EDTA
soft detergents, 236
soil salinity, 222
solar
  evaporation of sea water, 231
  ponds, 87
  radiation, reception, reflection, and penetration, 87–91
solubility of oxygen, 20, 21
solution, 138
  lake, 56
sources
  and mechanisms of ion supply, 8–11
  of atmospherically supplied ions, 9
  of domestic water, 215
South Africa, 172
South America, 81, 161, 167, 171, 225, 259
South Australia, 49, 50, 51, 58, 60, 144–6, 147, 149, 165, 195, 196, 199–201, 202, 204, 206, 207, 231, 233, 259
Southern Alps, 210
Southern Highlands, 79
southern trout, 155
South Island (New Zealand), 54, 55, 56, 58, 72, 79, 81, 93, 97, 99, 106, 111, 153, 154, 173, 211, 213, 222, 233
Southland (New Zealand), 56, 100
South Pole, 118
spangled perch, 226
species composition and size differentiation in calanoids, 107
specific conductance, 12
spectral
  composition, 90
  composition of solar radiation, 89
  distribution of radiant energy, 89
spectrum, 90
Speed River, 45
Spelaeogriphacea, 172
sphaerid, see Sphaeriidae
Sphaeriidae, 127, 154
Spongillidae, 127, 154
spotted barramundi, 155
spring and bore waters in Australia, 173
springs, 173–4
spruce, 258, 259
standard composition of waters, 1
standing crop, 26, 29, 66
  biomass, 26
  values for, 28
standing stock, 26
staphylinid beetles, 162
State Electricity Commission, 239
State Rivers and Water Supply Commission of Victoria, 238
stock, 26
  water standards, 220
stoneflies, see Plecoptera

storage reservoirs, 217, 232
Strait of Portita, 16
Stratford, 99
Stratiomyidae, 167, 175
stratiomyids, see Stratiomyidae
stream
  and river beds, relation of flow rates and nature, 143
  order, 138
  piracy, 139
  salinities, 146
streams, 49, 133, 135–69
  and rivers as geomorphological agents, 138
Stygiocaridacea, 171
Subantarctic, 79
subartesian
  bores, 220
  waters, 216
sublittoral, 64
submerged
  depression individuality, 63
  macrophytes, 36
subsequent streams, 137
subsistence fishing, 225
subsurface water basins in Australia, 216
suitability of water for stock, 220
sulphides, 249
sulphur bacteria, 86, 242
summer stagnation, 82
superimposed drainage, 137
surface seiche, 95
Sutherland Falls, 56
Sutton, 202, 210, 213
swamps, 71
Sweden, 29, 30, 45, 107, 165
Switzerland, 257
Sydney, 10, 113, 240
  Basin, 216
Synbranchidae, 172
Syncarida, 153, 171, 172, 173, 260
syncarid(s), see Syncarida
synergism, 252
Syracuse, 86
syrphids, 175
systemic insecticides, 238

Tabanidae, 167, 175, 247
tabanids, see Tabanidae
tachyauxesis, 123
tadpoles, 155, 159, 160, 178, 185
Tanganyika, 15
Tangiwai, 18
tanks, 232
Tanypodinae, 127, 247
tanypodine, see Tanypodinae
tardigrades, 131
Tarim River, 59
tarns, 179
Tasmania, 50, 54, 55, 56, 81, 97, 98, 99, 101,

GENERAL INDEX

Tasmania *(continued)*
  102, 106, 113, 115, 130, 131, 132, 136,
  140, 146, 147, 148, 153, 155, 172, 173,
  182, 187, 222, 229, 231, 232, 233, 239
Tasmanian
  Central Plateau, 55, 56
  lakes, 100
  whitebait, 226
Taupo, 223
T.D.S., *see* total dissolved solids
tectonic lakes, 49, 50
temperate reed-swamps, 36
temperature-depth curve, 78
temperature
  in Melbourne pond, 180
  of ponds, 179
  of running waters, 143
  of springs, 173
temporary ponds, 179, 184, 185, 186, 187
  littoral insect fauna of, 185
temporary pools, 176
tench, 227, 228
Tennant Creek, 147, 149
Tennyson, 239
terminal
  lakes, 193
  production, 29
terminology of the major ecological regions and biological communities of a lake, 63–5
terrestrial isopods, 209
tertiary
  production, 28
  treatment, 236
thalweg, 138
Thames River, 40, 45, 216
Theraponidae, 226
thereimictic lakes, 81, 85
thermal
  classifications, 81
  pollution, 239, 253
  pollution, effect on rivers, 253
  springs, 173, 174, 175
  springs and bores in Australia, 175
  stratification, 78–82, 119
  stratification, consequences of, 82–5
  stratification in ponds, 179
  stratification in running waters, 146
thermally stratified lakes, 64, 78, 94
thermocline, 64, 78, 83, 95, 198
Thiobacteria, 29
Tibet, 59
Tipulidae, 154, 162, 206, 207
tipulids, *see* Tipulidae
tortoise, 155
total
  dissolved solids, 11, 12
  incident radiation, 90
  ionic concentration, 1

total *(continued)*
  phosphorus, 20
totality of inorganic ions and various measures of it, 11–12
Tower Hill, 51
toxic algae, 257
toxicity of various insecticides, 238
transparency of lakes, 91
transverse seiches, 94
travertine, 173
treatment of raw water, 219
tree-holes, 174
Trent River, 245
triangular diagrams, 6
trichloroacetic acid, 238
Trichoptera, 127, 153, 154, 157, 158, 159, 160, 161, 164, 174, 206, 207, 241, 246, 247, 253
trichopterans, *see* Trichoptera
triclads, 159
trinodal seiches, 93
tripton, 27
troglobites, 172
troglophiles, 172
trophic
  levels, 25, 26
  relationships, 168, 169
  structure, 24
trophogenic layer or zone, 64, 66, 123
tropholytic layer or zone, 64, 123, 124
trout, 166, 172, 234, 259, 260
Truncatellidae, 210
tube samplers, 28
Tubificidae, 186, 244, 246, 248
tubificids, *see* Tubificidae
Tungatinah, 223
tunnel formation, 139
Turbellaria, 127, 131, 154, 172, 206, 248, 253
turbellarians, *see* Turbellaria
turbidity of lake waters, 93
turbulence, 105
turbulent flow, 143
turnover
  rate(s), 25, 43
  ratio, 43
tychoplanktonic, 103
types of osmotic regulation, 13–14
typhoid fever, 254

underground
  production of biomass, 37
  streams, 171
  waters, 171–4
underwater telescope, 91
uninodal
  internal seiche, 95
  seiches, 93, 94
Unionidae, 45, 153

United States, 59, 60, 85, 126, 136, 230, 235, 239, 241, 258
urine, human, 236
U.S.A., *see* United States
U.S.S.R., *see* Russia
Utah, 16, 202, 204

vadose water, 171
Valley Creek, 45
valley transverse profiles, 139
vascular hydrophytes, *see* macrophytes
Veliidae, 133, 154, 162, 247
veliid bugs, *see* Veliidae
Venice System of brackish water classification, 17
vertebrates in running water, 155
vertical
　distribution of periphyton biomass, 209
　migration of zooplankton, 117–20
　profile of lake edge, 123
　zonation of benthos, 123
　zonation of lake regions, 125
Victoria, 9, 16, 30, 51, 52, 53, 54, 56, 58, 60, 87, 92, 97, 99, 106, 109, 113, 126, 127, 128, 136, 146, 147, 148, 149, 160, 172, 181, 185, 193, 197, 198, 199, 201, 202, 203, 204, 205, 206, 208, 209, 217, 222, 230, 231, 232, 233, 237, 239, 253, 258, 259, 260
Victorian rivers, 148
viruses, 258
volcanic
　lakes, 51–4
　springs, 87
volcanoes, 8, 10
volcano-tectonic lakes, 52
volume development, 62
Volvocales, 189

Waiau Syncline, 100
Waikato River, 107, 222, 223
Waikoropupu, 174
Waikoropupu Springs, 173
Wairarapa, 58
Wairau, 58
Waitaki River, 222
Waitangi River, 53
Waitangi valley, 53
Waitomo, 58
wallum, 59, 97
Wannon River, 146, 149
Warburton River, 144, 145
warm monomictic lakes, 23, 79, 80, 81
warm thereimictic lakes, 81, 113
Warrah Creek, 152
Warrandyte, 148
Washington State, 33, 59, 85, 86, 202, 239, 241
water
　associated with terrestrial vegetation, 174–6

water *(continued)*
　fern, 259
　for agricultural purposes, 220
　hyacinth, 152, 259
　level, changes in, 234
　pollution, 234
　renewal as a factor in lake productivity, 76–7
　supply engineering, 217
water-skiing, 230
water-table, 171
weedicides, 237
weirs, 232
wells, 172
Western Australia, 11, 50, 60, 106, 109, 130, 136, 146, 147, 148, 172, 178, 180, 189, 190, 193, 207, 210, 216, 229, 231, 259
Western Australian rivers, 148
Western District of Victoria, 79, 92
wetlands, 230
wet weight, 27
Whangaehu River, 18
whitebait, 226
whitefish, 241
Wiluna, 60
wind-generated currents, 93–6
Windorah, 140
windrows, 75
Winkler method of oxygen determination, 20
Wisconsin, 47, 66, 68, 70, 257
world average fresh water, 1, 5, 6
World Health Organization, 218
Woronora River, 240

xanthophyceans, 131

Yallourn, 239
Yangtze River, 136
Yarra River, 136, 148, 255
yield, 29
Yorke Peninsula, 231

zinc, 72, 252, 253
zinc sulphate, 252
zonational patterns, 166
zonation of polluted rivers, 244, 245
zone
　of foul pollution, 245
　of mild pollution, 245
　of pollution, 245
zoobenthos, 65
zooplankton, 27, 45, 65, 74, 75, 82, 83, 85, 107–23, 203
　food of and effects of predation on, 107–11
　of rivers, 152
　of running waters, 152
　populations, 41
　seasonal aspects, 111–13
　vertical migration of, 113–20
Zygoptera, 127, 154, 247

postscript
LAKE PEDDER

As this book goes to press waters impounded behind a dam built by the Hydro Electric Commission of Tasmania have already merged with those of Lake Pedder. Unless some dramatic last-minute reprieve is granted the lake will be eventually submerged and destroyed beneath a reservoir. This now seems likely to occur despite the fact that Lake Pedder lies within a national park which was proclaimed in 1955. However, if the diversion tunnel at the Serpentine Dam were to be opened any time before the onset of next winter (say May 1973), it is unlikely that permanent damage to either the lake as a landform or its biota would result.

Lake Pedder is widely acknowledged as being one of the most beautiful, if not the most beautiful lake in Australia. Additionally all available evidence points to its having a more extensively developed psammon community than any other extant freshwater lake in southern Australia and to being the home of a number of endemic species (see pp. 65, 130, and 131 and Table 6:8). The plant species presently known only from Lake Pedder are two new species of *Centrolepis*, *Milligania* sp. nov. and *Triglochin* sp. nov. The animal species are: *Telmatodrilus multiprostratus* and *T. pectinatus* (Oligochaeta), two new phreatoicids (Isopoda), a new asellote (Isopoda), the crayfish *Parastacoides pulcher*, three new species of caddis-fly, a new corixid, a new notonectid, a new planorbid snail, and possibly the fish *Galaxias pedderensis*. Furthermore, the swampy moorland immediately to the east of Lake Pedder contains the largest known population of *Allanaspides helonomus*, a syncarid first discovered in 1969 and described in 1970.

Most of the species which on the basis of existing information are regarded as endemic to Lake Pedder seem to have an obligate relationship with lacustrine sand or sandy gravel. In the case of the endemic plants it appears that an additional requirement is that these sands be subject to winter flooding and summer drying and the period of flowering is appropriately geared to this regime. In other words, there is first a requirement for sand or gravel or mixtures of these as a substratum in which to grow, while oscillations in water level constitute a further essential aspect of their ecological requirements; they are true eulittoral plants (*sensu* Hutchinson 1967).

Some people have pointed out, with full justification, that in general terms the biology of southwestern Tasmania, like that of many other parts of Australia, is incompletely known. They also rightly indicate that although the biological exploration of Lake Pedder is anything but complete it has nevertheless been somewhat more intensive than that of some other lakes in southwestern Tasmania. However, there is no justification for concluding from these two truisms that most if not all of the species presently regarded as endemic to Pedder will eventually be found elsewhere in the southwest. To do so implies an inability to turn one's attention away from certain generalizations concerning this region and consider the evidence concerning the distribution within it of a certain quite specific habitat, lacustrine sand. While many uncertainties still exist with regard to the southwest we do know that there is no other sizeable lake with a sandy beach; all other lakes in the region are much smaller than Lake Pedder. None of these is known to have a well-developed sandy beach and so they are most unlikely to support a full complement of the distinctive Pedder species. The discrete nature of these lakes stands in marked contrast to the terrestrial continuum surrounding them. As a consequence there may well be more uncertainty concerning the distribution of certain terrestrial habitats in the southwest than there is regarding its lakes. In this respect it should be noted that many of the smaller lakes and tarns in this region have been photographed from the air and preliminary limnological surveys of several have been made, using a helicopter to gain access. As a result, a significant amount is known about the general nature of the lakes in this region.

When species have an obligate relationship with a special type of habitat and only one of that sort is definitely known within a circumscribed area such as the southwest, it would seem reasonable that indications of endemism should be taken seriously. This is especially so in a conservation issue in which the loss or gross alteration of such a unique habitat is threatened. To argue in such a case that apparent endemics will eventually turn up elsewhere may be regarded as somewhat of a mixture of wishful and irresponsible thinking.